Tomasz Blachowicz, Andrea Ehrmann
Spintronics

Graduate Texts
in Condensed Matter

Series Editor
Prof. Dr. Christian Enss
Heidelberg University
Kirchhoff-Institute for Physics
Im Neuenheimer Feld 227
69120 Heidelberg
Germany

Tomasz Blachowicz, Andrea Ehrmann

Spintronics

Theory, Modelling, Devices

2nd Edition

DE GRUYTER

Authors
Prof. Dr. Tomasz Blachowicz
Institute of Physics
Center for Science and Education
Silesian University of Technology
Boleslawa Krzywoustego 2
44-100 Gliwice
Poland
tomasz.blachowicz@polsl.pl

Prof. Andrea Ehrmann
Faculty of Engineering and Mathematics – ITES
Bielefeld University of Applied Sciences and Arts
Interaktion 1
33619 Bielefeld
Germany
andrea.ehrmann@hsbi.de

ISBN 978-3-11-138249-4
e-ISBN (PDF) 978-3-11-138373-6
e-ISBN (EPUB) 978-3-11-138383-5

Library of Congress Control Number: 2024941929

Bibliographic information published by the Deutsche Nationalbibliothek
The Deutsche Nationalbibliothek lists this publication in the Deutsche Nationalbibliografie;
detailed bibliographic data are available on the Internet at http://dnb.dnb.de.

© 2024 Walter de Gruyter GmbH, Berlin/Boston
Cover image: iStock/Getty Images Plus
Typesetting: Integra Software Services Pvt. Ltd.

www.degruyter.com

Preface to the second edition

Spintronics is not stopping in its dynamic development. Over the past few years, surprising new results have been obtained and new technological solutions have been implemented. However, all this requires an understanding of the fundamentals of the physical issues of spintronics. That is why the need to reissue this book arose.

In the second edition of the book, minor errors in the text have been corrected. In addition, the references have been expanded and updated to facilitate access to the latest research developments. In the third chapter, new information about exotic magnetization states, the hopfions, is provided. Also, in the same chapter we extend the analysis of the Dzyaloshinskii–Moriya interaction (DMI) for new symmetry classes of materials. In Chapter 6, some new remarks about the second quantization perspective for the Kondo effect analysis are added.

We wish the reader a lot of patience in learning the basics of the issues presented in the book, hoping that it will be a support and a good guide in the search for more scientific discoveries.

Tomasz Blachowicz and Andrea Ehrmann

https://doi.org/10.1515/9783111383736-202

Preface to the first edition

The research area of spintronics, although under investigation for decades, is still frequently excluded from physics in school and even in university. Although the mathematical formalisms used to describe spintronic processes are not exactly simple, the basic concepts and applications can often be understood without them – and teaching these should be part of physics education. While some of the possible future applications, such as quantum computing, are still dreams of the future, other applications have been used for years now, for example, in our computers' hard disk read/write heads, and should thus be understood at least by physicists.

Our aim is to give you an insight into this emerging field of theoretical and experimental physics and to guide you from a basic understanding of processes related to spintronics to the recent applications and future possibilities opened by ongoing research. We tried to explain as detailed as possible the physical background for the related phenomena combined with the underlying mathematical equations, guiding you from basic physics to the state of the art in the difficult area of spintronics.

Further, this book will present an overview of new trends and research results, as well as ideas in spintronics, at the beginning of the twenty-first century. It is suitable for students, industrial laboratory workers, and for interested scientists who were once discouraged from the theoretical quantum formalism up to now.

Starting with a short description of the spin and mathematical possibilities to describe it, some fundamental physical effects are introduced, taking into account different materials and structures. Micromagnetic modeling is used to illustrate magnetic effects in the typically very small spintronics devices. Their description is extended to contact effects in diverse interface pairs, followed by transport processes in spintronics structures. After depicting real spintronics devices and applications, the influence of external physical factors on spintronics devices is examined. Finally, materials nowadays used in commercial applications are described in detail.

In general, the presentation here goes from some theoretical backgrounds to many applied topics, suited mainly for less experienced scientists and students who want to know more about this exciting topic in recent physics.

We wish you an inspiring reading and hope the fascination experienced by researchers working in the field of spintronics will kindle your interest in this topic, too.

Tomasz Blachowicz and Andrea Ehrmann

https://doi.org/10.1515/9783111383736-203

Contents

1 Introduction

The term "spintronics" refers to electronic effects and material properties induced by the spin of an electron. Nowadays, this trivial sentence contains so many meanings, spreading every day to so many regions of material science and physics, including applied and theoretical perspectives, that all this requires significantly more exploration and understanding. This is the main reason to present the reader with another book on spintronics.

What is a spin? It is another level of freedom, an element of the quantum mechanics operator algebra, the angular momentum of an electron, a quantum number, a quantity that can be transferred in real devices, a quantity that can be back-reflected at the interface between two materials, a quantity that can be transferred in space without spatial movement of electrons, and also a quantity involved in spin-waves.

This chapter will give an overview of possible answers to this question, their mathematical description, and the consequences for spintronics applications.

1.1 What is a spin?

A spin is the angular momentum. It seems that the electron reveals the fundamental property of self-rotation, while it is a point-like elementary particle and it has no classically understood diameter. Thus, the angular momentum is a quantum–mechanical property. The electron spin was introduced by Wolfgang Pauli (Pauli 1926).

The length of the electron spin, being a quantized value, is expressed similarly to other atomistic angular momenta as proportional to the Planck constant h, namely,

$$L_s = \sqrt{s(s+1)}\,\frac{h}{2\pi},\qquad(1.1)$$

where L_s is the angular momentum and $s = 1/2$ is the spin quantum number, which always keeps this single value. The common idea about electron-spin is that it can have two possible states, usually called "up" and "down." Thus, the number $s = 1/2$ cannot be used to resolve these two states. The physical origin of the two states comes from the interaction between the spin and the magnetic field, including coupling with the field resulting from the electron orbital movement in an atom and also including the fields imposed externally or the field of magneto-crystalline origin due to the material. In other words, a spin vector can be projected onto an effective magnetic field direction while the two projections are of the same length, since there is no reason to emphasize any orientation.

As a matter of fact, the aforementioned magnetic interaction takes place between the magnetic field and the electron magnetic moment, being directly proportional to the angular momentum L_s. The magnetic moment of the free electron equals $\mu_B = e\,\hbar/2m$, where e and m are the electron electric charge and the electron mass, respectively. The magnetic

https://doi.org/10.1515/9783111383736-001

moment is named by physicists as the Bohr magneton; hence, $\mu_B = 9.27 \cdot 10^{-24}$ J/T. The magnetic moment of an electron involved in a real material can differ from the value $\mu_B = e\,\hbar/2m$. More details about this fact are provided in the next sections.

1.2 Single spin algebra

The absolute value of the electron angular momentum equals

$$L_s = \sqrt{s(s+1)}\hbar \text{ or } L_s = \sqrt{\frac{1}{2}\left(\frac{1}{2}+1\right)}\,\hbar, \tag{1.2}$$

where $\hbar = h/2\pi$ and where obviously $s = 1/2$. The value of the angular momentum $L_s = \sqrt{3/4}\,\hbar$ is a real physical quantity. On the other hand, in the presence of a magnetic field, the angular momentum can be projected – or better said, can physically exist – in two possible states, up and down; namely, it can be expressed as a quantity proportional to the atomic-scale angular momentum unit $\hbar$. This length of the projection equals $L_{zs} = m_s\hbar$, where the newly introduced magnetic spin quantum number m_s can obtain two equally possible values, $m_s = \pm 1/2$. This is why people sometimes say that an electron spin can be directed up or down, or equals 1 or −1, or sometimes more precisely that electron spin equals $1/2\hbar$ or $-1/2\hbar$. This simple physical meaning can be easily explained geometrically (Fig. 1.1).

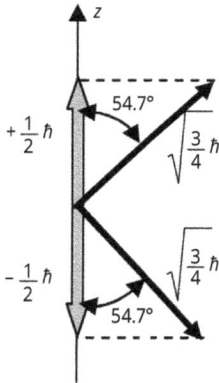

Fig. 1.1: The possible states for a free electron spin, $\pm\hbar/2$, under the influence of an externally applied magnetic field oriented along the z-axis direction.

The algebra describing the above situation is based on the so-called operators, that is, matrices and two-component vectors, representing the two possible physical spin states

$$|\text{up}\rangle = \begin{bmatrix} 1 \\ 0 \end{bmatrix}, \quad |\text{down}\rangle = \begin{bmatrix} 0 \\ 1 \end{bmatrix}. \tag{1.3}$$

The mentioned operator, equivalent to the externally applied magnetic field – here, for simplicity, acting along the chosen z-axis – is represented by the following 2×2 Pauli matrix:

$$\sigma_z = \begin{bmatrix} 1 & 0 \\ 0 & -1 \end{bmatrix}. \tag{1.4}$$

This operator, acting on the spin-state vector in the mathematical sense of a matrix multiplication, provides the possible physical values of the spin projected onto the magnetic field direction. This is the so-called eigenvalue equation or problem, namely,

$$\frac{\hbar}{2} \begin{bmatrix} 1 & 0 \\ 0 & -1 \end{bmatrix} \begin{bmatrix} 1 \\ 0 \end{bmatrix} = m_{\text{up}} \frac{\hbar}{2} \begin{bmatrix} 1 \\ 0 \end{bmatrix}, \tag{1.5}$$

$$\frac{\hbar}{2} \begin{bmatrix} 1 & 0 \\ 0 & -1 \end{bmatrix} \begin{bmatrix} 0 \\ 1 \end{bmatrix} = m_{\text{down}} \frac{\hbar}{2} \begin{bmatrix} 0 \\ 1 \end{bmatrix}, \tag{1.6}$$

where $m_{\text{up}} = +1$ and $m_{\text{down}} = -1$, and the measurable physical quantities are equal to

$$L_{zs} = m_{\text{up}} \frac{\hbar}{2} \text{ and } L_{zs} = m_{\text{down}} \frac{\hbar}{2}. \tag{1.7}$$

As mentioned before, the magnetic spin quantum number is equal to

$$m_s = m_{\text{up}}/2 = 1/2 \text{ and } m_s = m_{\text{down}}/2 = -1/2, \tag{1.8}$$

respectively, or the projections are just equal to $L_{zs} = \pm \hbar/2$.

At the end of this section, there is a short explanation about the origin of the characteristic $\sqrt{s(s+1)}$ factor met in many issues related to angular momentum problems in quantum formalism algebra. The origin can be understood if we extend our consideration to three dimensions, where we have an operator $\widehat{\sigma} = [\sigma_x; \sigma_y; \sigma_z]$ consisting of the three spin components along with the additional condition of

$$\sigma_x^2 = \sigma_y^2 = \sigma_z^2 = \begin{bmatrix} 1 & 0 \\ 0 & 1 \end{bmatrix} = \widehat{I}. \tag{1.9}$$

The total spin angular momentum operator is a three-dimensional quantity and equals

$$\widehat{L} = \frac{\hbar}{2} \widehat{\sigma}, \tag{1.10}$$

while its squared value has the following form

$$(\widehat{L})^2 = \left(\frac{\hbar}{2} \widehat{\sigma} \right)^2 = \frac{\hbar^2}{4} \left(\sigma_x^2 + \sigma_y^2 + \sigma_z^2 \right) = \frac{\hbar^2}{4} (I + I + I) = \frac{3\hbar^2}{4} \widehat{I}. \tag{1.11}$$

Hence, it is worth emphasizing that the squared angular momentum is a measurable quantity, while the nonsquared angular momentum cannot be measured directly. Thus, the eigenvalue problem for the squared spin angular momentum can be written as

$$\frac{3\hbar^2}{4}\,\widehat{I}\,|\alpha\rangle = \frac{3\hbar^2}{4}\,|\alpha\rangle = \frac{1}{2}\left(\frac{1}{2}+1\right)\hbar^2|\alpha\rangle \tag{1.12}$$

with the arbitrary physical spin-wave state $|\alpha\rangle$ represented by the single-column vector. Remember that on the right-hand side of the above eigenvalue equation, we have a really measurable physical quantity of positive algebraic sign, here equal to $1/2(1/2+1)\hbar^2$. This is why the length of the real spin angular momentum is expressed by the characteristic square-root taken from quantum numbers $\sqrt{1/2(1/2+1)}\hbar = \sqrt{s(s+1)}\hbar$.

1.3 A spin in a real material

The free electron spin magnetic moment $\vec{p}_s$ is a vector quantity, being directly proportional and antiparallel to the angular momentum vector $\vec{L}_s$ (comp. information in Section 1.1), namely,

$$\vec{p}_s = -\frac{e}{m}\vec{L}_s \tag{1.13}$$

or

$$p_s = \frac{e}{m}L_s = \frac{e}{m}\sqrt{\frac{3}{4}}\hbar, \tag{1.14}$$

and the magneton Bohr value can be derived from the spin projection $L_{zs.}$

$$p_{zs} = \frac{e}{m}L_{zs} = \frac{e}{m}\sqrt{\frac{3}{4}}\hbar \cdot \cos(54.7°) = \mu_B. \tag{1.15}$$

Thus, p_{zs} can have two possible values, $p_{zs} = \pm\mu_B$. However, in general, this simple formula does not hold for conduction electrons in real solid states. In such a case, the magnetic moment is influenced by interactions with fields produced by the environment via spin–orbit coupling (a relativistic interaction between spin and particle motion in a potential), and the expression takes the form

$$p_{zs} = g_s m_s \mu_B = \pm\frac{g_s}{2}\mu_B, \tag{1.16}$$

where g_s is the so-called spin-electron g-factor. It is exactly equal to 2.0023 only for the free electron case. Hence, experimentally measured or calculated values can significantly differ from the ideal value of $g_s \cong 2$ (Tab. 1.1). The most exotic situations can take place in semiconductors, thus in nonmagnetic materials.

Tab. 1.1: Values of the spin-electron g-factor for chosen semiconductors and metals.

Material	g_s factor
Ge doped by Sb (g_s – the tensor quantity)	0.820–1.922 (Pontinen et al. 1960)
InAs	−14.9 (Konopka 1967)
InSb	−51.3 (Isaacson 1968)
GaAs	−0.5 (White et al. 1972)
GaSb	−7.9 (Reine et al. 1972)
GaP, AlAs, AlSb	2.0 (Kaufmann et al. 1976)
Si quantum well	1.9944 (Wilamowski et al. 2007)
Ge quantum well, tuned by external magnetic field	0.905–1.915 (Giorgioni et al. 2016)
Cu	2.033 (Grechnev et al. 1989)
Ag	1.983 (Grechnev et al. 1989)
Au bulk	2.11 (Grechnev et al. 1989)
Au nanoparticle	0.3 (Davidović et al. 1999)
Co nanoparticle	7.3 (Garland et al. 2013)

It should also be clarified that the formula connecting magnetic and angular momenta can be written separately for orbital movement (case of an orbital g-factor) or for the total angular movement, including coupling between spin and orbital movement (case of the total angular momentum g-factor). This is, at the moment, out of the main scope of this book.

1.4 The 3d electrons in ferromagnetic atoms

Where does ferromagnetism stem from, the property of solid-state matter to keep magnetization even without the presence of an external magnetic field? The question is quite fundamental since the answer bases on fundamental principles of nature. Namely, the natural state for each physical system is the one of minimum total energy.

The most significant components of solid-state energy come from Coulomb-like electric interactions between atoms, between the atoms' electrons, and between electrons and nuclei, on the one hand, as well as from the exchange, a quantum-originated interaction between electrons, on the other hand. In other words, electric interaction controls the space distribution of atoms and atom dimensions themselves, while the exchange interaction is responsible for magnetic properties. To be even more precise, ferromagnetic properties are mostly influenced by the exchange between electron spins and do not result from orbital magnetic moments. Nevertheless, ferromagnetism happens since it is energetically favorable.

Thus, let us briefly review five elements from the fourth period of the periodic table of the elements: Mn, Fe, Co, Ni, and Cu, located in neighboring groups: 7, 8, 9, 10, and 11, respectively. In iron, for example, the distance between atoms is such that four electrons located at the third orbit and the d-subshell have spins oriented in the same direction. In this case, we can say that the magnetic contribution to energy exceeds the electric component. The other subshells of the iron atom, including the outer 4s sub-shell with two electrons, have even numbers of electrons with antiparallel coupled spins. Roughly speaking, this situation, advantageous for ferromagnetism, takes place if the atom diameter is greater than 3 radii of the 3d shell diameter. Precisely, the ratio of the diameter to the 3d subshell diameter, for ferromagnetic iron, cobalt, and nickel, equals 3.26, 3.64, and 3.94, respectively (Murray et al. 2007). Similarly, for manganese, the ratio is equal to 2.94 and the Coulomb interaction dominates over the magnetic ones. The greater ratio means larger distances, smaller Coulomb-type electric interactions, and enhanced exchange coupling between 3d-shell electrons. For copper, all 3d electrons are compensated and ferromagnetism is not accessible. To provide more direct intuitions into this important point, Fig. 1.2 shows a graphical representation of the spin configurations in atomic shells for the cases commented on earlier.

It is worth noticing that strong ferromagnetic properties can be achieved in alloys of metals that are not ferromagnetic alone or exhibit only weak ferromagnetism. Nevertheless, from the aforementioned perspective, making alloys means control of the spatial distribution of atoms to achieve favorable domination of spin-originated ferromagnetism over the Coulomb electric interactions. A good example of such a strong ferromagnetic material is the alloy $Nd_2Fe_{14}B$, commonly known as neodymium magnet. Neodymium alone is ferromagnetic, due to a favorable, four-parallel spin configuration at the last 4f subshell; however, owing to thermal factors, it loses ferromagnetic properties over 19 K temperature, very far below room temperature (Jiang et al. 2017).

Thus, to give the reader some sense of temperature influence on ferromagnetic order, we can compare the energy kept by a single-spin magnetic moment, represented by the Bohr magneton, immersed in the magnetic field induction B, that is $\mu_B B$, with the competing thermal energy, represented by kT, the product of the Boltzmann constant and ambient temperature. Assuming $T = 300$ K, $k = 1.38 \cdot 10^{-23}$ J/K, and $\mu_B = 9.27 \cdot 10^{-24}$ Am^2, we have

$$B = \frac{kT}{\mu_B} = 446.6T, \qquad (1.17)$$

a field value exceeding the values achievable in typical electromagnets in standard experimental situations by far. In other words, spin-originated magnetic fields in ferromagnetic materials are huge.

The temperature at which a material loses its ferromagnetic properties is called the Curie temperature. For the representative elements of Fe, Co, and Ni, it is equal to 1,043, 1,388, and 627 K, respectively (Heller 1967).

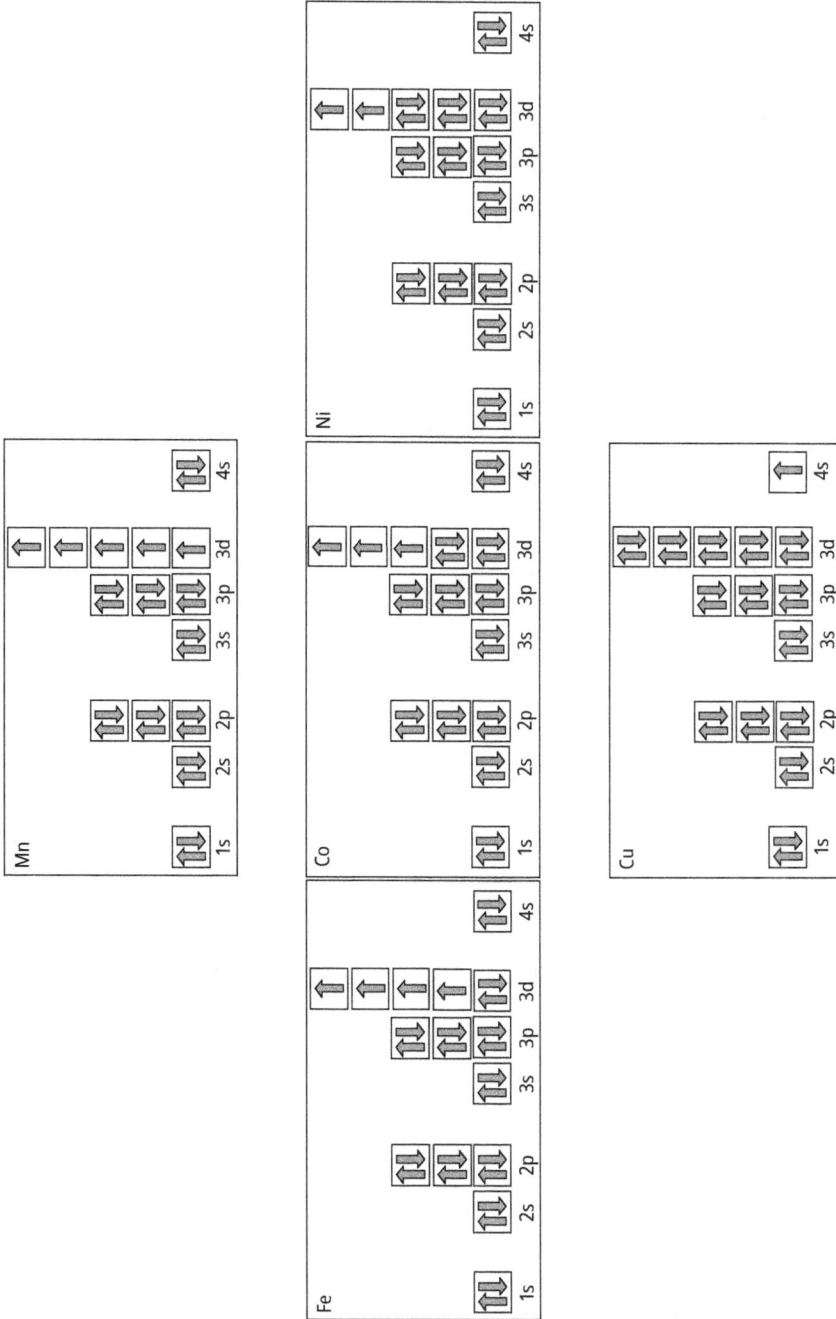

Fig. 1.2: Electronic shell configurations for neighboring elements of the fourth period of the periodic table of the elements. Fe, Co, and Ni are ferromagnetic.

1.5 An electron in a phase space: the density of states

The notion of phase space is widely applied by solid-state physicists and material scientists. It tells how much space is available for electric current carriers, and the availability is counted from the perspective of energy or momentum scales. This section is mathematically the most advanced within the introduction chapter; however, it enables a useful presentation of spintronics practical devices, exactly from the energetic perspective.

An electron, which is a quantum particle, is a point-like object. The smallest amount of space accessible for it is governed by the Heisenberg principle of uncertainty for each spatial direction and momentum values accessible. The infinitesimal region where a particle can be found equals

$$dx \cdot dp_x = h,$$
$$dy \cdot dp_y = h, \tag{1.18}$$
$$dz \cdot dp_z = h.$$

Thus, the product of infinitesimal spatial and momentum volumes equals $dV \cdot dV_p = h^3$, while for a single spin state, one from the two possible ones, the available volume is half of it.

$$dV \cdot dV_p = \frac{h^3}{2}. \tag{1.19}$$

The momentum part of this volume, as it is of interest to us, equals

$$dV_p = \frac{h^3}{2 \cdot dV}. \tag{1.20}$$

In the momentum space, we have the total volume, ranging from 0 to any maximum value of p, and it equals

$$V_p = \frac{4}{3}\pi p^3, \tag{1.21}$$

and the volume of the thin layer of thickness dp is $dV_p = 4\pi\,p^2\,dp$ – it was obtained by calculating the first derivative of p. Within the small range $\langle p; p + dp \rangle$, we have the following number of phase-space elements:

$$\frac{4\pi p^2 dp}{dV_p} = \frac{8\pi p^2 dp}{h^3}dV. \tag{1.22}$$

However, the number of states per spatial volume unit equals

$$\frac{8\pi p^2}{h^3} dp = g(p)\ dp, \tag{1.23}$$

where $g(p)$ is the function of the momentum density of states. Now, finally, expressing the same on the energy scale and remembering the obvious relations $E = p^2/2m$ and $dE = (p/m)dp$ since this energetic perspective is used by electronic engineers, we have

$$\frac{8\pi p^2 dp}{h^3} = \frac{8\pi p^2}{h^3} \cdot \frac{1}{2}\sqrt{\frac{2m}{E}} dE = \frac{8\pi 2mE}{h^3} \cdot \frac{1}{2}\sqrt{\frac{2m}{E}} dE = \frac{4\pi (2m)^{3/2}}{h^3}\sqrt{E} dE = g(E)dE, \tag{1.24}$$

thus, finally, we obtained the function of the energy density of states with its characteristic shape

$$g(E) = \text{const} \cdot \sqrt{E}, \tag{1.25}$$

while the number of available states in a whole available energy range equals

$$n = \int_{0}^{E_F} g(E)dE. \tag{1.26}$$

The number of states n is expressed in units of $1/m^3$, which means it represents the spatial concentration of available states.

The most important point of the above dependence is that the density of states is proportional to the square root of energy, and what is even more important, the relation $g(E) \sim \sqrt{E}$ is cut off at the maximum available energy named the Fermi energy level E_F, this parameter being an intrinsic property of a given material.

At the end of this section, one point has to be emphasized. Since in nonmagnetic conductors both subsets of electrons, possessing spin-up and spin-down states, are indistinguishable, the density of state function can be split into two independent regions of equal areas, as seen in Fig. 1.3a. Hence, a question arises: Are there any materials where this balance can be broken? The answer is positive. It can happen in ferromagnetic materials; however, even in such nonmagnetic metals like Cu, a set of electrons placed in an external magnetic field behaves paramagnetically, meaning the number of electrons keeping magnetic moments in the direction of the external field exceeds the number of electrons with magnetic moments oriented against the field (Figs. 1.3b, 1.3c). For ferromagnetic materials, this situation can happen even if no external magnetic field is applied. This is an example of a favorable spin-orbit interaction between an electron's magnetic moment and the magnetic moments coming from orbital movements of electric carriers. To summarize these facts shortly, it is worth noticing that the effect can be fully explained by the inclusion of electrostatic energy components of real multiatomic systems, and the point is that the physical system always finds the minimum of the total energy.

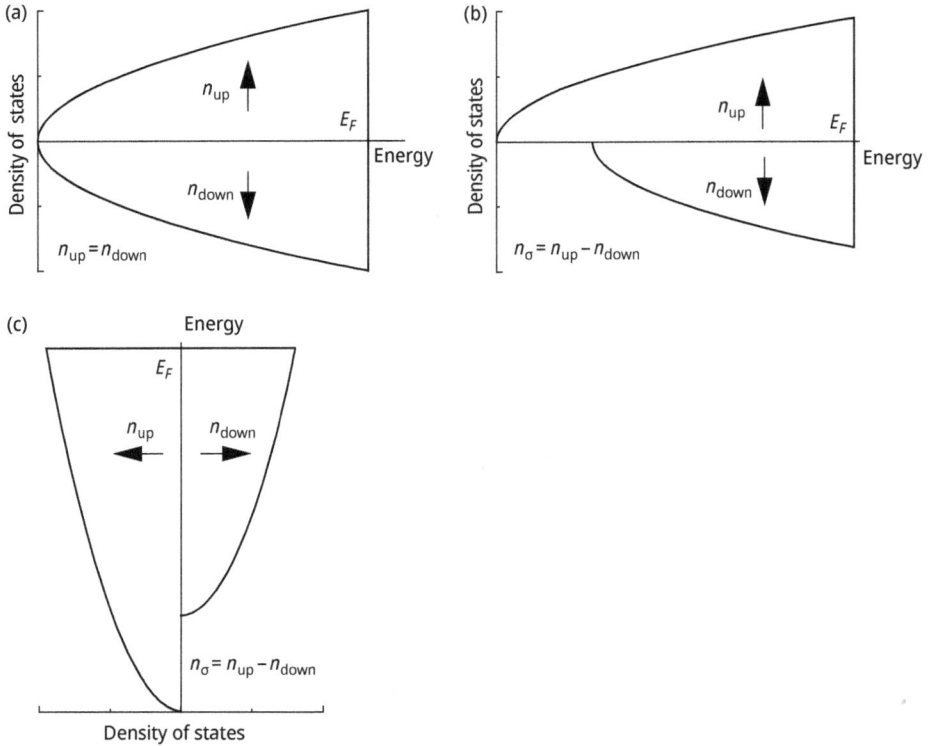

Fig. 1.3: Energy density of states functions for electrons: with the same number of particles with spin-up and spin-down states (a), being in the state of spin polarization $n_\sigma = n_{up} - n_{down}$ (b), and another graphical variant of the spin-polarization state, rotated by 90°, as often seen in scientific publications (c).

In many material phases, in ferromagnetic metals and also in nonmagnetic materials, the imbalance between spin-up and spin-down systems is fundamental for spin-dependent phenomena explanations. Especially, the effect of spin tunneling in magnetic tunnel junctions can be understood from the perspective of the mentioned imbalance of available state described by the density of states function.

1.6 Full quantum perspective for the electron spin description: the second quantization

At the beginning of this section, some basic definitions will be provided. For the reader, it might be confusing to look at the completely new terms and expressions without extensive, many-pages introductory explanations. However, the reason for such an approach is quite rational. Namely, it is an example of new knowledge based on axioms. Axioms are logical sentences, supported by physical observations, while they do not require strict mathematical proofs based on previous mathematical models. Similar situations took place, for example, in ancient Greece where five geometry axioms were formulated by Euclid (fourth century BC), or quite recently within the first half of the previous age, when postulates of quantum mechanics were formulated by Heisenberg, Schrödinger, Dirac, Born, Bohr, Jordan, Weyl, and von Neumann (Fedak et al. 2009). Thus, the formulations given here are based on the quantum mechanics axioms, which introduce specific algebra of operators and physical states.

Firstly, for a system with n electrons, the physical situation – that is, the existence or nonexistence of different electrons counted by the index i – is represented by the field operators $\Psi(\vec{r}, \sigma)$ and $\Psi^+(\vec{r}, \sigma)$. They are linear combinations of single-electron states, φ_i or φ_i^+,

$$\Psi(\vec{r},\sigma) = \sum_i^n c_i \varphi_i(\vec{r},\sigma) \tag{1.27}$$

and

$$\Psi^+(\vec{r},\sigma) = \sum_i^n c_i^+ \varphi_i^+(\vec{r},\sigma), \tag{1.28}$$

where c_i^+ and c_i are the creation and annihilation operators of the ith electron. Each electron's physical state is represented by the wave function $\varphi_i(\vec{r}, \sigma)$ being a function of the spatial position $\vec{r}$ and the spin variable σ.

To understand the physical sense of the operators above, let us assume we have no electrons in a given region of interest; thus, we are dealing with a vacuum:

$$\Phi_{\text{vac}} = |0\rangle = \{0, 0, \ldots, 0_i, \ldots, 0_n\}. \tag{1.29}$$

The above expression contains three kinds of symbols used by theoreticians: the text symbol Φ_{vac}, the Dirac so-called ket-vector $|0\rangle$, and the normal sequential form $\{0, 0, \ldots, 0_i, \ldots, 0_n\}$ used in algebra of sets.

Thus, first, using the creation operator acting on the ith electron, we obtain

$$c_i^+ |0\rangle = \{0, 0, \ldots, 1_i, \ldots, 0_n\}, \tag{1.30}$$

or acting again onto the right-hand side, using the annihilation operator, we go back to the vacuum state, that is,

$$c_i c_i^+ |0\rangle = \{0, 0, \ldots, 0_i, \ldots, 0_n\}. \tag{1.31}$$

Secondly, an important expression used in many practical situations is the operator of the number of particles written as

$$\hat{N} = \sum_i^n c_i^+ c_i = \int \Psi^+(\vec{r},\sigma)\Psi(\vec{r},\sigma)d^4V, \tag{1.32}$$

where integration goes over the Cartesian volume and the spin variable. Thus, it is constructed from the field operators, and the interior of the integral above can be interpreted as the operator of density of particles $\Psi^+(\vec{r}, \sigma)\Psi(\vec{r}, \sigma)$.

Thirdly, as mentioned, the term $c_i^+ c_i$ (equivalent to the density of particles allocated around a given point) can be used to describe the energy of a physical system (the energy is obviously proportional to the number of electrons) expressed by the so-called Hamiltonian in energy units. This statement will be slightly explained using the concrete example of the three-terminal spin device, the conceptual prototype of a spin transistor. It consists of two leads and an island placed between them, that is, two tunneling interfaces on the left and right sides of the island (Fig. 1.4) (Barnaś et al. 2008). The device energy (Hamiltonian) consists of four components for each interface, namely,

Fig. 1.4: The concept of the spin transistor. It is a three-terminal device, where the island is tuned by capacitance coupling.

$$H = H_L + H_I + H_Q + H_C, \tag{1.33}$$

where H_L is the lead energy, H_I is the island energy, H_Q is the extra electrostatic energy needed to transfer electrons into the island, and H_C is the coupling (interaction) energy between a lead and the island obtained due to this transfer. All these terms are constructed from the $c_i^+ c_i$ operators. For example, the island part of the Hamiltonian equals

$$H_I = \sum_{\vec{k}\sigma} \varepsilon_{(I)\vec{k}\sigma} c_{(I)\vec{k}\sigma}^+ c_{(I)\vec{k}\sigma}. \tag{1.34}$$

The careful reader possibly has noticed some differences in summation indexes in comparison to the formulas presented before. Indeed, the commonly used custom in literature is based on summation in the momentum space and not in the spatial one, where $\vec{r}$ is replaced by the wave vector $\vec{k}$. An additional reason for this is quite rational since the quantum formalism is based on wave-functions φ involved in momentum space. Hence, $\varepsilon_{(I)\vec{k}\sigma}$ is the energy of a single electron located in the island – see the index (I).

More details about physical effects in such spin devices, regimes of tunneling, underlying electrostatic effects, and other electromagnetic parameters will be provided in the next chapters.

1.7 Spin transport and random-walk

The nature of spin transport is stochastic since the physics of underlying processes at atomic and meso-scales is not classical, or, in other words, is not fully deterministic.

A good preliminary intuition of the spin transport in one dimension can be provided by the expression describing the concentration of electrons at a given location x and at a given time t, $n = n(x, t)$, which is modified stochastically from the left and the right (Žutić et al. 2004), namely,

$$n(x,t) = P_{\rightarrow} \cdot n(x - \Delta x, t - \Delta t) + P_{\leftarrow} \cdot n(x + \Delta x, t - \Delta t), \tag{1.35}$$

where $P_{\rightarrow}$ is the movement probability of electrons with concentration $n(x - \Delta x, t - \Delta t)$ from the left to the point of interest (x, t) and $P_{\leftarrow}$ is the movement probability from right to left, assuming the equivalent concentration $n(x + \Delta x, t - \Delta t)$ at the earlier moment of time $t - \Delta t$. Because of the probabilistic approach, such a model is sometimes called a random-walk method (Fig. 1.5). As a matter of fact, the expression above is classical, and it has no electron-spin included in an evident way.

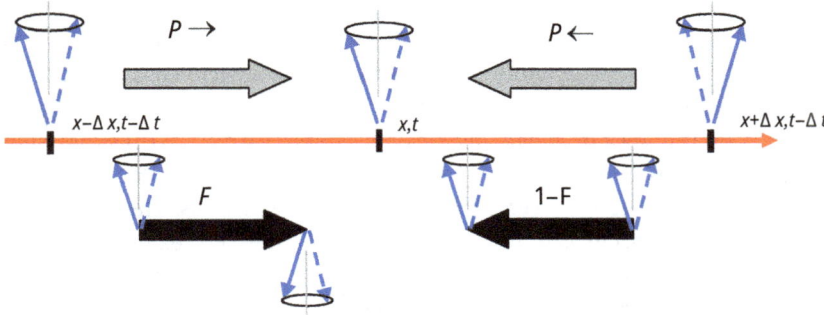

Fig. 1.5: Stochastic nature of spin random-walk transfer in space and time (one-dimensional model). Explanations for the symbols used are provided in the text above.

The next step in our approach is to include the spin transport quantitatively in the random-walk terms given above. This can be achieved by the introduction of the spin density or spin imbalance $n_\sigma = n_{up} - n_{down}$ (cf. Fig. 1.3). Now, the amount of quantities is somehow doubled since during spin transport from place to place, a spin can flip to the opposite orientation with a given probability F. Thus, writing adequate equations for up and down spins separately, we get

$$n_{up}(x,t) = P_{\rightarrow} \cdot (1-F) \cdot n_{up}(x - \Delta x, t - \Delta t) + P_{\leftarrow} \cdot (1-F) \cdot n_{up}(x + \Delta x, t - \Delta t) +$$
$$+ P_{\rightarrow} \cdot F \cdot n_{down}(x - \Delta x, t - \Delta t) + P_{\leftarrow} \cdot F \cdot n_{down}(x + \Delta x, t - \Delta t) \tag{1.36}$$

and

$$n_{\text{down}}(x,t) = P_{\rightarrow} \cdot (1-F) \cdot n_{\text{down}}(x-\Delta x, t-\Delta t) + P_{\leftarrow} \cdot (1-F) \cdot n_{\text{down}}(x+\Delta x, t-\Delta t) +$$
$$+ P_{\rightarrow} \cdot F \cdot n_{\text{up}}(x-\Delta x, t-\Delta t) + P_{\leftarrow} \cdot F \cdot n_{\text{up}}(x+\Delta x, t-\Delta t)$$

$$(1.37)$$

where F is the probability of spin flipping, while $1-F$ is the probability of the spin not flipping. Now, calculating the difference between n_{up} and n_{down}, we obtain

$$n_{\text{up}}(x,t) - n_{\text{down}}(x,t) = P_{\rightarrow} \cdot (1-F) \cdot n_{\sigma}(x-\Delta x, t-\Delta t) + P_{\leftarrow} \cdot (1-F) \cdot n_{\sigma}(x+\Delta x, t-\Delta t) +$$
$$+ P_{\rightarrow} \cdot F \cdot n_{\sigma}(x-\Delta x, t-\Delta t) + P_{\leftarrow} \cdot F \cdot n_{\sigma}(x+\Delta x, t-\Delta t) =$$
$$= P_{\rightarrow} \cdot n_{\sigma}(x-\Delta x, t-\Delta t) + P_{\leftarrow} \cdot n_{\sigma}(x+\Delta x, t-\Delta t),$$

$$(1.38)$$

thus an equation exactly similar to the equation for the normal electron concentration $n = n(x,t)$. This important result enables the derivation of the spin-transport equation, which is, as a matter of fact, the equation of diffusion. This step will be shown more precisely in another chapter of this book.

The results obtained here are still immersed at the classical physics level where, for example, the spatial coordinates are continuous variables. Then, a problem of description related to a more atomistic point of view is slightly different. First, it is based on the notion of the probability of finding a given particle concentration n within a small region around a given point, that is $P(n)$. Second, it is based on the spin-dependent rate (number of events per second) of spin carriers $y_{\sigma}(n)$ expressing the transport of particles into and out of a region of interest. These rates, in a fully quantum formalism, substitute concentrations in classical equations. The concentrations are now hidden as variables in the rates and probabilities. For example, $y_{\sigma}(n)$ is the rate of tunneling through a barrier at the interface between two different materials. The reader should pay attention that the symbol $y_{\sigma}(n)$ and the others below have the subscript σ as the index, informing about the dependence of these phenomena on the spin. Thus, the quantum rate equation written equivalently to the classical case, for a given interface, can appear as follows

$$P(n) \cdot \left(y_{\sigma}^{\rightarrow}(n) + y_{\sigma}^{\leftarrow}(n) \right) = P(n-1) \cdot y_{\sigma}^{\rightarrow}(n-1) + P(n+1) \cdot y_{\sigma}^{\leftarrow}(n+1), \qquad (1.39)$$

where $y_{\sigma}^{\rightarrow}$ is the rate of electrons tunneling from left to right and $y_{\sigma}^{\leftarrow}$ is the rate of electrons tunneling from right to left, as the upper arrows show. Note that the equation counts single-electron events by the equivalent probabilities $P(n)$, $P(n-1)$, and $P(n+1)$.

The intrinsic property of the quantum rate equation is clearly revealed in the description of transport between different materials in a spintronics device, represented by the three spatially separated regions (the left lead, the island, the right lead, Fig. 1.4), where $P(n)$, $P(n-1)$, and $P(n+1)$ are now probabilities of finding the adequate number of excess electrons moving through the two interfaces. In our classical case, the quantity $n_\sigma(x, t)$ was counted from the left $(x - \Delta x)$ and the right $(x + \Delta x)$. Here, a similar equation, expressed in the quantum style for the two-interface system, reads

$$P(n) \cdot \left(\overrightarrow{\gamma_{L\sigma}}(n) + \overleftarrow{\gamma_{L\sigma}}(n) + \overleftarrow{\gamma_{R\sigma}}(n) + \overrightarrow{\gamma_{R\sigma}}(n) \right) = P(n-1) \cdot \left(\overrightarrow{\gamma_{L\sigma}}(n-1) + \overleftarrow{\gamma_{R\sigma}}(n-1) \right) +$$
$$+ P(n+1) \cdot \left(\overleftarrow{\gamma_{L\sigma}}(n+1) + \overrightarrow{\gamma_{R\sigma}}(n+1) \right) \tag{1.40}$$

where $\overrightarrow{\gamma_{L\sigma}}$ is the rate of electrons tunneling from the left lead to the island, $\overleftarrow{\gamma_{L\sigma}}$ is the rate of electrons tunneling from the island to the left lead, $\overleftarrow{\gamma_{R\sigma}}$ is the rate of electrons tunneling from the right lead to the island, and $\overrightarrow{\gamma_{R\sigma}}$ is the rate of electrons tunneling from the island to the right lead. All the symbols are equivalent to the states possessing $n - 1$, n, or $n + 1$ electrons, respectively.

What is worth being emphasized is that the rates are directly related to quantum material parameters, such as a change in the electrostatic energy of the system due to single electron tunneling, tunneling probabilities, or the density of states at interfaces. This fundamental fact can be derived from the so-called Fermi golden rule (Dempsey et al. 2011). For example, for electrons tunneling from the left lead to the island, at the single interface (Fig. 1.5), the rate is equal to

$$\overrightarrow{\gamma_\sigma}(n) = \frac{2\pi}{\hbar} \cdot \rho_{L\sigma} \rho_{I\sigma} |t_{L\sigma}|^2 \cdot \frac{\Delta \overrightarrow{E_\sigma}(n)}{\exp\left(\Delta \overrightarrow{E_\sigma}(n)/kT\right) - 1}, \tag{1.41}$$

where $\rho_{L\sigma}$ is the density of states at the left lead side, $\rho_{I\sigma}$ is the density of states at the island side, $t_{L\sigma}$ is the tunneling coefficient, k is the Boltzmann constant, T is the struc-

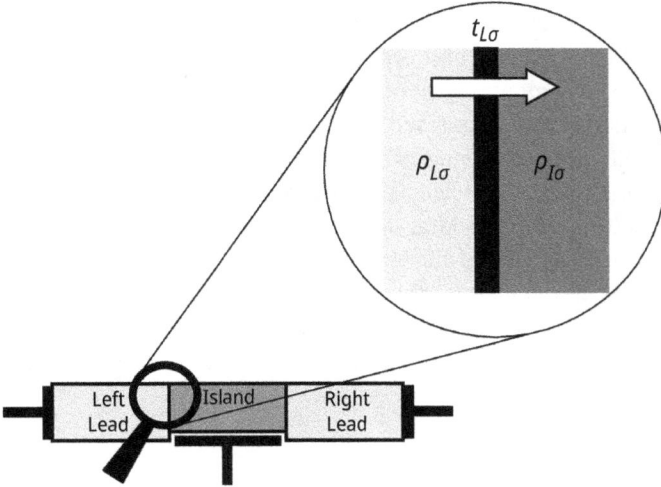

Fig. 1.6: A tunneling event at the interface between the left lead and the island depends on ρLσ – the density of states on the left lead side, ρIσ – the density of states on the island side, and tLσ – the tunneling coefficient.

ture temperature, and $\Delta E_\sigma^{\rightarrow} \rightarrow (n)$ is the energy change as the electron tunnels across the barrier, while on the island side there are n electrons. As mentioned before, the rate is logically dependent on n, the number of excess electrons transferred from one material to the other one (Fig. 1.6).

More details related to transport effects will be presented in the next chapters of this book, providing equations for rates and currents as well as some practical remarks about spin-dependent devices.

1.8 Magnetization and micromagnetism

This section stays somehow in contradiction to the previous ones, or more precisely, it introduces micromagnetism as the continuous, classical function of spatial variables with the most fundamental quantity being the magnetization vector. The magnetization is an average value of magnetic moments per unit volume,

$$\vec{M}(x,y,z) = \frac{\sum\limits_{i=1}^{N} \vec{p}_m^{(i)}}{V},$$ (1.42)

where $\vec{p}_m^{(i)}$ is the ith contribution to the magnetization vector $\vec{M}$ coming from an electron spin or from the electron's orbital movement, and V is the sample volume. Thus, the micromagnetic approach enables easy simulations of magnetization dynamics under the influence of external fields of variable orientation and intensity in materials. In laboratories, measurable magnetic effects are usually sensed by magnetization dynamics. Interchangeably, the vector $\vec{M}$ can be substituted by the magnetic polarization vector $\vec{J} = \mu_0 \cdot \vec{M}$ or by the normalized unit vector $\vec{m} = \vec{M}/M$.

Usually, for micromagnetic computations, four fundamental contributions to the sample energy are taken into account: one resulting from interactions between magnetization and the externally applied magnetic field, the Zeeman term E_{ext}; the exchange energy treated here at the macroscale as the isotropic background E_{exch}; the anisotropy component resulting from the interatomic coupling E_{ani}; and the demagnetizing energy caused by uncompensated magnetic moments at surfaces and edges E_{demag}. All contributions together constitute the total energy of the system E_{tot}, expressed in energy per volume unit. It is worth emphasizing that for the micromagnetic approach, the total effective magnetic field is the negative functional derivative of the total energy, namely,

$$H_{\text{tot}} = -\frac{\delta E_{\text{tot}}}{\delta \vec{J}},$$ (1.43)

and the time evolution of the polarization vector is the apparent function of the polarization vector and time

$$\frac{\partial \vec{J}}{\partial t} = f\left(\vec{J}, t\right).$$ (1.44)

Next, an elementary type of spin and magnetization motion is the rotation around the direction of the effective magnetic field – this is the so-called precession. The classical equation of the angular momentum $\vec{L}$ evolution, based on Newton's second law of dynamics in this case, is expressed by the time derivative of the angular momentum:

$$\frac{d\vec{L}}{dt} = \vec{\omega}_0 \times \vec{L},$$ (1.45)

where $\vec{\omega}_0$ is the angular velocity vector of the precession around the local field direction. However, since the magnetic moment is antiparallel to the angular momentum vector, $\vec{s} \sim -\vec{L}$, and the angular velocity of spin precession is proportional to the magnetic field $\vec{B}$ induction, $\vec{\omega}_0 = \gamma\vec{B}$, with γ being the electron gyromagnetic ratio, each spin-dynamics equation contains the term

$$\frac{d\vec{s}}{dt} = -\gamma\vec{s} \times \vec{B}.$$ (1.46)

Hence, the simplest equation of motion for the magnetic polarization dynamics can now be shown by inclusion of damping effects. The equation reads (Landau et al. 1935)

$$\frac{\partial \vec{J}}{\partial t} = -\frac{\gamma}{1+\alpha^2}\vec{J} \times \vec{H}_{\text{tot}} - \frac{\alpha\gamma}{J_s \cdot (1+\alpha^2)}\vec{J} \times \left(\vec{J} \times \vec{H}_{\text{tot}}\right).$$ (1.47)

This is the so-called Landau–Lifshitz–Gilbert equation, being commonly used alternatively to the similar Gilbert equation of the following shape (Gilbert 2004):

$$\frac{\partial \vec{J}}{\partial t} = -\gamma\vec{J} \times \vec{H}_{\text{tot}} + \frac{\alpha}{J_s}\vec{J} \times \frac{\partial \vec{J}}{\partial t}.$$ (1.48)

In both equations, the second term on the right is responsible for the precession damping. The damping is characterized by the phenomenological Gilbert damping constant α. More details about micromagnetic equations are provided in the third chapter.

To give the reader some flavor of the micromagnetic method used for simulations of small magnetic objects, graphical results for spatial distributions of magnetization are shown in Figs. 1.7 and 1.8 as well as for the demagnetizing field distribution in Fig. 1.9 for some nanosamples. The simulations were carried out using the MAGPAR simulator (Scholtz et al. 2003). Figures 1.8 and 1.9 were prepared using the ParaView software (Ahrens et al. 2005). In the figures, "snapshots" of the magnetization for a certain external magnetic field, swept with a defined speed along a distinct direction, are shown. Depicted by color codes and/or arrows, the orientations of the magnetization

(Figs. 1.7 and 1.8) in each separate point of the model are given. By integration over local magnetic moments, calculated on a 3D grid, the overall hysteresis loop can be calculated consequently.

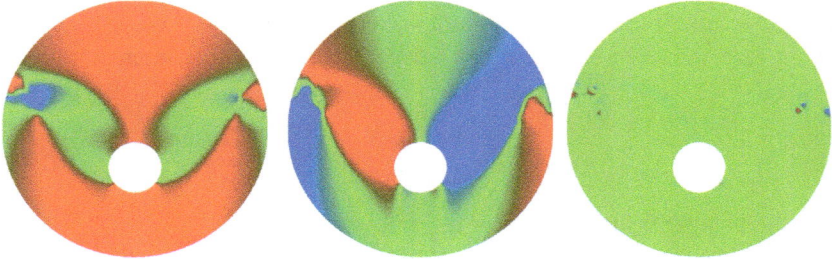

Fig. 1.7: Magnetization vector distributions on a flat nanodisk of 100 nm diameter, equipped with an asymmetrically located hole, at a given moment of magnetization evolution. The three figures show the three components of the magnetization vector M_x, M_y, and M_z, from left to right, respectively. The meaning of the colors is as follows: red – the component is parallel to the respective axis; blue – the component is antiparallel to the axis; and green – the component is perpendicular to the axis.

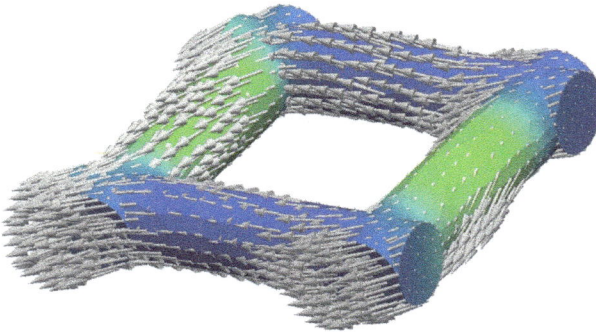

Fig. 1.8: Vector field of magnetization in a four-wire square sample (Błachowicz et al. 2013). The figure presents the so-called onion state of magnetization.

Micromagnetic simulations enable, in this way, not only analyzing "macroscopic" values, such as the hysteresis curve, but also identifying magnetic states and magnetization reversal processes, allowing for tailoring desired magnetic properties for different basic research and technical applications.

We hope this prolog-like chapter gave the reader some first impressions about the nature of spin as well as some hints of underlying physical effects.

More details about the micromagnetic approach, numerical requirements, and procedures in applications of spin-related effects will be provided in the next chapters of this book.

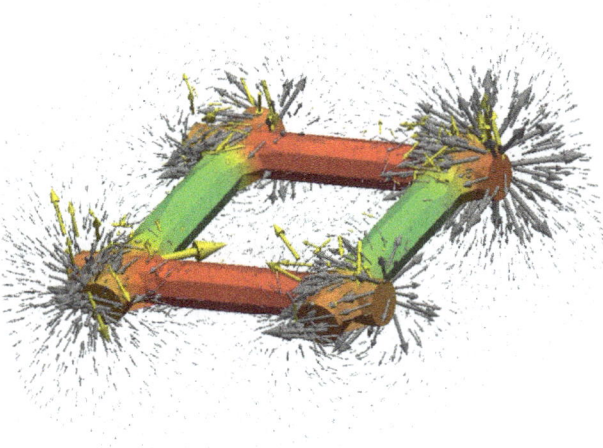

Fig. 1.9: Demagnetizing vector field distribution in a four-wire square sample obtained from micromagnetic simulations.

2 Physical effects in spintronic structures

2.1 Classical dipole interactions versus quantum exchange interaction

Forces between classical magnets can be described, among others, by their dipole–dipole interaction. The idea behind this description is quite simple: Magnets are – besides some very special cases – dipoles, without the possibility to form perfectly separated magnetic monopoles. It is well known that cutting a magnet with north and south pole results in two magnets with a north and south pole each. Imagining the magnetic fields around such a macroscopic magnet, they will form closed loops from the north pole through the open space outside the magnet and back to the south pole.

Making the magnet smaller and smaller will finally reveal the magnetic field around a magnetic dipole, which is defined as infinitesimally small, while its magnetic moment $\vec{m}$ is kept constant.

The forces between two magnetic dipole moments $\vec{m}_1$ and $\vec{m}_2$ at a distance $\vec{r}$ can be calculated as follows:

$$\vec{F_1} = \vec{\nabla}\left(\vec{m_1} \cdot \vec{B_2}\right) \tag{2.1}$$

with the force $\vec{F}_1$ working on the dipole moment $\vec{m}_1$ and the magnetic flux density $\vec{B}_2$ of the second dipole moment $\vec{m}_2$. The symbol ∇ defines the nabla operator, which is a vector consisting of partial derivations:

$$\vec{\nabla} = \left(\frac{\partial}{\partial x}, \frac{\partial}{\partial y}, \frac{\partial}{\partial z}\right) = \frac{\partial}{\partial x}\vec{e_x} + \frac{\partial}{\partial y}\vec{e_y} + \frac{\partial}{\partial z}\vec{e_z} \tag{2.2}$$

with the unity vectors $\vec{e_i}$.

Further evaluation of the aforementioned equations results in the following approximation of the force exerted on a dipole with moment $\vec{m}_1$ by a dipole with moment $\vec{m}_2$ (Yung et al. 1998):

$$\vec{F_1} = \frac{3\mu_0}{4\pi r^4}\left(\vec{m_1}\left(\vec{r}\cdot\vec{m_2}\right) + \vec{m_2}\left(\vec{r}\cdot\vec{m_1}\right) + \vec{r}\left(\vec{m_1}\cdot\vec{m_2}\right) - 5\vec{r}\left(\vec{r}\cdot\vec{m_1}\right)\left(\vec{r}\cdot\vec{m_2}\right)\right). \tag{2.3}$$

This means that the force is proportional to the product of the involved magnetic moments and inversely proportional to the distance between them to the fourth power. The overall force has vectorial components parallel to the magnetic moments $\vec{m}_1$ and $\vec{m}_2$ as well as to the distance $\vec{r}$. This approximation was shown to give results sufficiently near to the exact solution for not too small distances from the dipole, that is, in the far-field.

The dipole–dipole interaction has a quite practical application in nuclear magnetic resonance spectroscopy to distinguish geometrical shapes of molecules or of the

https://doi.org/10.1515/9783111383736-002

distances between different molecules by comparing the measured coupling forces with theoretically calculated values.

While this classical approach to calculate forces between magnetic dipoles is easily understandable, switching from macroscopic magnets to electrons or other quantum mechanical particles results in a completely new and not directly intuitive effect. First, it must be mentioned that such particles are subdivided into bosons and fermions, with the first having spins of integer numbers (0, 1, 2, . . .) and the latter having half-integer spins (1/2, 3/2, 5/2, . . .). Electrons have spin ½ and thus belong to the fermions. For them, the Pauli exclusion principle is valid, stating that it is not possible for two or more identical fermions (i.e., fermions with identical spin) to be in the same state at the same time. More precisely, two electrons in the same atom must differ at least in one of the following quantum numbers: the principal quantum number n, the angular momentum l, the magnetic quantum number m_l, or the spin quantum number m_s.

This means, on the other hand, that an exchange interaction must occur between completely identical fermions. This interaction is no force in the common sense, such as the electro-magnetic, weak or strong force, since there are no force carriers involved.

Integrating the spin into the electron wave functions, overall wave functions are gained either including a symmetrical orbital wave function and an antisymmetric spin wave function or vice versa. Depending on the sign of the exchange constant that defines the energy difference between both possible overall wave functions, the exchange energy can either favor parallel spin orientation – supporting ferromagnetism – or antiparallel spin orientation – supporting antiferromagnetism. It should be mentioned that the exchange constant and the overall wave functions are mostly based on electrical repulsion of the electrons and Pauli's exclusion principle, that is, on electrical interactions and not on magnetic ones – while the results of these interactions are collective magnetic phenomena. On the other hand, the above-described quantum effects cannot explain all ferromagnetic materials quantitatively, in some cases not even qualitatively. In metallic solids and other materials, special approaches for the electron wave functions have to be used, which depend on the orbits of the valence electrons. Some of them will be described in the following sections.

2.2 Double-exchange interaction

The double-exchange interaction can occur between ions with different oxidation states, that is, with different numbers of electrons. The oxidation state – or oxidation number – defines the number of missing electrons, as compared to the neutral atom.

The double-exchange interaction theoretically describes how easily an electron can be exchanged between two ions. In the first description by Zener (1951), it was used to show a quantitative relationship between the electrical conductivity and the

Curie temperature of a ferromagnet. The idea of the double-exchange interaction is as follows: When two ions with mixed valence are bonded via an intermediate O^{2-} anion, for example, Mn^{4+} and Mn^{3+} in Mn–O–Mn, electrons can move between the metal ions. This is energetically easier when during this process the electron can keep its spin. Since this possible delocalization of electrons reduces the overall energy, neighboring ions are preferred ferromagnetically coupled if the exchange splitting between both spin states is large enough.

While the double-exchange interaction is known to be the basis for ferromagnetic order in several materials, it can also create an antiferromagnetic state under special circumstances. Ono and Ishihare (2017), for example, recently found that a normal metallic ferromagnet could be photoexcited into an antiferromagnetic state, with the switching time between both states being correlated with the light amplitude and wavelength. This antiferromagnetic interaction that could only be detected for a nonequilibrium electron distribution was attributed to the double-exchange interaction. Similarly, Palii et al. (2020) showed that antiferromagnetic spin alignment could be produced by double-exchange interaction in a tetramer model-system $d^2 - d^2 - d^1 - d^1$ where two electrons were shared over four spin cores. On the other hand, Hortensius et al. (2023) found a photo-induced macroscopic ferromagnetic moment in usually antiferromagnetic $CaMnO_3$ which was attributed to carrier-mediated double-exchange between the Mn ions in this manganite.

On the other hand, double-exchange models have been extended from manganites for which they were first established to several other systems. Recently, Hussein et al. (2018) created a model for a Monte-Carlo simulation of copper-based superconductors with high critical temperature. By replacing the original strong Coulomb repulsion between copper d electrons with an exchange coupling between itinerant electron spins and localized spins at the copper lattice positions and adding a small antiferromagnetic coupling between the latter, they could reproduce several experimental features of doped and undoped systems.

In the so-called diluted magnetic semiconductors, for example, aluminum arsenide doped with transition metal (vanadium, chromium, or manganese) impurities, different spin states can show different physical character. In the low-doped system $Al_{0.9375}(TM)_{0.0625}As$, the majority spins show metallic behavior, while the minority spins behave like those in a semiconductor. This half-metal shows perfect spin polarization of 100%, which makes it well suitable for diverse spintronics applications, in combination with ferromagnetism due to double-exchange coupling (Boutaleb et al. 2018).

Similarly, ZnTe doped with chromium was found to show half-metallic behavior, while the most stable state in the system is ferromagnetic. Depending on the amount of chromium, the spin polarization at the Fermi level was near 100%. The Curie temperature, the ferromagnetic ordering temperature, was higher than 800 K. In this system, the ferromagnetic order was also attributed to double-exchange interaction (Goumrhar et al. 2018).

Another half-metallic ferromagnet is CrO_2, which was investigated in the form of a nanorod powder (Seong et al. 2018). Here, the mixed-valence states are Cr^{3+} and Cr^{4+}, giving rise to the double-exchange interaction. Half-metallic ferromagnets were also found in Janus FeXY (Li et al. 2021) or Janus Mn_2XSb monolayers (Ma et al. 2022) as well as half-Heusler alloys (Behera et al. 2022).

It should be mentioned that the double-exchange interaction is excluded in systems like $Fe_3O_4{}^+$ or $FeFe_5O_7{}^+$ so that here the super-exchange dominates the exchange interaction. The super-exchange describes the strong coupling between next-to-nearest neighbor cations via a nonmagnetic anion (Anderson 1950, Goodenough 1955). The super-exchange is typically antiferromagnetic, but can become ferromagnetic if the next-to-nearest neighbor ions are oriented at 90° with respect to the nonmagnetic anion between them (Weihe and Güdel 1997).

In some systems, combinations of the (ferromagnetic) double-exchange interaction with the (antiferromagnetic) super-exchange interaction have to be used to explain the experimental findings. Lu et al. (2018), for example, investigated rare-earth nickelates with a noncollinear magnetic ground state and site-dependent magnon excitations and could model both effects by taking into account a strong competition between the aforementioned interactions. Similarly, Apostolov et al. (2018) attributed the magnetic properties of $La_{1-x}Sr_xMnO_3$ nanoparticles to the competition between the ferromagnetic double-exchange and the antiferromagnetic super-exchange interaction. Such nanoparticles could be used for magnetic hyperthermia treatment or drug delivery to tumor cells.

In the next section, another possible interaction between the electrons inside solid materials will be described.

2.3 The RKKY interaction

The RKKY interaction is a coupling mechanism of the localized d- and f-shell electron spins in metals. This exchange interaction is intermediated by the conduction electrons and works over relatively large distances.

First, the RKKY interaction was proposed by Ruderman and Kittel (1954) as an indirect exchange interaction to explain unexpectedly broad nuclear spin resonance lines in silver. Kasuya (1956) suggested a similar mechanism for the coupling of inner d-electron spins, before Yosida (1957) set up a calculation of all interactions that had to be taken into account. The RKKY interaction is thus named according to the four researchers who developed this theory.

Interestingly, the mathematical description of the RKKY interaction contains a term oscillating between positive and negative coupling, that is, between ferromagnetic and antiferromagnetic interaction, depending on the distance between the involved electrons. This effect was indeed found in experiments dealing with the giant magnetoresistance, which will be described in detail later (Parkin and Mauri 1991).

The RKKY interaction can be found in many different materials. A special case occurs in Nd-doped ZnO where the exchange interaction was found to be based on the hole carriers instead of the electrons, resulting in significantly increased magnetic properties for doping conditions supporting the RKKY interaction (Hou et al. 2018). The same effect was found in ultrathin ferromagnetic $CrTe_2$ films (Wang et al. 2024). In V-doped ZnO, however, the RKKY interaction is based on electrons, as in most other systems (Mamouni et al. 2018).

Recently, the RKKY interaction was also often used to explain the effects in 2D or 1D materials. Phosphorene, for example, a 2D hexagonal lattice of phosphorus atoms, can be used to create zigzag nanoribbons. Opposite to graphene and other 2D materials, here the so-called edge modes occur, which are completely isolated from the bulk states. If two magnetic impurities are now positioned on the nanoribbon, they can be used to examine the edge states. The distance between the impurities only slowly changes the sign of the bulk RKKY interaction, while the sign of the edge modes oscillates much faster with varying distance (Islam et al. 2018). In hole-doped phosphorene, the relative orientation of the impurities was shown to strongly influence the RKKY interaction, with the maximum values for the impurities oriented along the zigzag direction and the smallest ones for the armchair direction (Zare et al. 2018). In graphene nanoribbons in armchair geometry, a strong dependence of the RKKY interaction on doping was found, as compared to metallic samples, as well as an increase of the effect with the ribbon width (Hoi and Yarmohammadi 2018).

Such direction-dependent effects are visible in most solid-state materials due to their crystal structure, outer shape, external influences, and so on. These so-called anisotropies are described in the next section.

2.4 Examples of anisotropy effects in low-dimensional objects

Ferromagnetic materials typically have several anisotropies describing the magnetic properties along different orientation. In a crystal, the crystal axes are typically either "easy" axes along which the magnetization is preferably oriented, or "hard" axes along which the magnetization avoids being oriented due to energetic reasons. Figure 2.1 shows some axes in a cubic crystal.

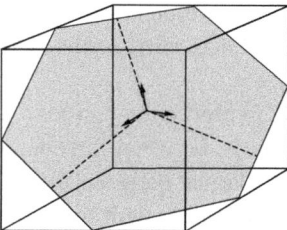

Fig. 2.1: Crystal axes in a cubic crystal. The light-gray plane is a (111) plane, the upper and lower square planes are (001) planes, and the sides are (100) and (010) planes, respectively.

While the exact definition of the plane or the axis – for example, (100) or (010) or (001) plane – depends on the chosen coordinate system, from the crystallographic perspective these plane and the corresponding normal vectors are identical. This can be shown by the notation of the axes: While [100] means the direction along the positive x-axis, <100> denotes all equivalent directions along positive or negative x-, y-, or z-axis.

Depending on the magnetic material, the easy axes can be found in different orientations. For the antiferromagnet CoO, for example, the spins are slightly canted with respect to the (111) planes (cf. Fig. 2.1) and oriented in the (110) planes (in Fig. 2.1, the intersection lines of the depicted (111) plane with the (110) planes are marked by dotted lines) (Miltényi 2000). In magnetite, <111> is the easy magnetization direction, while <100> is the hard direction of the magnetization.

This impact of the crystal orientation on the magnetic properties is called "magnetocrystalline anisotropy."

A completely different anisotropy occurs when external mechanical influences act on a magnetic material. The "magnetoelastic" anisotropy is related to the spin–orbit coupling and the resulting magnetostriction. This effect describes the dependence of the anisotropy constants on an externally applied strain that changes the dimension of the crystal in one direction, resulting in a change of the magnetization. The inverse effect also occurs: By magnetizing a sample, its dimensions are changed.

In small magnetic elements, there is another anisotropy becoming more and more important with decreasing sample size, the "shape anisotropy." Generally, the shape anisotropy tries to orient the magnetic moments in a sample along the borders of this sample to minimize stray fields. From the physical point of view, demagnetization fields occur due to surface charge distributions, working against the magnetization that produces them. The demagnetization fields are thus responsible for the reduction of stray fields.

The shape anisotropy is typically the dominant anisotropy for samples smaller than some micrometers, partly up to 10 or 20 μm, depending on the respective material's saturation magnetization. This means that for typical spintronics elements, the shape anisotropy is always important and in most cases the dominating anisotropy.

To give the reader a certain feel for the influence of shape anisotropy in small magnetic elements, the following sections provide an overview of magnetization reversal processes and corresponding magnetic hysteresis curves for low-dimensional objects such as nanodots, nanowires, and thin films. All simulations were performed using the micromagnetic simulator Object Oriented Micro Magnetic Framework (Donahue and Porter 1999) and working with typical parameters of iron.

2.4.1 Dots (0D)

Magnetic nanodots are of high technical interest because they can be produced relatively easily, opposite to shapes with corners, and the magnetic moments in them can create the so-called vortex states, which nearly completely eliminate stray fields. The latter is of technological importance since without stray fields, the magnetic state of one nanodot does not influence the magnetic states of neighboring dots. This enables using close-packed dot arrays to store information in the single dots.

A vortex state can be imagined like a swirl, with the magnetization being oriented along the edges of the dot either clockwise or counterclockwise, and in the middle pointing "up" or "down," like in a twister. This state is usually quite stable, meaning that it can be taken even without the necessity to switch the magnetic field to values with opposite signs than the previous saturation, and that it often takes relatively large fields to destroy this state and reach opposite saturation.

Figure 2.2 depicts a typical magnetization reversal process of a relatively large dot with a diameter of 400 nm and a height of 5 nm as a basis for comparison (images loosely based on Ehrmann and Blachowicz 2019). Starting with no. 1 in positive saturation (i.e., with the magnetic field oriented from the left to the right), the external magnetic field is decreased to zero (no. 4) and swept further to negative values (i.e., now pointing from the right to the left). For small negative magnetic fields (no. 5), a sudden change in the magnetization state occurs, until in no. 6, a vortex is formed. Here, the arrows clearly show that the magnetization in the nanodot is oriented clockwise around the middle of the vortex, parallel to the edges. With more and more negative external field, this vortex moves to the top, resulting in more magnetic moments being oriented along the external magnetic field (here indicated by the blue areas). Finally, when the vortex nearly reaches the border of the nanodot, it is dissolved, and the magnetization in the whole dot is oriented approximately from right to left (no. 9) until for negative saturation (not shown here), most magnetic moments are oriented parallel to the external magnetic field. Only along the borders, the strong shape anisotropy still counteracts the external magnetic field, as can also be recognized in no. 1 for positive saturation.

While these "snapshots" of the magnetization inside the nanodot during magnetization reversal can easily be understood and clearly depict the processes occurring here, a quantitative description of the magnetic properties of a system under examination can better be given by its hysteresis loops and the coercive field, which are easily accessible by measurements. These properties are depicted in Fig. 2.3. Here, the x-axis shows the external magnetic field, measured in milli-Tesla (mT), while the magnetization M inside the sample is depicted on the y-axis, scaled by the saturation magnetization M_{sat}.

The longitudinal hysteresis loop (black line) shows how the magnetization behaves with respect to the orientation of the external magnetic field. A value of approximately 1 corresponds to positive saturation (all magnetic moments in Fig. 2.2 oriented

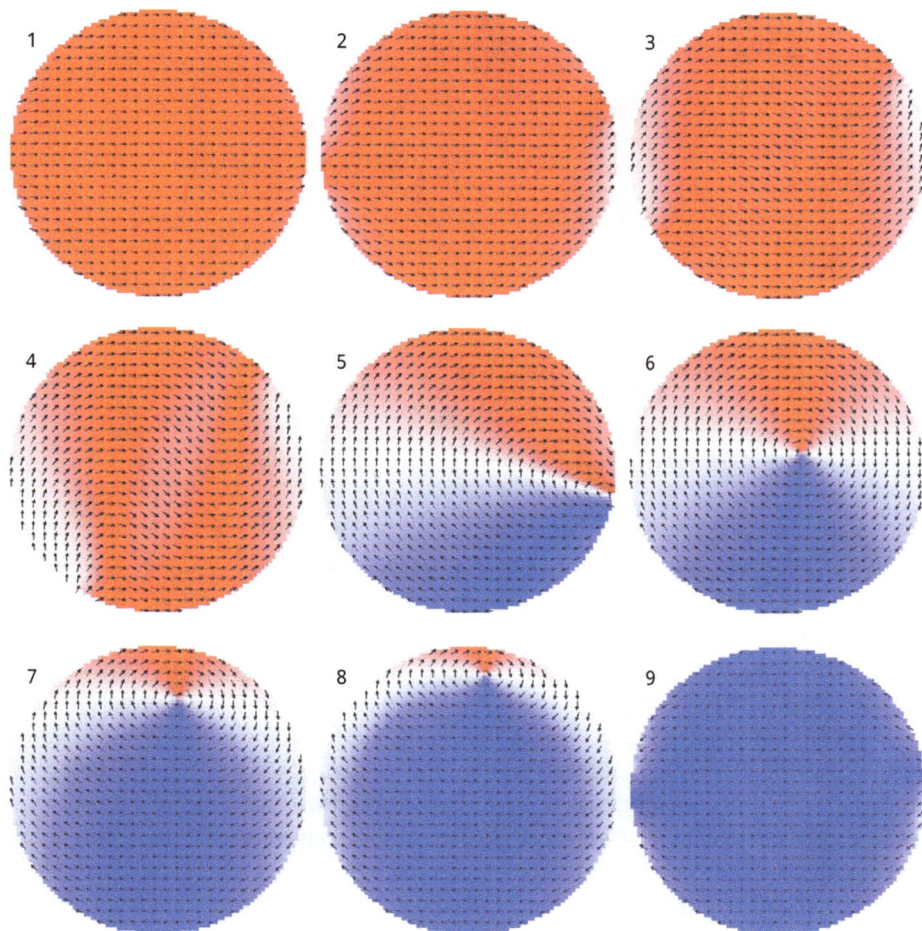

Fig. 2.2: Magnetization reversal process in a nanodot of dimension 400 nm × 400 nm × 5 nm from positive to negative saturation.

to the right side) and a value of −1 to negative saturation, respectively. Starting from positive saturation, the "jump" in the longitudinal hysteresis loop at about −2 mT corresponds to the nucleation of the vortex visible in Fig. 2.2 (no. 5), while the second "jump" at about −52 mT occurs when the vortex vanishes again.

While the longitudinal hysteresis loop could be calculated from the "snapshots" of the magnetization in the dot, as depicted in Fig. 2.2, by summing over all magnetic moments in the sample along the magnetic field direction, the transverse hysteresis loop corresponds to the overall magnetization perpendicular to the magnetic field direction. This loop is depicted in Fig. 2.3 by a red line.

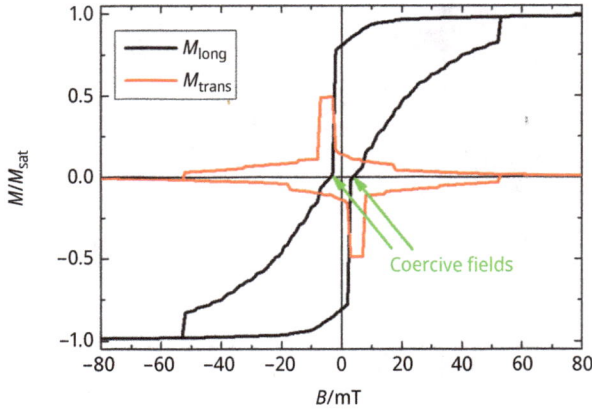

Fig. 2.3: Longitudinal and transverse hysteresis loops of the magnetization reversal processes depicted in Fig. 2.2.

Opposite to the longitudinal loop, the transverse hysteresis is approximately zero in saturation. Large values occur only between approximately −2 and −8 mT, which is the situation depicted in Fig. 2.2 (no. 5). Combining both hysteresis loops, several conclusions can be drawn about the magnetization reversal process; however, the "snapshots" in Fig. 2.2 still give more information. In an experimental investigation, nevertheless, such snapshots are very often not available, and the longitudinal and – sometimes – transverse magnetization components are the only sources that can be used as a base for the analysis of the whole magnetization reversal process. This is one of the reasons why micromagnetic simulations, as the one used here, are often supportive in investigating new systems.

Another important property that is often used to define magnets is the coercive field, as depicted by the green arrows. Usually, the hysteresis loops are symmetrical around zero, and the absolute values of both coercive fields are identical and defined as *the* coercive field. In special systems, this symmetry may be broken, and left and right coercive fields are different. Depending on the system and the reason for the asymmetry, sometimes an average coercive field is defined in such cases, together with the shift of the hysteresis loop. One possible reason for such a symmetry break is the so-called exchange bias, which will be described more in detail in Chapter 5.

After this short introduction, Fig. 2.4 shows a comparison of hysteresis loops, simulated for nanodots of height 5 nm and four different diameters. The black line is identical to the one in Fig. 2.3. Reducing the diameter from 400 to 200 nm leads to a strong increase in the coercive fields and at the same time a different shape of the hysteresis loop. Decreasing the diameter again to 100 or 50 nm results in another shape of the hysteresis loop, this time again with much smaller coercive fields.

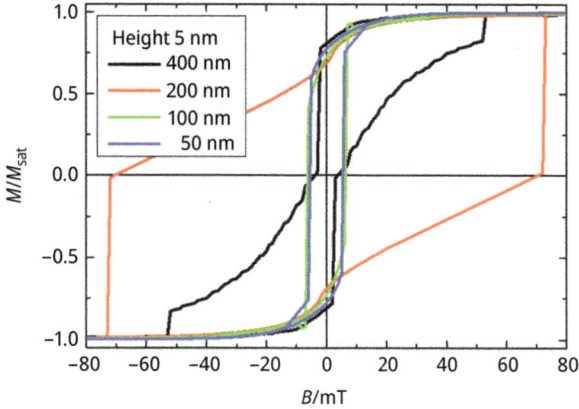

Fig. 2.4: Longitudinal hysteresis loops of iron nanodots with a height of 5 nm and different diameters.

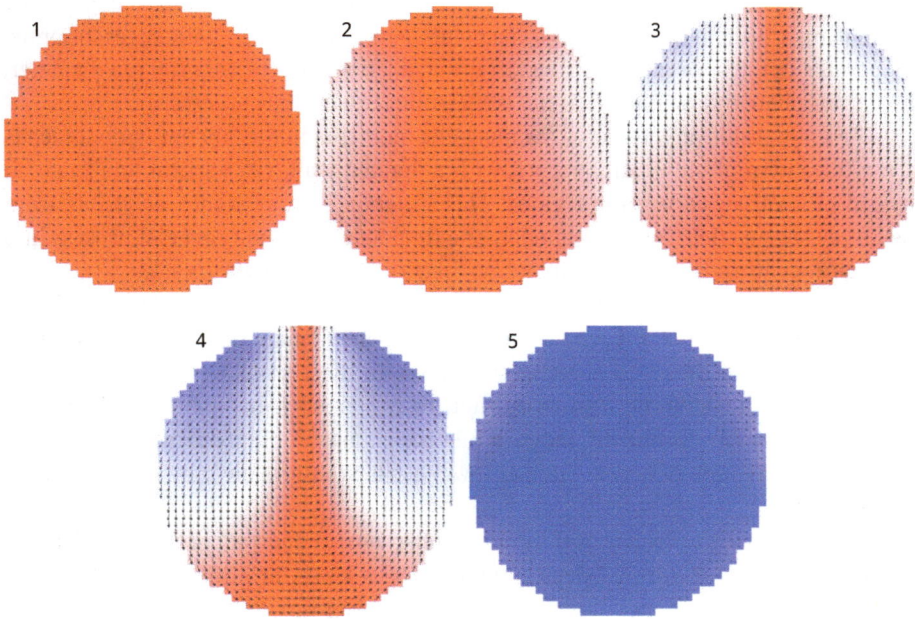

Fig. 2.5: Magnetization reversal process in a nanodot of dimension 200 nm × 200 nm × 5 nm from positive to negative saturation.

To investigate the reasons for these changes, Figs. 2.5 and 2.6 depict the snapshots of the magnetization in the nanodots with diameters of 200 and 100 nm, respectively.

For a nanodot diameter of 200 nm (Fig. 2.5), something completely different happens compared to the 400 nm nanodot depicted in Fig. 2.2. Here, the first approach to

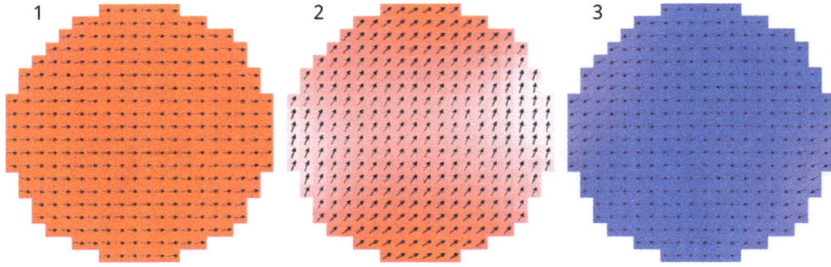

Fig. 2.6: Magnetization reversal process in a nanodot of dimension 100 nm × 100 nm × 5 nm from positive to negative saturation.

start magnetization reversal shows antiparallel magnetic moments at the left and right edges of the dot (no. 2). Conversely, in the larger nanodot, both edges have parallel magnetic moments (Fig. 2.2, no. 4). The difference is based on the minimization of stray fields, which is energetically favorable and can be supported in the larger nanodot by introducing a third area with canted magnetic moments between both edges, oriented antiparallel to the edge magnetization. In the smaller nanodot of only 200 nm diameter, the distance between both edges is not sufficiently large for this process.

The antiparallel alignment of the magnetic moments in the 200 nm nanodot leads to the formation of only half a vortex, with a domain wall (red vertical line in no. 4) instead of a vortex core being created. This situation is quite stable, as visible from the large coercive field for this nanodot. The final magnetization reversal process into the negative saturation occurs without the formation of a real vortex state.

In the smaller nanodots, the situation is different again. Here, as depicted in Fig. 2.6 for a diameter of 100 nm, the whole particle builds one domain that first rotates coherently and then switches into negative saturation. In these small dots, no vortices are created, either. It should be mentioned that the hysteresis loops for the smallest particles have the most "common" shape, while the shapes of the loops in which vortex nucleation or creation of a stable domain wall occur are normally only found in nanoscale magnets.

After testing the influence of the dot diameter, next the impact of the dot thickness will be investigated. Figure 2.7 depicts the hysteresis loops for an increased dot thickness of 20 nm; Fig. 2.11 shows the corresponding loops for dots of 50 nm thickness.

The first point attracting attention in Fig. 2.7 is the large number of small "steps" in the hysteresis loops for the dots with diameters between 200 and 50 nm. For the dot with a diameter of 400 nm, instead, there is a small jump around ±30 mT. In addition, all curves start magnetization reversal now before reaching zero magnetic field. To understand the reversal processes better, Figs. 2.8–2.10 show snapshots of the magnetization again.

Figure 2.8 depicts the magnetization reversal of the nanodot with diameter 50 nm and thickness 20 nm. Opposite to the thinner nanodot with an identical diameter,

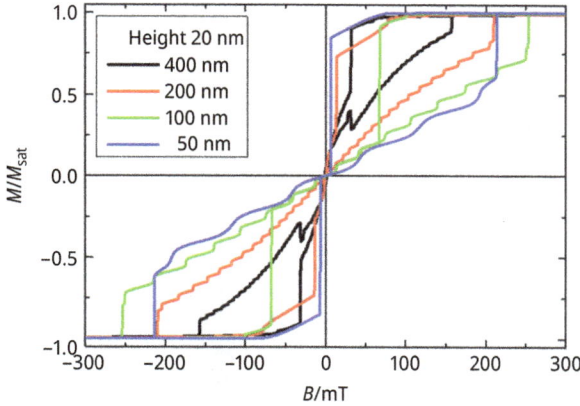

Fig. 2.7: Longitudinal hysteresis loops of iron nanodots with a height of 20 nm and different diameters.

here the reversal process occurs via vortex nucleation and propagation. Apparently, in the thicker dot, creating such a "point-like domain wall" is energetically favorable as compared to the coherent rotation of the complete magnetization. This may be attributed to the fact that in the thinner dot, the vortex core – which has to be oriented perpendicular to the vortex plane, that is, to the sample plane – is strongly suppressed by the shape anisotropy, while in the thicker dot, the energy reduction by forming the vortex throughout the whole sample thickness is larger than the necessary energy to overcome the shape anisotropy at the top and bottom layer.

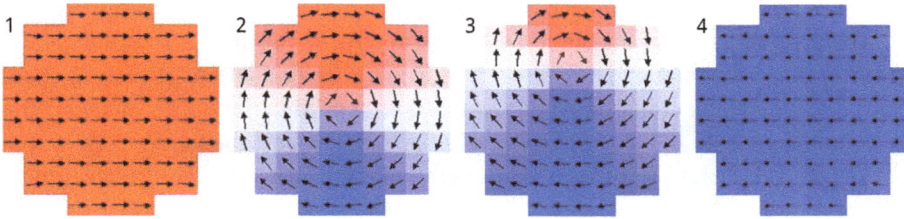

Fig. 2.8: Magnetization reversal process in a nanodot of dimension 50 nm × 50 nm × 20 nm from positive to negative saturation.

For a diameter of 100 nm, the process looks approximately identical. Figure 2.9 depicts the magnetization reversal for the dot with a diameter of 200 nm.

Here, something new happens: First, opposite to the thinner dot with diameter 200 nm (Fig. 2.5), the magnetization along the left and right borders is oriented parallel (nos. 2 and 3). Opposite to the dot with a diameter of 400 nm and a height of 5 nm (Fig. 2.2); however, this does not result in the formation of one vortex core at the right side, but in the creation of two vortices with opposite precessional orientation (clock-

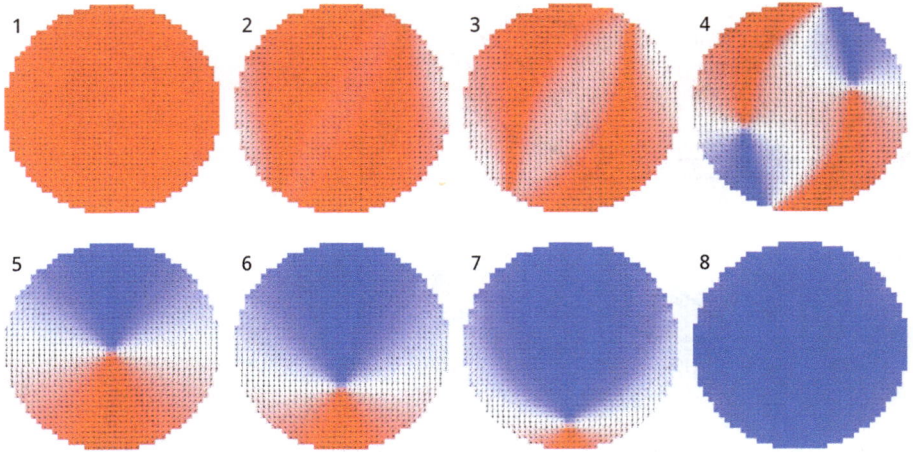

Fig. 2.9: Magnetization reversal process in a nanodot of dimension 200 nm × 200 nm × 20 nm from positive to negative saturation.

wise for the lower left vortex and counterclockwise for the upper right one). This state does not stay stable for long, but jumps into the usual state with one vortex core (no. 5), which propagates until it vanishes.

The small field range in which two vortex cores occur is hard to recognize in the respective hysteresis loop as shown in Fig. 2.7. For the magnetization reversal process from positive to negative saturation, it can be found between approximately +13 and +1 mT, that is, after the first clear jump, but before the reversible area starts. Tests show that the whole range in which the small "oscillations" are visible is reversible – by sweeping the external magnetic field back and forth within this field range, the vortex core is moved "up" and "down," with the little oscillations depicting small jumps of the core from one simulated cell to the next. Within this range, changing the external magnetic field is completely reversible. These reversible vortex core propagations can usually be identified from hysteresis loops since they have zero net magnetization for vanishing external magnetic field, that is, the hysteresis loops traverse the point of origin.

For the largest dot with a diameter of 400 nm and thickness of 20 nm, something new happens again (Fig. 2.10). Starting with the two vortex cores that were also established in the dot with diameter 200 nm (no. 3), this double-core structure here does not directly vanish, but stays stable for a broad field range. The cores start "dancing" (or in more physical terms: preceding) around each other (no. 3–no. 9), in this way allowing a larger area between them to rotate magnetization and align it with the (now negative) external magnetic field. Only in step no. 10, a jump into the usual vortex state occurs, with the core moving to the upper edge (no. 11) and finally dissolving to reach the negative saturation state.

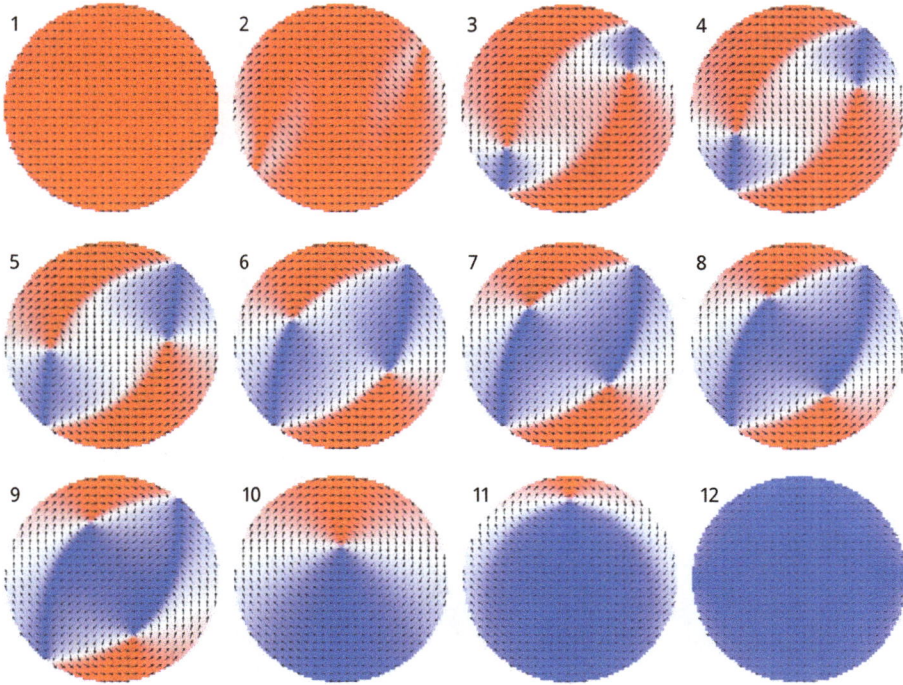

Fig. 2.10: Magnetization reversal process in a nanodot of dimension 400 nm × 400 nm × 20 nm from positive to negative saturation.

The visible "jump" in the corresponding black hysteresis loop shown in Fig. 2.7 around −16 mT corresponds to the switch from two vortex cores to only one vortex.

Finally, Fig. 2.11 shows hysteresis loops simulated for a dot thickness of 50 nm. It should be mentioned that for the smallest diameter, the thickness and diameter are equal.

The three smaller nanodots show hysteresis loops similar to the ones identified as magnetization reversal by vortex nucleation and propagation in Fig. 2.7, with a broad field range of reversible processes around the point of origin and the two separate hystereses that are typical for vortex phases. Investigating the snapshots again indeed shows the expected processes.

Only for the largest dot, a two-step magnetization reversal is visible. Comparing this curve with the ones in Fig. 2.7, this reversal process may be based on the nucleation of two vortex cores first, which afterward merge into only one vortex when the second jump of the magnetization segues into the reversible area. To test this idea, Fig. 2.12 depicts the snapshots of the magnetization reversal in the dot with a diameter of 400 nm and a height of 50 nm.

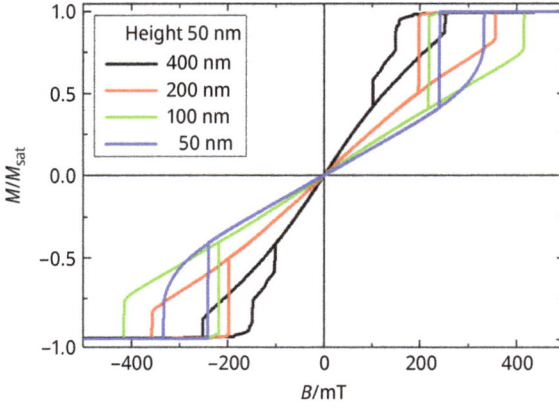

Fig. 2.11: Longitudinal hysteresis loops of iron nanodots with height 50 nm and different diameters.

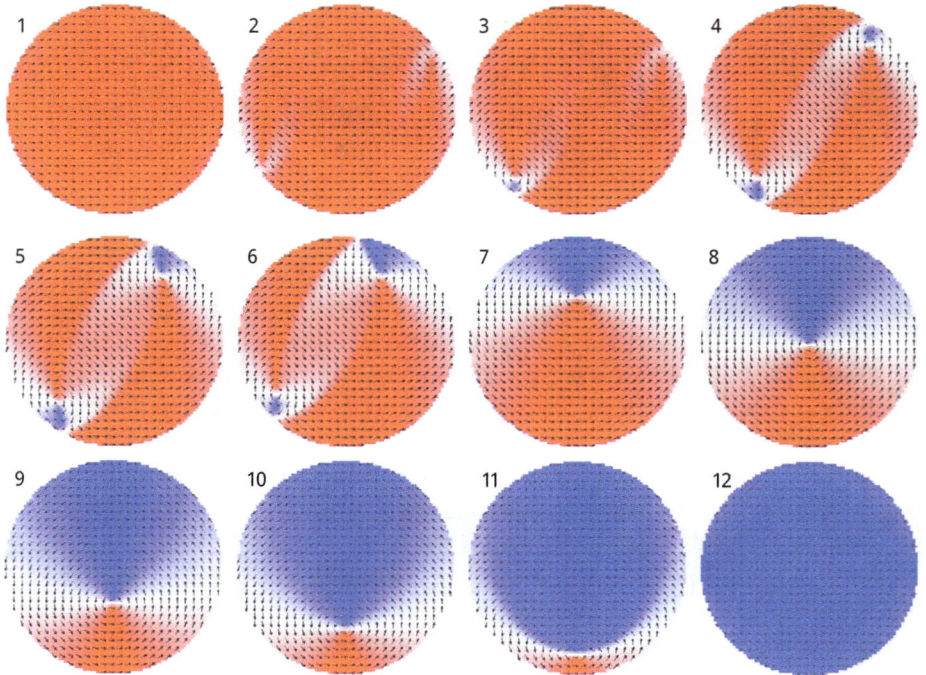

Fig. 2.12: Magnetization reversal process in a nanodot of dimension 400 nm × 400 nm × 50 nm from positive to negative saturation.

As expected from the hysteresis loop, Fig. 2.12 clearly shows that indeed the magnetization reversal process starts with the nucleation of two vortex cores (nos. 4–6). Opposite to the situation in Fig. 2.10, but similar to the magnetization reversal depicted in Fig. 2.9, this double-core structure is not stable for long, but soon switches into a single-vortex state without previous precession of both vortex cores around each other.

A more detailed overview of magnetization reversal processes in round and square nanodots can be found in the recent works by Ehrmann and Blachowicz (2018, 2019). After this first introduction of possible magnetization reversal processes in nanodots and about how to interpret hysteresis loops of these systems, the next section will introduce possible magnetic states in nanowires of different cross-sections and lengths-to-diameter ratios, oriented in different angles with respect to the external magnetic field.

2.4.2 Wires (1D)

Wires can be imagined as strongly extended dots, typically with round (in case of wires growing from bottom to top through a template or in a nonmagnetic matrix) or square cross-sections (in case of wires growing parallel to the substrate, for example, by sputtering or molecular beam epitaxy (MBE)). Since they can be grown in different ways, starting from template-assisted or lithographic methods to self-organized processes to textile methods like electrospinning, they are of high technological interest. Besides, they can be used in diverse applications, such as biomedicine (Mukhtar et al. 2020), energy storage, microwave electronics, catalysis (Moreno et al. 2021), sensors, and actuators (Alam et al. 2020).

Magnetic nanowires can show various magnetic properties, depending on material and dimensions. In polycrystalline gadolinium nanowires of 20–100 nm prepared by direct current electrodeposition, for example, a mixture of ferromagnetic and superparamagnetic clusters was found (Khan et al. 2018). Creating nickel nanowires with a wet-chemical process (hydrazine hydrate reduction method) in the presence of Tween 80 as a capping agent resulted in pure nickel nanowires without undesired oxygen, showing soft ferromagnetic properties (Deepa and Therese 2018). Magnetron sputtering was used to prepare multishell $Ni_{80}Fe_{20}$ nanowires with Cu separating the magnetic shells, showing giant magneto-impedance, which could be tailored by changing the anisotropy in the magnetic shells (Lv et al. 2018). In CoNi and CoNiFe nanowire arrays, a vortex domain wall could be created for tailoring the iron content (Noori et al. 2018). In Ni, Fe, and Co nanotubes of average diameter 250 nm obtained by chemical electrodeposition in porous polycarbonate membranes as templates, no specific reversal modes were found to be dominant, opposite to results of simulations (Paulo et al. 2018).

As this short introduction shows, the simulations depicted here cannot give a comprehensive overview of all possible magnetization reversal processes, but aims at letting the reader get a feel for possible reversal processes in dependence of the wire morphology. It should be mentioned that here only wires with square cross-section are depicted, while round wires may partly show different magnetization reversal processes, and the effect of bending (Blachowicz and Ehrmann 2018) or interconnected nanofiber networks (Blachowicz et al. 2020, Blachowicz et al. 2021) is ignored since recently most nanowires are grown more or less linearly.

Figure 2.13 depicts hysteresis loops simulated for nanowires of 400 nm length and different diameters. Opposite to the hysteresis loop simulated for the nanodots, here the curves loop mostly like squares. Only for the largest diameter, a small step is visible.

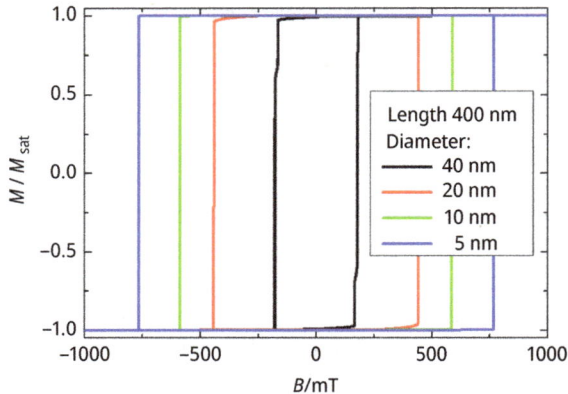

Fig. 2.13: Longitudinal hysteresis loops of iron nanowires with length 400 nm and different diameters.

Comparing the snapshots of the magnetization reversal process, the different magnetization reversal processes can be identified. Figure 2.14 shows the reversal of the nanowire with diameter 20 nm; Fig. 2.15 shows the reversal of the nanowire with diameter 40 nm.

Fig. 2.14: Magnetization reversal process in a nanowire of dimension 400 nm × 20 nm × 20 nm from positive to negative saturation.

While in both situations, magnetization reversal starts with a rotation of the magnetic moments at both ends of the wire. An intermediate state is created afterward only in the thicker wire, consisting of two vortices with opposite rotational orientation, thus minimizing stray fields. This state corresponds to the small step in the 40 nm loop in Fig. 2.13. In the thinner wires, no such vortices or other structures can be created during magnetization reversal.

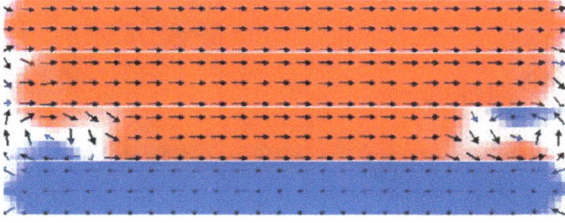

Fig. 2.15: Magnetization reversal process in a nanowire of dimension 400 nm × 40 nm × 40 nm from positive to negative saturation.

After comparing wires with identical length and different diameters, the length-to-diameter ratio is kept constant, and the wire dimensions are increased or decreased proportionally. Figure 2.16 depicts the results.

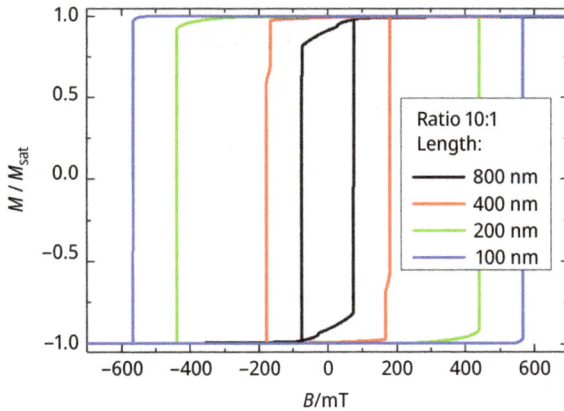

Fig. 2.16: Longitudinal hysteresis loops of iron nanowires with length:diameter ratio 10:1 and different lengths.

The curve for 400 nm is already known; the hysteresis loops of the smaller systems do not offer any interesting features. Apparently, magnetization reversal occurs here again in one step, without the nucleation of a domain wall or a vortex. The largest nanowire, however, shows a new feature with two small steps, suggesting possibly interesting reversal processes. Figure 2.17 shows the snapshots of magnetization reversal in this nanowire.

Again, magnetization reversal starts with a rotation of the magnetic moments at both ends, leading to vortex formation. Similar to the sample with half the dimensions, in the next step, a second vortex is created with a rotational direction opposite to the one at the respective end. Finally, the complete magnetization reversal happens

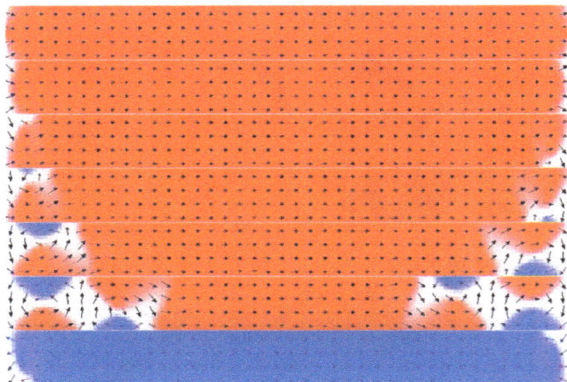

Fig. 2.17: Magnetization reversal process in a nanowire of dimension 800 nm × 80 nm × 80 nm from positive to negative saturation.

by dissolving the vortices and switching into a state similar to negative saturation, with only the magnetization at both ends of the wire still being slightly rotated.

Opposite to the investigations of the nanodots, in a nanowire the orientation of the external magnetic field may play an important role. Figure 2.18 thus shows the hysteresis loops simulated for a magnetic field orientation of 45° with respect to the wire axis.

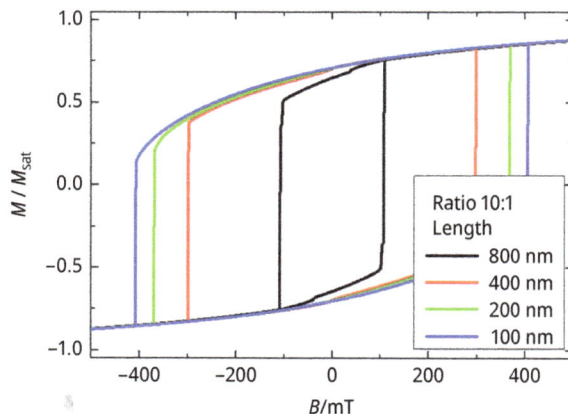

Fig. 2.18: Longitudinal hysteresis loops of iron nanowires with length:diameter ratio of 10:1 and different lengths, simulated for an angle of 45° between the external magnetic field and the wire axis.

While the hysteresis loops have a different shape now since they are depicted along the external magnetic field, that is, also under an angle of 45° with respect to the wire axes, no new interesting features are visible. The snapshots of the magnetization reversal process reveal that for 800 nm, there are again two vortices, while for 400 nm length, only one vortex is created, and in the smaller wires, no such intermediate states occur.

It should be mentioned again that the features shown here are not typical for all materials. As described at the beginning of this chapter, the magnetic materials as well as possible crystal orientations inside the nanowires will strongly influence the widths for which vortices or other forms of domains can be built. This statement is also valid for the next section in which thin layers will be investigated.

2.4.3 Layers (2D) and spatial systems (3D)

Thin films are a typical form of magnetic samples since they can be produced relatively easily by different techniques. Thin layers of Ni–Mn–Sn with Cu/Fe/Co substitutions, for example, were grown by magnetron sputtering and afterward annealed, resulting in a highly crystalline ordered phase (Modak et al. 2018). Besides magnetron sputtering, which is very often used nowadays, normal sputtering can also be applied to create magnetic thin films (Huminiuc et al. 2018).

When pulsed laser deposition is used to prepare magnetic thin films, not only the substrate and the deposition temperature but also the laser wavelengths strongly influence the composition and structure of the grown materials, thus affecting the magnetic properties (Oujja et al. 2018).

Besides other diverse possibilities to produce thin films, molecular beam epitaxy (MBE) allows for growing epitaxial thin films, that is, films with a defined crystal orientation and monolayer by monolayer (O'Hara et al. 2018; Uchitomi et al. 2018, Saha et al. 2022).

Plasma-assisted chemical vapor deposition was used to prepare a thin film of TiO_2 that after reduction to TiO_x (with $x < 2$) became ferromagnetic (Ren et al. 2018). Besides chemical vapor deposition, it is also possible to use chemical solution deposition to create polycrystalline, phase-pure magnetic films (Zhang et al. 2018). Another chemical method that is especially suited for oxide thin films is polymer-assisted deposition, resulting in epitaxial thin films from diverse materials (Vila-Fungueirino et al. 2018, Ren et al. 2023).

Simulating thin films has a major disadvantage: They have a very high ratio of the spatial dimensions with respect to the thickness, that is, they should be simulated as being extended in x- and y-direction to not only some micrometers but also some millimeters – which makes simulations last unreasonably long if normal personal computers are used. Here we show only one simulation of a thin film of thickness 5 nm and spatial dimensions 1 μm × 1 μm. Figure 2.19 depicts a series of snapshots for the magnetization reversal from positive to negative saturation.

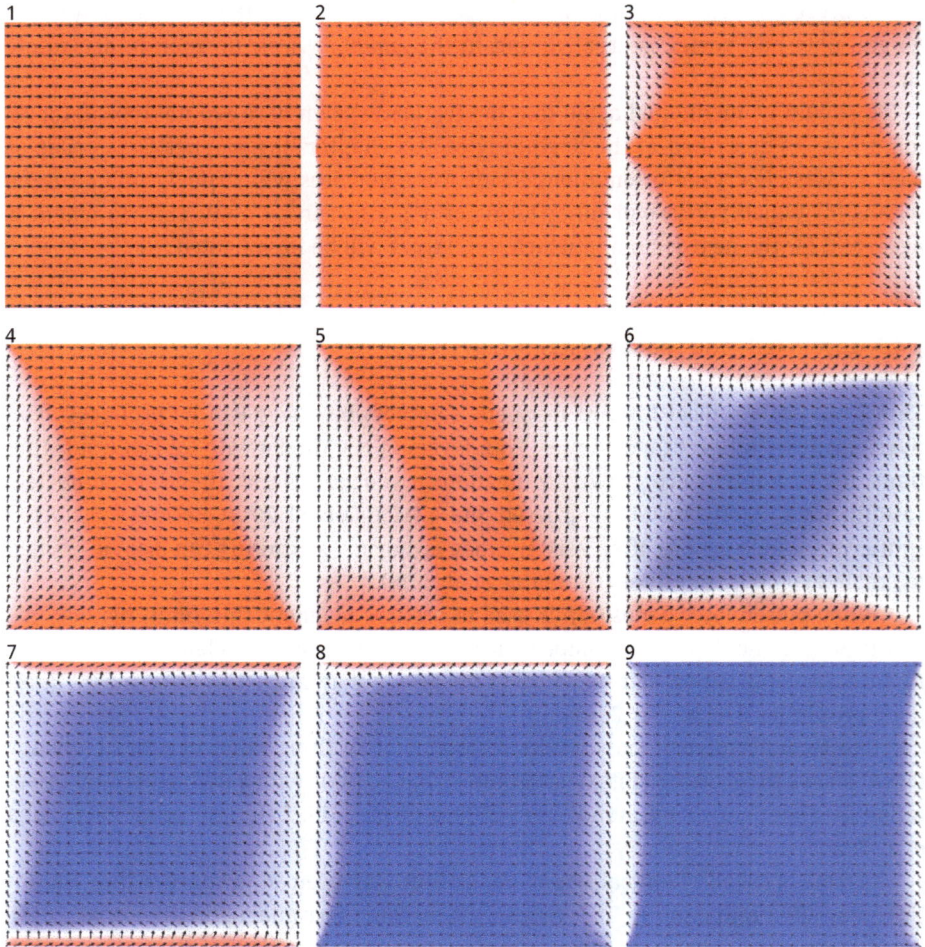

Fig. 2.19: Magnetization reversal process in a thin film of dimension 1,000 nm × 1,000 nm × 5 nm from positive to negative saturation.

First, magnetization reversal starts with the usual canting of the magnetic moments at the edges that are perpendicular to the external magnetic field, following the shape anisotropy, to reduce stray fields (no. 2). This process proceeds from the edges into the inner part of the thin layer (no. 3) until in the next step a more global energy minimum is found (no. 4).

Developing this situation further, finally the middle area between the 90°-canted areas along the left and the right edges switches its magnetization (no. 6). While the situation can still be described as one large meander, the main direction of this meander has now changed. With a further increasingly negative field, the areas along the upper and the lower border with the "wrong" magnetization direction get thinner and thinner (no. 7), until the first of them reverses (no. 8) and finally also the second one (no. 9). Approaching negative saturation, the magnetization at the borders will be oriented along the external magnetic field opposing the shape anisotropy in these regions.

It should be mentioned that the example depicted here is only one possibility among a broad variety, including coherent rotation of the complete magnetization with or without stable intermediate states, domain wall nucleation and propagation, formation of domains in different numbers and shapes, and so on. In a realistic thin film with much larger spatial dimensions than used for the simulation here, the shape anisotropy along the edges would be more or less negligible, while only the shape anisotropy due to the thin layer would play a role. In addition, the crystallinity plays an important role, inducing additional anisotropies. The simulation shown here can thus only give the first idea of how magnetization reversal processes in thin films may work.

2.5 Magnetization dynamics in low-dimensional magnetic systems

In nanodots or small disks, the most interesting and typically examined dynamics are those of a vortex core. Such gyrotropic dynamics or precessions of vortex cores are, for example, influenced by out-of-plane magnetic fields, with the frequency of this gyrotropic mode being nonlinearly dependent on the external magnetic field (Fried et al. 2016). Vortex dynamics can be influenced by an AC magnetic field or an electrical current, both enabling switching the vortex core under suitable conditions – which allows for using the polarity of vortex states, that is, the vortex core pointing "up" or "down" as "bits" in data storage (Gaididei et al. 2008).

Theoretical examinations based on the Landau–Lifshitz–Gilbert equation showed that with increasing disk thickness, additional exchange-dominated gyrotropic "flexure" modes occurred with nodes along the disk thickness. Such 3D eigenmodes were found to be important for the explanation of experimental findings in thick permalloy disks. Additionally, the eigenfrequencies decreased with increasing disk thickness (Noske et al. 2016). Similarly, Ding et al. (2014) investigated relatively thick circular permalloy nanodots with ferromagnetic resonance without applying an external magnetic field. They found flexure oscillations of the vortex core string with 0–2 nodes along the dot thickness, with the $n = 1$ mode sometimes showing higher intensity than the uniform $n = 0$ mode. Flexure oscillations of the vortex core in permalloy nanodots of different thickness also showed that for thicker nanodots, the $n = 1$ mode had a higher intensity than the $n = 0$ mode (Kakazei et al. 2020).

For small dot dimensions, Lyberatos et al. (2011) found meta-stable vortex states in micromagnetic simulations, which spiraled toward either the center or the borders of a dot, depending on the initial position.

The dependence of the resonance motion of vortices in nanodots with round or parallel-epiped shape was investigated by Kim et al. (2015). They found an increase in the gyrotropic frequency with increasing nanodot thickness and could also show that the aspect ratio of the dot influenced the oscillation frequency.

Extending these examinations from pure magnetic nanodots to trilayer nanopillars from ferromagnetic/nonmagnetic/ferromagnetic layers, vortices, or single domains were found as ground states, depending on nanopillar radius and thickness. In this system, the magnetostatic interlayer coupling was shown to influence the magnetic ground state in both ferromagnetic layers. The stability of the vortices was additionally found to depend on the spacer thickness (Sukhostavets et al. 2012). In an earlier study, Aranda et al. (2010) used a spin-polarized current perpendicular to the dot layer to shift the vortex to the border of the dot and let it vanish, which was followed by oscillations of the single domain state and afterward the returning of the vortex with opposite polarization of the core. The authors underlined the strong dependence of the gyrotropic motion of the vortex core on the simulation conditions, especially the chosen damping and the cell sizes used in the micromagnetic model. Comparing single-layer and trilayer permalloy nanodots of diameter 120 nm, Kuchibhotla et al. (2021) found two separated FMR modes near the saturation field for single-layer nanodots as well as a high-frequency center mode and low-frequency edge mode, while the trilayer nanodots showed two center modes with the spins in the ferromagnetic layers precessing in-phase or out-of-phase.

Besides these investigations of magnetization dynamics in single nanodots, diverse research groups also focused on the collective behavior of nanodot arrays. While such arrays are natural in experimental investigations, they need significantly increased calculation times in micromagnetic simulations, which is why they are usually limited to relatively small amounts of nanodots. Nevertheless, they are highly interesting due to the possibility of using them as elemental switches in magnetic memories or nanomagnetic logic gates (Madami et al. 2018).

In arrays of permalloy dots with a diameter of 50 nm and different spacing, Rana et al. (2011) found one dominant edge mode for large distances and an increase of damping and precession frequency with decreasing interdot distances as well as a jump in damping combined with a mode splitting, when the interdot distance became identical with or smaller than the dot diameter.

Mondal et al. (2018, 2016), for example, investigated the collective behavior of nanodot arrays in honeycomb or octagonal forms in experiment and theory. They varied the distances between the dots with diameter 100 nm and thickness 20 nm between 30 and 390 nm and found significant changes in the magnetization dynamics, probed with time-resolved magneto-optical Kerr microscopy. While for the largest distance, the dots showed isolated precessional dynamics, they preceded collectively for the smallest distance. Additionally, different precessional modes were found for both lattice types. These effects were attributed to the dipolar interaction between neighboring magnetic moments.

Using permalloy ($Ni_{80}Fe_{20}$) nanodots of two different alternating diameters in an array, collective modes were found that depended on the applied magnetic fields and field directions, with new interaction modes becoming visible in addition to the original modes of the small and the large nanodots (De et al. 2017).

The influence of a temperature gradient on the magnetization dynamics of an NiFe nanodot array was investigated by vector network analyzer measurements (Asam et al. 2018). Here the temperature gradient resulted in modulations of the ferromagnetic resonance frequency, stemming from the damping-like and the field-like components of the spin transfer torque.

Using a multivector formalism of magnetization dynamics, Lisenkov et al. (2016) modeled semi-infinite and finite magnetic nanodot arrays. The edge modes occurring along the borders of the array as well as along potential domain walls between different array regions could be simulated. In this way, they could calculate the ferromagnetic resonance of a triangular nanodot array, the spin-wave edge modes at both sorts of aforementioned borders, and the influence of the modes along the internal domain walls on the ferromagnetic resonance spectrum of a nanodot array in a chessboard AFM ground state.

Different dot shapes were investigated by Mahato et al. (2015). They found two ferromagnetic resonance modes in arrays of round nanodots and even up to four modes in square, diamond-shaped, and triangular dots. Additionally, the symmetry of the lattice and the array boundaries influenced the results. Depending on the dot shape, different rotational symmetries were found, which could be attributed to different collective modes, based on varying internal magnetic fields and stray fields. Interestingly, the collective behavior was strongest in the round dots in which the smallest stray fields could be expected, and weakest in the diamond-shaped dot arrays. In an earlier study, they found even eight modes in a lattice of triangular dots including a broad frequency band (Mahato et al. 2014).

After these 0D magnetic systems, next we will focus on the 1D systems, that is, magnetic nanowires. Such nanowires can be interpreted as geometrically confined regions in which domain walls can be trapped. The geometric position of a domain wall in a magnetic nanowire strongly depends on the geometry and the material properties. For well-adjusted properties, domain walls can be kept oscillating with a single defined frequency, depending only on the DC current density in the nanowire, in this way enabling DC/AC conversion and the creation of a nano-oscillator with defined frequency (Sbiaa et al. 2018).

Permalloy nanowires with continuous width modulation were investigated by Xiong et al. (2017b). Opposite to the single defined frequency mentioned earlier, here three to five different resonance modes were found, with their number increasing with increasing wire diameter. These modes could be described by a standing spin-wave quantization. Furthermore, the coercive fields measured along the wire axis were significantly increased in comparison with nanowires of identical, but homogeneous thickness.

Herranen and Laurson (2017) have shown by micromagnetic simulations that the velocity of the domain wall propagation in nanowires of approximately rectangular cross-section is proportional to small external fields, while for larger fields additional internal degrees of freedom occur in the domain wall, leading to a sudden decrease in the velocity. These threshold fields as well as the mechanism of higher degrees of freedom depend on the thickness of the strips. For thick strips, horizontal Bloch lines occur in the domain walls, while for thin strips either vertical Bloch lines occur, or the domain wall magnetization rotates uniformly. For an intermediate regime, circulating Bloch lines along the domain wall perimeter were found.

Generally, domain walls even in relatively simple nanowires can be of complex nature with complicated dynamic evolution, as Hayward and Omari (2017) have shown in permalloy nanowires experimentally and by micromagnetic simulations. Especially defect sites make quantitative predictions challenging, while general qualitative statements are possible with micromagnetic simulations.

The injection, pinning, and depinning of domain walls along a geometrical constriction in a permalloy nanowire were also investigated by Mohanan and Kumar (2017). Interestingly, they found different depinning fields for vortices with opposite chiralities. Correspondingly, the pinning probability at the constriction also depended on the chirality. This result underlined the stochastic nature of pinning and depinning at such a constriction, opposite to the reliable deterministic switching, which is usually desired.

In several applications, it is necessary to move domain walls through a nanowire to transport information in magnetic logic devices. Unfortunately, higher velocities of the domain wall motion result in instabilities in the domain wall and thus a sudden decrease in the velocity, the so-called Walker breakdown. To understand this effect better, and finally to find solutions to overcome this problem, Vandermeulen et al. (2016) used a semianalytical ansatz to model the internal degrees of freedom in such a domain wall, resulting in the definition of different variables describing the domain wall as well as the opposing and reinforcing effects on the domain wall.

A new possibility to move domain walls along a defined path instead of having them oscillate was recently suggested by Liu et al. (2017). Opposite to the usual technique of applying an external magnetic field along the magnetization of a magnetic domain, they used a perpendicular field pulse to move domain walls in magnetic nanostrips. By rotating the external field with a suitable frequency, the perpendicular orientation can be obtained, allowing for moving the domain wall continuously with maximum velocity.

Transverse field pulses were also suggested by Li et al. (2017). They used uniform transverse magnetic fields to increase the domain wall speed. At the same time, however, this resulted in a twisting in the azimuthal distribution of the transverse domain wall, which is not desired for high-density nanowire devices. This is why they realized a fully planar transverse domain wall with only arbitrary tilting in a biaxial nanowire in which the polar and the azimuthal degree of freedom are fully decoupled.

Another result was found by Kuncser et al. (2017) investigating soft magnetic Ni–Cu nanowires with a high aspect ratio. They observed nucleation and propagation of corkscrew domain walls for an external magnetic field applied along the wire axis, but a reversible coherent spin rotation in case of a magnetic field perpendicular to the wire axis. Applying the magnetic field at an arbitrary angle – with the axial component being large enough for domain wall nucleation – resulted in more complex superposed reversal mechanisms.

Besides domain walls, vortices are also an important aspect here. Magnetization reversal typically starts at the ends with the formation of vortex domains, that is, vortices with the core oriented along the cylinder axis. The vortex nucleation fields depend strongly on the angle between the wire axis and the external magnetic field (Mehlin et al. 2018).

In nanotubes, the length of the domain walls formed at the ends depends on the aspect ratio and the anisotropy constant – for relatively short vortex domain walls in comparison to the whole nanowire length, the magnetization in the middle flips uniformly, while for larger vortex domain walls, these couple and build a Néel domain wall in the middle of the nanotube due to the opposite rotation directions of both vortex domain walls, before the complete magnetization is suddenly aligned along the external magnetic field as soon as the switching field is reached, and the Néel domain wall collapses (Chen et al. 2018).

Another important aspect, influencing domain wall dynamics in nanowires, is the temperature. While micromagnetic simulations are typically carried out at 0 K, Broomhall and Hayward (2017) performed them at a finite temperature. Taking into account the doping of permalloy nanowires with rare-earth elements such as holmium, they found that even small amounts of holmium were sufficient for domain walls to maintain defined magnetization structures during propagation and thus to avoid stochastic pinning or depinning effects, making the process of domain wall nucleation more reliable.

Ohmic heating, actually a problem in down-scaling attempts, can create thermal gradients that can be used to excite coherent auto-oscillations of the magnetization in nanowires. Safranski et al. (2017) investigated yttrium iron garnet ($Y_3Fe_5O_{12}$)/Pt bilayer nanowires and found that Ohmic heating of the platinum layer resulted in spin current injection into the $Y_3Fe_5O_{12}$ layer, resulting in the mentioned auto-oscillations and coherent microwave radiation generation. In this way, spin caloritronic devices for magnonics applications could be created. In the same system, Jungfleisch et al. (2017a) found field-like and antidamping-like torques influencing the precession of the magnetization, which was excited by spin-Hall effects. The spin-Hall effect describes the finding that spin accumulation of opposite sign occur on the opposite lateral surfaces of a specimen which carries an electric current. At high microwave powers and corresponding heating due to microwave absorption, leading to decreasing effective magnetization, they also observed a strong modification of the resonance field. The spectra also revealed quantized spin wave modes across the wire width, which depended on the wire geometry.

Another external impact on magnetization reversal in nanowires is given by the strain on the wire. This factor can be used to tailor the magnetic properties of the deposited nanowires. First calculations showed for cobalt wires grown on strained platinum or gold surfaces that the magnetic anisotropy increased under compressive strain and correspondingly decreased under tensile strain in the wire. In addition, a surface strain resulted in a decrease in the magnetic field necessary for switching the magnetization in the cobalt wire (Polyakov et al. 2017).

Similar to the trilayer nanopillars investigated by Sukhostavets et al. (2012), magnetic/nonmagnetic/magnetic trilayer nanowires were examined by Lupo et al. (2016). In relatively thin nanowires, they found a completely different spin-wave spectrum than in broader wires, depending on the interlayer coupling. This change of the magnetization dynamics was attributed to a competition of the shape anisotropy with the RKKY interaction and additional exchange interaction across the nonmagnetic spacer layer.

Similar to the aforementioned nanodots, nanowires can also be arranged in an array. One possibility to do so is the creation of the so-called artificial spin-ice, a network of ferromagnetic nanowires with geometrically frustrated magnetic domain structure. In a square spin-ice system, spin-torque ferromagnetic resonance measurements have revealed that the collective magnetization – especially the angular orientation of the vortex regions – in this system influences the magnetization dynamics as well as the magnetoresistive properties. In this way, reconfigurable microwave oscillators and magnetoresistive devices can be lithographically produced (Jungfleisch et al. 2017).

While magnetization dynamics in the 0D and 1D magnetic systems described before are mostly related to domain walls and vortices, a broader spectrum of possible magnetization dynamics will be found in the 2D systems described next.

Thin magnetic films can be grown by diverse techniques, as described earlier. The quality of their crystal structure – if the layer is not planned to be polycrystalline – and the surface roughness, however, depend not only on the growing technique but also on the substrate. Sometimes buffer layers or seed layers are first grown on the substrate before the magnetic layer is added. Akansel et al. (2018), for example, have investigated five different seed layers below $Fe_{65}Co_{35}$ magnetic thin films and found a significant influence of the chosen material on the coercivity of the magnetic films. In the magnetic properties, they found a decrease in the damping parameter, which is antiproportional to the lifetime of precessions of the magnetic moments.

In polycrystalline yttrium iron garnet thin films, a platinum buffer layer was used. In this way, the damping parameter could be reduced by approximately a factor of 3 for an yttrium iron garnet film of thickness 100 nm (Pati et al. 2017).

The damping parameter can also be modified by the layer thickness and a possible heat treatment. Akansel et al. (2018a) used ion beam sputtering of Co_2FeAl thin films with different thicknesses between 8 and 20 nm. While the saturation magnetization and the coercive field varied slightly with the film thickness, the damping parameter was antiproportional to the magnetic film thickness. These findings were attributed to two-magnon scattering and spin pumping into the nonmagnetic capping layer (which is usually applied only to avoid oxidation of the magnetic thin film).

In sputtered polycrystalline iron–gallium thin films between 20 and 80 nm, broadband ferromagnetic resonance measurements were used to detect spin-wave resonances perpendicular to an external magnetic in-plane field. The damping constants for films of 40–80 nm thickness were found to be approximately 0.04, which was similar to the values found in epitaxially grown layers, making the simpler sputtering process also feasible for industrial-scale production of thin FeGa films for microwave applications (Gopman et al. 2017).

A possibility to influence the damping factor without modification of the growth conditions or a thermal after-treatment of the samples is given by microwave irradiation. For permalloy grown on a SiO_2 substrate, relatively short irradiation times of a few minutes – which did not change the sample structure – could significantly reduce damping by smoothing the sample surface, while longer irradiation times destroyed the surface by cracks, resulting in oxidation of the permalloy layer and correspondingly severely increased damping factor (Azadian et al. 2017).

More complicated magnetic properties can be found in FeN thin films where weak stripe-domains can occur. Investigations of the dynamic properties showed a broad spectrum of dynamic eigenmodes that were influenced by the orientation of the external magnetic field parallel or perpendicular to the stripes (Camara et al. 2017).

In $Co_{49}Pt_{51}$ films of only 5 nm thickness, deposited at different temperatures between room temperature and 350 °C, the dynamic magnetic properties were measured by Brillouin light spectroscopy. The Brillouin light spectroscopy measures inelastic scattering between incident laser light and different waves in a crystalline lattice, that is, phonons (acoustic modes), polarons (charge displacement modes in dielectric materials), or magnons (magnetic oscillation modes). In magnetic materials, it is typically applied to detect spin waves. Here, magnetic modes were found that could be described by coexisting hard and soft magnetic areas in the sample. The perpendicular magnetic anisotropy that is typical for Co/Pt systems was shown to depend on the deposition temperature similar to the coercive fields (Abdallah et al. 2018).

In nearly the same system, CoPt, Polley et al. (2018) used time-resolved magneto-optical Kerr effect measurements to investigate the demagnetization. Here, the laser-based THz (terahertz) generation of larger magnetic fields may result in a very fast precessional magnetization reversal, an effect that is about one order of magnitude stronger in CoPt than in easy-plane anisotropy systems and which may be utilized for ultrafast switching.

Besides the precession of magnetic moments, the direct switching of the magnetization in a thin film is of large technological interest. One possibility to achieve coherent control of the magnetization is by ultrashort light pulses of terahertz frequencies. These light pulses can create strong electromagnetic fields in ferromagnetic layers. Using a ferromagnetic semiconductor thin film, terahertz responses were recently achieved, which was attributed to spin-carrier interactions in the ferromagnetic semiconductor (Ishii et al. 2018).

Light can also be used to trigger spin precessions in an external magnetic field. For this purpose, the so-called pump-probe experiments are performed in which the strong pump-pulse excites the precession and the significantly weaker probe pulse measures it using the time-resolved magneto-optical Kerr effect (cf. Chapter 4). Bonda et al. (2018), for example, investigated spin precessions in a ferromagnetic Heusler alloy. For this epitaxially grown thin-film system, they found damping parameters between 0.04 and 0.11, which is approximately one order of magnitude larger than found in a nearly identical, but stoichiometric Heusler alloy film (Dubowik et al. 2011).

This technique can also be used to investigate magnetization dynamics in multi-layer systems. Berk et al. (2018), for example, used the time-resolved magneto-optical Kerr effect to investigate the exchange coupling of thin layers of Fe and FePt over a Pt layer of varying thickness between 0 and 5 nm. They found that the exchange coupling started at an interlayer thickness of 1.5 nm and became maximum at 0 nm, with a ferromagnetic coupling between both layers in the exchange coupled regime.

Even two pump pulses with varying time between them were used by Shibata et al. (2018) to induce spin precession in permalloy thin films. The measured spin precession strongly depended on the time interval between both pump pulses and thus on the phase of the precession when the second pulse arrives. Here, thermal excitation can be recognized as the source of the oscillation, which was underlined by a numerical simulation.

Two pump pulses with varying helicity of the laser light were used by Medapalli et al. (2017) to trigger all-optical switching of ferromagnetic Co/Pt multilayer films with perpendicular anisotropy. They found that the first ultrashort laser pulse demagnetized the film significantly, independent of the orientation of the circular polarization of the light. The second pulse triggered domain wall propagation in the films in a preferred direction, which could be tailored by the orientation of the circular polarization. In this way, all-optical switching of the multilayer stack could be realized in a two-step process.

While such precessional switching experiments are typically carried out in CoPt or other thin film systems with perpendicular magnetic anisotropy, the question arises whether this technique can also be used in systems with in-plane anisotropy. Theoretical examinations have shown that this is in principle possible, but necessitates longer light pulses and better defined laser parameters. Especially for iron nanoparticles, the canting of magnetic moments – instead of the desired switching – could be attributed to the laser-induced spin–orbit torque. This clearly shows that spin switching in systems with perpendicular magnetic anisotropy is much more robust and thus reliable for technological applications (Zhang et al. 2017).

After this short overview of magnetic states in diverse magnetic materials and shapes, the next chapter will give an introduction to micromagnetic calculations that are typically used to understand magnetization dynamics especially in low-dimensional systems, as already used for the examples in this chapter.

3 Micromagnetic equations in magnetization dynamic studies

Dynamics of magnetization are a key issue in magnetic materials research. Magnetization is a vector quantity that evolves in time around a given point of material. The evolution can be analyzed both from the local and the global perspectives. Thus, taking into account the first, point-like approach, the evolution can be presented as the change of the magnetization vector in time, from the state at the moment t to the state at the infinitesimally later moment $t + dt$ (Fig. 3.1).

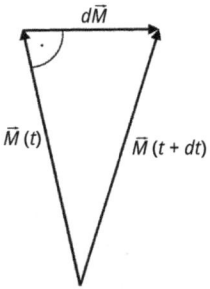

Fig. 3.1: The evolution of the magnetization vector $\vec{M}$, at a given point in space, during an infinitesimally small time interval dt. The infinitesimally small magnetization change vector $d\,\vec{M}$ indicates the direction of evolution and, importantly, is perpendicular to $\vec{M}$.

It is easy to apply now the following approximation, using the Taylor expansion series around the value $\vec{M}(t)$ in order to evolve to the next state $\vec{M}(t + dt) = \vec{M}(t) + d\vec{M}$ (cf. Fig. 3.1), namely

$$\vec{M}(t + dt) = \vec{M}(t) + \frac{\partial \vec{M}}{\partial t}\, dt, \tag{3.1}$$

leading us to the obvious conclusion that the change vector $d\vec{M}$ and the first time derivative $\partial\vec{M}/\partial t$ are directly proportional. This simple remark is of high importance from the mathematical formalism perspective met in differential equations describing magnetization evolution. This class of equations always has the quantity $\partial\vec{M}/\partial t$, for example, on the left side, and the physical reasons for the evolution on the right side. Taking advantage of this simple consideration, the meaning of four possibilities of the magnetization evolution is presented in Fig. 3.2. They are precession, damping, slowing of precession, and reversal. The physical reasons for the first two are as follows: the externally applied magnetic field $\vec{H}$ seen in the figure as the vector perpendicular to the page, and the interaction with the magnetic moment in a ferromagnetic sample, respectively. The latter two factors might result from the influence of the external magnetic moments, imposed by an external magnetic terminal, and transferred into the given material by a spin-polarized electron current. A more detailed discussion of all these factors will be provided later.

https://doi.org/10.1515/9783111383736-003

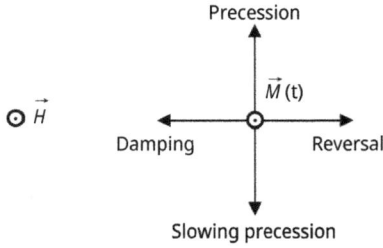

Fig. 3.2: The scheme of the main four physical factors influencing the magnetization vector $\vec{M}$ in the presence of the effective (inside ferromagnetic material) magnetic field $\vec{H}$.

At first glance, the equation of evolution should look as follows:

$$\frac{\partial \vec{M}}{\partial t} \sim \text{precession} + \text{damping} + \text{other factors}. \qquad (3.2)$$

This class of equations is named based on the spatial scale involved (local single moment vs. global material magnetization, and single material vs. multilayered structure): the Landau–Lifshitz (LL), the Landau–Lifshitz–Gilbert (LLG), or Landau–Lifshitz–Gilbert–Slonczewski equation.

Thus, the precessional movement is the fundamental type of dynamical behavior at the atomic scale. Also, looking at mesoscale effects and going beyond the microdimensions, a precession is something natural. What exerts precessions is the angular momentum vector from the mechanical point of view or the magnetic moment vector from the electrodynamical perspective, respectively. In this sense, both vectors are equivalent, and hence, both vectors are proportional to each other, and for both vectors, a precession means the occurrence of the time-dependent vector change $\partial \vec{M}/\partial t$. This change is enforced by the external magnetic field via the $(\vec{M} \times \vec{H})$ term. The precession is not everlasting – it is damped due to material factors. In one of the known forms met in the literature, which will also be discussed in detail a few pages later, the damping is expressed by a factor proportional to the doubled vector product $\vec{M} \times (\vec{M} \times \vec{H})$, where $\vec{H}$ is the external magnetic field intensity. Both the mentioned components are always perpendicular to $\vec{M}$ (Fig. 3.3).

The external field and the damping are not the only two factors influencing the precession. A very specific situation takes place in magnetoelectronic devices prepared, for example, in thin-layer technology. One of the layers, for example, an external terminal, can be made from ferromagnetic material kept all the time at saturation. Electrons passing through the ferromagnetic terminal are spin-polarized, and in this way, the magnetic moment, adopted from the ferromagnetic terminal, is transferred into the next, free magnetic layer, which is susceptible to magnetization evolution. This phenomenon is known as the spin-transfer torque (STT) effect since the electrons mediate (transfer) the magnetization state from one material to another. The effect is used to overcome damping and the effective field in order to reverse from one magnetization state to another. The STT effect can be analyzed from the microscopic perspective, which can provide the reader with much practical intuition. One important point is

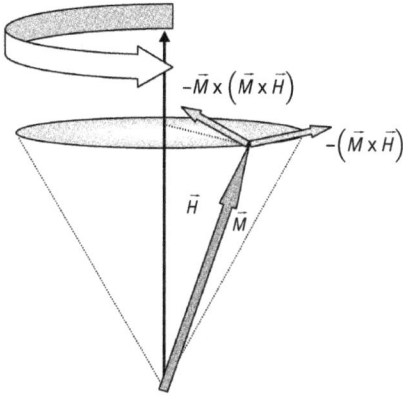

Fig. 3.3: A precession phenomenon. The two fundamental factors of magnetization evolution are shown: the precession $\vec{M} \times \vec{H}$ and the damping $\vec{M} \times (\vec{M} \times \vec{H})$. (original figure was rearranged)

that in the underlying mathematical expressions, the external magnetic field vector $\vec{H}$ is substituted by the fixed magnetization $\vec{M}_p$, resulting from an external terminal in two equivalent terms, similarly to the normal precession and the damping, namely $(\vec{M} \times \vec{H}) \rightarrow (\vec{M} \times \vec{M}_p)$ and $\vec{M} \times (\vec{M} \times \vec{H}) \rightarrow \vec{M} \times (\vec{M} \times \vec{M}_p)$, respectively. Of course, in order to put all these components to the right-hand side of eq. (3.1), the appropriate amplitudes should be introduced and their meaning should be clarified. Before that, let us introduce four basic facts correlated with the STT effect. They are as follows:

1) The STT is the vector $\vec{N}_S$ (where "S" stands for "Slonczewski," the name of the person who predicted this effect) and is another reason for the occurrence of $\partial \vec{M}/\partial t$ which, as already mentioned, is perpendicular to $\vec{M}$ at a given point inside the magnetic material (Fig. 3.2).
2) The $\vec{N}_S$ vector has two components, parallel and perpendicular to the plane (say to the page of this book, for simplicity), where both the vectors, the magnetization $\vec{M}$ of the free layer and the fixed magnetization $\vec{M}_p$ of the external ferromagnetic terminal, determine this plane at a given moment of time (cf. Fig. 3.3).
3) The parallel component is proportional to $\vec{M} \times (\vec{M} \times \vec{M}_p)$.
4) The perpendicular component is proportional to $\vec{M} \times \vec{M}_p$.

The overall effect can then be treated as the superposition of parallel and perpendicular effects.

In order to understand more deeply the mathematical terms listed in the last two points, there is a need to clarify first the simpler situation, meaning the situation when there is no STT. Thus, when we have no external polarizing terminal and no damping, the equation of motion for the magnetic moment $\vec{M}$ of the free layer is managed by the following equation:

$$\frac{\partial \vec{M}}{\partial t} = -\gamma \left(\vec{M} \times \vec{H} \right) \tag{3.3}$$

for the pure, everlasting precession, where $\gamma > 0$ is the gyromagnetic ratio of the electron and $\vec{H}$ is the effective magnetic field intensity being a sum of externally (eventually) applied field and the internal (material originating) fields. However, for the more realistic damped, that is, dissipative case, the equation obtains an extra term, and thus we have

$$\frac{\partial \vec{M}}{\partial t} = -\gamma \left(\vec{M} \times \vec{H} \right) - \lambda \left[\vec{M} \times \left(\vec{M} \times \vec{H} \right) \right], \tag{3.4}$$

where λ is the so-called Landau–Lifshitz damping parameter. This equation is valid for a simple, single-atomic magnetic moment or for a single-electron magnetic moment (Landau et al. 1935). The reason for this comes from the physical meaning of the γ and λ constants being related to a single atomic-scale object. The reader should carefully check (comp. Figs. 3.1–3.3) the mutual orientations of both the vector products. Besides, before we follow the discussion in order to include the Slonczewski term, it must be said that there are three types of precessional equations in use in magnetoelectronics. They are the Gilbert, the LL, and the LLG types. This needs some explanations; where do all these come from?

In classical mechanics and, what is important for the micromagnetic approach, from a classical micromagnetic perspective, the basic equation of motion is Newton's second law (eqs. (3.3) and (3.4)). This well-known equation can be derived, however, from the more general formalism based on the so-called Lagrangian and the Lagrangian equation. The Lagrangian L refers to the energy of the physical system; precisely, it equals $L = T - U$, that is, the difference between the kinetic energy T and the potential energy U. The Lagrangian equation, on the other hand, looks as follows:

$$\frac{d}{dt}\left(\frac{dL}{dv}\right) - \frac{dL}{dx} = 0 \tag{3.5}$$

Thus, it is a combination of derivatives with respect to time, velocity, and position. For example, for the most elementary case of the mass m connected to an undamped spring having the coefficient of elasticity k, the subsequent expressions are as follows:

$$L = T - U = \frac{mv^2}{2} - \frac{kx^2}{2} \tag{3.6}$$

and

$$\frac{dL}{dv} = mv, \quad \frac{d}{dt}\left(\frac{dL}{dv}\right) = m\frac{dv}{dt} = ma, \quad \frac{dL}{dx} = -kx. \tag{3.7}$$

This is why, finally, the Lagrangian equation takes the simple form of $ma + kx = 0$, or even better, $F = -kx$, if we introduce the force F, a very fundamental quantity in physics.

We know that damping can be represented in classical mechanics by a force proportional to the velocity (for a slow movement), that is: $F_d = -bv$ (the minus sign results from the fact that the force is inversely directed to the velocity vector). So how will an equivalent Lagrangian look in this situation? The answer is relatively simple.

Since damping forces are proportional to the velocity, and $dL/dv = -bv$, the damping force can be introduced as the first velocity-derivative, from the quadratic term of $0.5\,bv^2$. In other words, if position is equivalent to magnetization, then the damping effect should be proportional to velocity or the first time-derivative of magnetization (see eqs. (3.8–3.9) below).

Now, we will concentrate on magnetic issues. The Lagrangian equation can be applied here as well if we simply exchange the classical position with the magnetization vector and the classical velocity with their first time derivative (we can name it as generalized velocity), that is

$$x \rightarrow \vec{M}, \tag{3.8}$$

$$v \rightarrow \frac{\partial \vec{M}}{\partial t}, \tag{3.9}$$

Then, we can similarly introduce the magnetization damping term as proportional to the generalized velocity

$$-bv \rightarrow -\eta \frac{\partial \vec{M}}{\partial t}, \tag{3.10}$$

This means that the magnetic effective field intensity is now reduced due to material damping, and the equation of motion, mirroring Newton's second law, can be written in the modified form as

$$\frac{\partial \vec{M}}{\partial t} = -\gamma \left[\vec{M} \times \left(\vec{H} - \eta \frac{\partial \vec{M}}{\partial t} \right) \right] \tag{3.11}$$

or as

$$\frac{\partial \vec{M}}{\partial t} = -\gamma \vec{M} \times \vec{H} + \gamma \eta \vec{M} \times \frac{\partial \vec{M}}{\partial t}. \tag{3.12}$$

Now, some remarks about η, the energy loss parameter, which should provide a better understanding of the topic. The effective energy loss parameter is, in general, the spatiotemporal quantity $\eta = \eta(\vec{r},t)$ as it can result from local defects, moving domain walls, local magnetization fluctuations, and other factors. This is why it gives us the possibility to also treat η as an effective average material parameter for a sample as a whole, the same for all local parts. This assumption was made by Gilbert (2004), who introduced the so-called damping constant α by the use of the following simple substitution:

$$\alpha = \eta \gamma M_S, \tag{3.13}$$

where M_S is the magnetization at saturation, the quantity again representing the sample as a whole. Moreover, a, which is the simplicity of the Gilbert model, is expressed by this saturation magnetization M_S – the only parameter for the sample as a whole. This is the intrinsic difference between the LL and the Gilbert equations. The first one is related to a single atomic object, the latter uses the global parameter in the above meaning found in eq. (3.13).

Thus, the subsequent equation of motion takes the following form:

$$\frac{\partial \vec{M}}{\partial t} = -\gamma \vec{M} \times \vec{H} + \frac{a}{M_S} \vec{M} \times \frac{\partial \vec{M}}{\partial t}. \tag{3.14}$$

A different type of equation, as mentioned before (eq. (3.4)), was introduced by Landau and Lifshitz, where λ is the so-called Landau–Lifshitz damping parameter. The Gilbert and LL equations are only partially equivalent (the issue of global vs. local descriptions); however, the unified equation, compromising both approaches, can usually be written in a form similar to the LL equation while, importantly, it can be defined now for a given material (global approach) in which the saturated magnetization M_S can be introduced due to the same reasons as Gilbert did for his approach. Thus, the LLG looks as follows:

$$\frac{d\vec{M}}{dt} = -\gamma' \left(\vec{M} \times \vec{H} \right) - \lambda' \left[\vec{M} \times \left(\vec{M} \times \vec{H} \right) \right], \tag{3.15}$$

where

$$\gamma' = \frac{\gamma}{1 + \gamma^2 \eta^2 M_S^2} = \frac{\gamma}{1 + \alpha^2} \tag{3.16}$$

and

$$\lambda' = \frac{\gamma^2 \eta}{1 + \gamma^2 \eta^2 M_S^2} = \frac{\gamma}{M_s} \frac{\alpha}{1 + \alpha^2}. \tag{3.17}$$

In this sense, the Gilbert and the LL equations are equivalent, which is materialized in the form of the LLG equation.

Now, finally, we can switch back to the important topic of the Slonczewski spin-torque effects (Slonczewski 2005). Thus, let us sketch some elementary scenarios for a typical magnetoelectronic situation of a three-layered system composed of a ferromagnetic layer with fixed magnetization (FM$_1$), a nonmagnetic spacer enabling tunneling of electrons (NM), and a second ferromagnetic layer (FM$_2$) capable of changing its magnetization state. Remember the meaning of $(\vec{M} \times \vec{M}_p)$ and $\vec{M} \times (\vec{M} \times \vec{M}_p)$ terms. Figure 3.4 provides the results for several representative situations from the vector formalism point of view for the spin-transferred torque in this trilayer system.

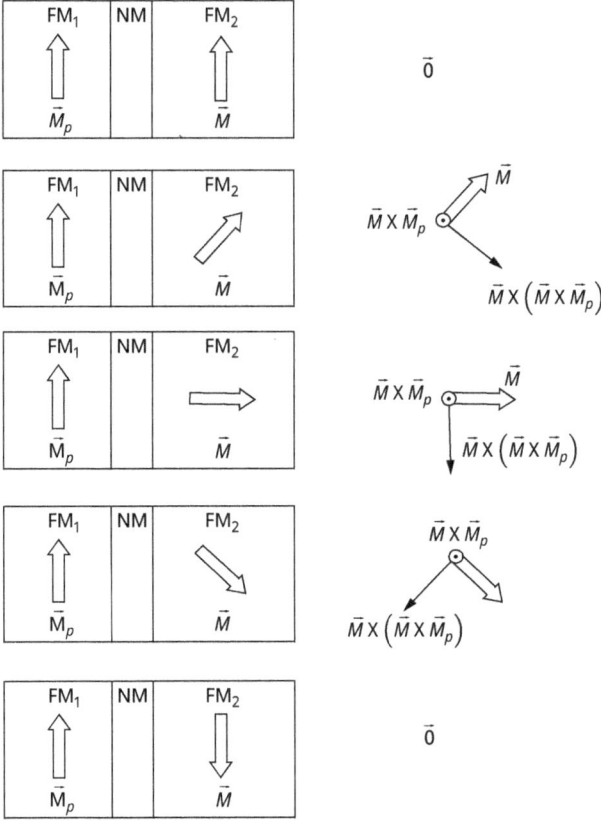

Fig. 3.4: The five characteristic situations of the Slonczewski-type STT for a three-layered system, with the ferromagnetic material with fixed magnetization FM_1 and the free-layer ferromagnetic material FM_2 having the ability to switch. The nonmagnetic spacer NM between both ferromagnetic layers has a properly adjusted thickness enabling the tunneling of electrons from FM_1 to FM_2. On the right side, the equivalent parallel and perpendicular components of the spin-transfer torque are shown.

Thus, the commonly used form of the LL equation, referencing only individual single-atomic magnetic moments, including all terms known so far, is usually written as

$$\frac{d\vec{M}}{dt} = -\gamma\left(\vec{M}\times\vec{H}\right) - \lambda\left[\vec{M}\times\left(\vec{M}\times\vec{H}\right)\right] - \gamma a_j\vec{M}\times(\vec{M}\times\vec{M}_p) - \gamma b_j\left(\vec{M}\times\vec{M}_p\right), \qquad (3.18)$$

where a_j is the amplitude of the in-plane contribution to the STT effect and b_j is the amplitude of the out-of-plane component. Thus now, using similar arguments that guided us from a single magnetic moment case (local approach) to a real magnetic material (global approach), the final version of the LLG equation can be written as follows:

$$\frac{d\vec{M}}{dt} = -\gamma'(\vec{M} \times \vec{H} - \lambda'\left[\vec{M} \times \left(\vec{M} \times \vec{H}\right)\right] -$$

$$-a_j\left[\lambda' M_s\left(\vec{M} \times \vec{M}_p\right) - \frac{\gamma'}{M_s}\vec{M} \times \left(\vec{M} \times \vec{M}_p\right)\right] + b_j\left[\gamma' M_s\left(\vec{M} \times \vec{M}_p\right) + \frac{\lambda'}{M_s}\vec{M} \times \left(\vec{M} \times \vec{M}_p\right)\right]$$

$$(3.19)$$

or in the equivalent form as

$$\frac{d\vec{M}}{dt} = -\gamma'\left(\vec{M} \times \vec{H}\right) - \lambda'\left[\vec{M} \times \left(\vec{M} \times \vec{H}\right)\right] - \gamma'\left(\alpha a_j - b_j\right)\left(\vec{M} \times \vec{M}_p\right) + \lambda'\left(\frac{a_j}{\alpha} + b_j\right)\left[\vec{M} \times \left(\vec{M} \times \vec{M}_p\right)\right].$$

$$(3.20)$$

Here, the vector direction of both STT components can be easily recognized. Namely, since α is usually a small number ($\sim 10^{-3}$), the first, perpendicular $(\vec{M} \times \vec{M}_p)$ term is oriented in the opposite direction to the undamped precession term $- (\vec{M} \times \vec{H})$ (the $(\alpha a_j - b_j)$ term is negative), while the term with the doubled vector product, the in-plane $\vec{M} \times (\vec{M} \times \vec{M}_p)$ contribution, can be oriented against the material damping $-\vec{M} \times (\vec{M} \times \vec{H})$. In other words, the STT reduces precession and works against damping and finally can be a reason for the magnetization reversal in real magnetoelectronic devices like spin valves or tunnel junctions.

The mentioned effects can also be understood from a general physical perspective. Since the torque is an external factor to the material of interest, the physical system reacts to the changes, causing the effect, and the precession is slower.

Complementing all the above considerations, eq. (3.20) can be rewritten using again the Gilbert substitution ($\alpha = \gamma \eta M_s$)

$$\frac{\partial \vec{M}}{\partial t} = -\gamma'\left(\vec{M} \times \left(\vec{H} + (\gamma \eta M_S a_j - b_j)\vec{M}_p\right)\right) - \lambda'\vec{M} \times \left[\vec{M} \times \left(\vec{H} - \left(\frac{a_j}{\gamma \eta M_S} + b_j\right)\vec{M}_p\right)\right],$$

$$(3.21)$$

which can be a reason for finding the specific conditions when the Slonczewski terms can be completely reduced. This can happen if

$$\gamma \eta M_S a_j = b_j.$$

$$(3.22)$$

Thus, the first term in eq. (3.21) reveals the normal precession proportional to $-\gamma'$ $(\vec{M} \times \vec{H})$. Hence, there is no perpendicular STT. Also, if

$$\vec{H} = \left(\frac{a_j}{\gamma \eta M_S} + b_j\right)\vec{M}_p,$$

$$(3.23)$$

the precession is stable, in the sense that the damping is compensated by the in-plane STT contribution. The second term in eq. (3.21), on the right-hand side, is zero.

The STT effects can also be nicely visualized using micromagnetic simulations (Fig. 3.5).

(a)

(b)

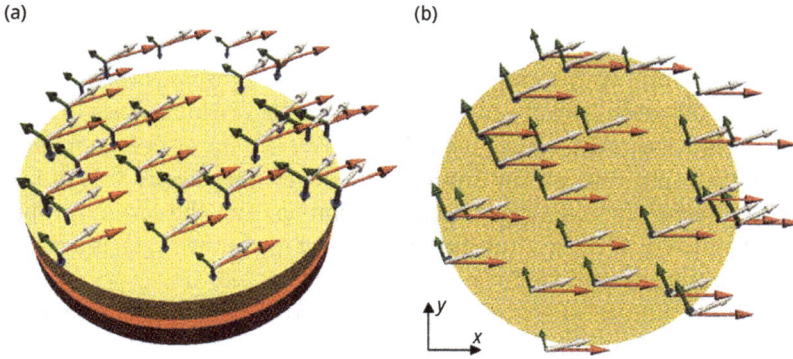

Fig. 3.5: Perspective (a) and top view (b) of spin-transfer torque in a ferromagnetic (Py) disk of 100 nm diameter and 10 nm thickness. The magnetization of the bottom terminal layer (brown) equals $\vec{M}_p = \vec{i}$ and is parallel to the x-axis. Description: magnetization vectors (gray), externally applied magnetic field (red), out-of-plane component of STT ($\vec{M} \times \vec{M}_p$) blue), in-plane component of STT $\vec{M} \times (\vec{M} \times \vec{M}_p)$ (green). Positions of vectors are shown at randomly chosen points. (original figure was rearranged)

Now, finally, after the above discussion, a natural question arises – which practical mechanism is responsible for the transfer of the fixed $\vec{M}_p$ magnetization into a free magnetic layer? The answer is simple: an electrical current flow or, more precisely, a spin-polarized current. For this reason, the LLG equation can now be modified in order to show explicitly the electric current density.

For the reader's convenience, we provide almost the same shape of the equation as eq. (3.21), using the Gilbert damping parameter α for simplicity:

$$\frac{\partial \vec{M}}{\partial t} = -\gamma'\left(\vec{M} \times \left(H + (\alpha a_j - b_j)\vec{M}_p\right)\right) - \lambda'\vec{M} \times \left[\vec{M} \times \left(\vec{H} - \left(\frac{a_j}{\alpha} + b_j\right)\vec{M}_p\right)\right]. \qquad (3.24)$$

The right-hand side of the equation consists of, going from left to right, "precessional" and "damping" terms, respectively. What may be striking is that the equivalent STT contributions have opposite algebraic signs: $+(\alpha a_j - b_j)\vec{M}_p$ and $-(a_j/\alpha + b_j)\vec{M}_p$, respectively. Also, the influence of damping stays in contradiction, namely, the first precessional amplitude equals $(\alpha a_j - b_j) = \alpha(a_j - b_j/\alpha)$, while the second damping amplitude is equal to $(a_j/\alpha + b_j) = 1/\alpha(a_j + \alpha b_j)$. Going further, thus considering only mathematical dependencies on the damping factor α and assuming for simplicity $a_j = b_j = 1$, we can conclude that the precessional SST contribution, also known as the perpendicular one, is just proportional to α, since $\alpha(a_j - b_j/\alpha) = \alpha(1 - 1/\alpha) = \alpha - 1 \sim \alpha$. Similarly, the damping STT contribution, also known as the in-plane one, is proportional to $1/\alpha$, since $1/\alpha(a_j + \alpha b_j) = 1/\alpha(1 + \alpha) = 1/\alpha + 1 \sim 1/\alpha$.

Now, from the spin-polarized electric-current perspective, the efficiency of STT should be proportional to the current density j, flowing from the fixed (pinned) magnetization layer via the nonmagnetic spacer into the free ferromagnetic layer. Next, the STT effect should be proportional to mutual orientations of the magnetizations of the pinned and free layers due to the vector product $(\vec{M} \times \vec{M}_p)$ found in eq. (3.24), thus proportional to the angular difference derived from the vector product $(\vec{M} \times \vec{M}_p) = M \cdot M_p \sin(\theta) \sim \sin(\theta)$. Finally, the STT is a complex quantum process related to exchange coupling between pinned and free layers and scattering processes during spin transport depending on mutual orientations of magnetization and the nonmagnetic layer conductance G. Also, it depends on the electrical resistances of the whole trilayer system, or even more precisely, on the level of spin polarization expressed by the electrical resistance difference between the more favorable orientation $R_\uparrow$ (spin-up) and the less favorable orientation $R_\downarrow$ (spin-down). The latter function, which is of quantum origin, is defined in the scientific literature as the STT efficiency $g(\theta)$ (Lehndorff 2008).

Thus, the precessional SST contribution, related to $(\vec{M} \times \vec{M}_p)$ and also known as the perpendicular one, is proportional to

$$j \cdot g(\theta)\sin(\theta) \cdot \alpha, \tag{3.25}$$

while the damping SST contribution, also known as the in-plane one, is proportional to

$$j \cdot g(\theta)\sin(\theta) \cdot \frac{1}{\alpha}, \tag{3.26}$$

and where the STT efficiency $g(\theta)$ equals

$$g(\theta) = \mu_B \left(\frac{m}{2e^2}\right) \left(\frac{R_\downarrow - R_\uparrow}{R_\uparrow + R_\downarrow}\right) \frac{\Lambda}{\Lambda \cos^2(\theta/2) + \frac{1}{\Lambda}\sin^2(\theta/2)}. \tag{3.27}$$

In the above formula, the spin-polarization level is calculated from the electrical resistance $R_\downarrow$ for the less favorable mutual orientations of the magnetization vector of the magnetic regions and the more favorable case of $R_\uparrow$. Hence, $R_\downarrow > R_\uparrow$ and the spin-polarization level is equal to

$$P = \frac{R_\downarrow - R_\uparrow}{R_\uparrow + R_\downarrow}, \tag{3.28}$$

while the geometry-dependent conductance factor, seen in eq. (3.27), is calculated from Lehndorff (2008)

$$\Lambda = \sqrt{A \cdot G \frac{R_\uparrow + R_\downarrow}{2}}. \tag{3.29}$$

Thus finally,

$$j \cdot g(\theta)\sin(\theta) \cdot \alpha = j \cdot \mu_B \left(\frac{m}{2e^2}\right) \left(\frac{R_\downarrow - R_\uparrow}{R_\uparrow + R_\downarrow}\right) \frac{\lambda \sin(\theta)}{\lambda \cos^2(\theta/2) + (1/\lambda)\sin^2(\theta/2)} \cdot \alpha \tag{3.30}$$

and

$$j \cdot g(\theta)\sin(\theta) \cdot \frac{1}{\alpha} = j \cdot \mu_B \left(\frac{m}{2e^2}\right) \left(\frac{R_\downarrow - R_\uparrow}{R_\uparrow + R_\downarrow}\right) \frac{\lambda \sin(\theta)}{\lambda \cos^2(\theta/2) + (1/\lambda)\sin^2(\theta/2)} \cdot \frac{1}{\alpha}. \qquad (3.31)$$

As a concrete example, let us consider a trilayer system of Fe/Ag/Fe nanopillars along with the assumed value of $\Lambda = 1.5$. Thus, the angular part of the STT efficiency

$$\frac{\Lambda\sin(\theta)}{\Lambda\cos^2(\theta/2) + (1/\Lambda)\sin^2(\theta/2)} \qquad (3.32)$$

multiplied by α or $1/\alpha$ (compare eqs. (3.25) and (3.26)) is presented in Fig. 3.6.

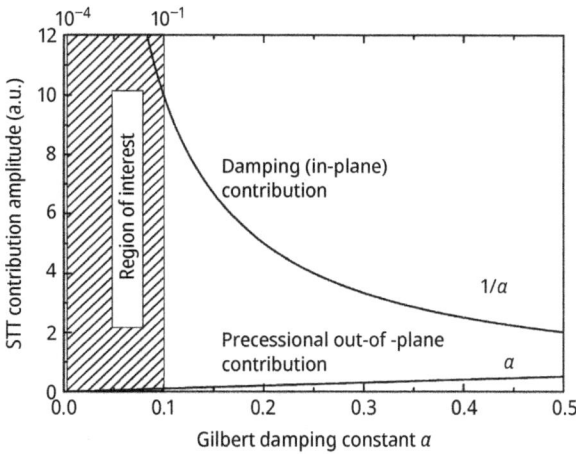

Fig. 3.6: The two contributions to the in-plane and out-of-plane STT efficiencies. The marked region of interest results from the accessibility of the Gilbert damping constant α in realistic materials.

The conclusion for that issue is quite obvious; the in-plane contribution is about one order of magnitude larger for relatively high damping ($\alpha = 0.1$), and even more significantly dominating for low-valued Gilbert constants ($\alpha = 0.001$) – in such cases, the precessional-like term can be neglected.

It is also interesting to analyze the angular part of the STT efficiency, thus the function given in eq. (3.32) (Fig. 3.7). The function is asymmetrical and has a single maximum value. The results in Fig. 3.7 were calculated for the geometry-dependent conductance factor $\Lambda = 1.5$.

The facts described in this chapter at least partially introduce the reader to the topic of equations used in micromagnetism. In order to extend knowledge about this interesting field of physics, more details can be found in the scientific literature.

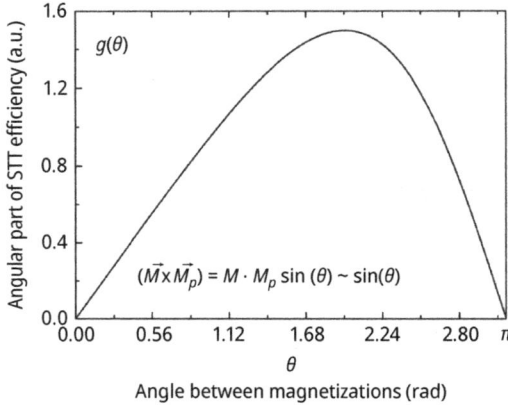

Fig. 3.7: The angular contribution to the STT efficiency (eq. (3.32) calculated for the geometry-dependent factor $\Lambda = 1.5$).

3.1 Magnetization dynamics in materials with strong spin–orbit coupling: the Dzyaloshinskii–Moriya interaction

The history of this relatively exotic type of interaction began in the middle of the previous century, thanks to the works of Dzyaloshinskii (1957) and Moriya (1960). The material under debate was, at that time, the hematite α-Fe_2O_3 crystal, generally an antiferromagnetic material exhibiting very weak ferromagnetism at high enough temperatures (>250 K) with an internal structure consisting of four sublattices, and correspondingly having four underlying subsets of spins (magnetic moments). The occurrence of weak ferromagnetism was explained as a result of the spin–orbit coupling and thus reorientations of spin moments with respect to the rhombohedral [111] crystal axis. It was observed that below 260 K, the so-called Morin temperature, the spins of two of the four sublattices were canted, which results in weak ferromagnetism. Hence, the bulk phase of α-Fe_2O_3 belongs to the trigonal crystal system and the space group $R\,\bar{3}c$, and these special effects always have to be related to specific underlying crystallographic symmetries.

The approach provided by Dzyaloshinskii was phenomenological. The Moriya approach was strict and quantitative and was developed on the basis of the anisotropic superexchange interaction induced by the spin–orbit coupling. It is worth underlining also that strong spin–orbit coupling can be induced in ferromagnetic materials combined with nonmagnetic ones, for example, in diamagnetic metals, through the interlayer spins coupling. This property is applied in many magneto-electronic devices nowadays.

What will be presented in this section is the approach tested with micromagnetic simulations, thus beyond the pure quantum mechanics scale. Before this, however, some strict remarks about spin–orbit coupling will be given.

The physical origin of the spin–orbit coupling results from the interaction between electrons moving in the presence of atomic nuclei; thus, it relies on the electron magnetic moment interaction with the magnetic field produced by the relative movement of the nuclei. Such an influence of the magnetic fields on electrons depends on the spatial distribution of atoms, thus on the general crystallographic symmetry. With the simplest description, we can say that the spin–orbit coupling can be described by the additive superposition of N pairs of electron magnetic moments, and it can be described effectively with the system energy (the so-called Hamiltonian with the unit energy per volume):

$$H = \frac{1}{N} \sum_{i,j} J_{ij} \vec{p}_i \cdot \vec{p}_j, \tag{3.33}$$

where $\vec{p}_i$ and $\vec{p}_j$ are electrons' magnetic moments and J_{ij} is the so-called exchange integral, the constant describing the strength of interaction between electrons, however, being dependent on the spatial distribution of atoms with their orbits in the sample under investigation. In order to switch from the above theoretical and quantum-like description to the scale which is more adequate for micromagnetic experiments, we should substitute the elementary magnetic moments $\vec{p}$ by the magnetization vector $\vec{M}$ (in units of magnetic moment per volume). In the Hamiltonian itself, this modification leads to the following transition:

$$\frac{1}{N} \sum_{i,j} J_{ij} \vec{p}_i \cdot \vec{p}_j \rightarrow J\vec{M} \cdot \left(\vec{\nabla} \cdot \vec{M} \right), \tag{3.34}$$

while, what is worth emphasizing, the scalar product is still in use, and J is now the effective constant valid for a whole piece of the material. The scalar product is obviously symmetrical. Dzyaloshinskii and Moriya noticed that for some materials with specific crystallographic symmetry (with broken inversion symmetry), the spin–orbit coupling should lead to an extra asymmetrical energy term expressed by the vector product

$$H = \frac{1}{N} \sum_{i,j} D_{ij} \vec{p}_i \times \vec{p}_j, \tag{3.35}$$

where D_{ij} is the Dzyaloshinskii–Moriya interaction (DMI) constant. Thus, similarly, doing the transition from quantum to continuous limit, we obtain

$$\frac{1}{N} \sum_{i,j} D_{ij} \vec{p}_i \times \vec{p}_j \rightarrow D\vec{M} \cdot \left(\vec{\nabla} \times \vec{M} \right), \tag{3.36}$$

with D being the single-valued effective parameter. The above expression is not universal and is used here to provide some introductory intuition. In general, different scenarios for crystal-symmetry-induced spin-originating DMI interactions are possible. Thus, the above approach is valid for bulk, macroscopic noncentrosymmetric crystals, where the underlying effects are relatively weak. The situation is different for two-dimensional systems, or even more so for zero-dimensional objects, where DMIs can overcome dipolar fields and significantly compete with the exchange energy terms.

Now, some mathematical details for representative systems will be provided. First, the $(\vec{\nabla} \times \vec{M})$ term, written in vector form, equals

$$\left(\vec{\nabla} \times \vec{M}\right) = [\nabla_x; \nabla_y; \nabla_z] \times [M_x; M_y; M_z] = \left[\vec{i}\frac{\partial}{\partial x}; \vec{j}\frac{\partial}{\partial y}; \vec{k}\frac{\partial}{\partial z}\right] \times [M_x; M_y; M_z]$$

$$= \left[\vec{i}\left(\frac{\partial M_z}{\partial y} - \frac{\partial M_y}{\partial z}\right) + \vec{j}\left(\frac{\partial M_x}{\partial z} - \frac{\partial M_z}{\partial x}\right) + \vec{k}\left(\frac{\partial M_y}{\partial x} - \frac{\partial M_x}{\partial y}\right)\right]. \tag{3.37}$$

Thus, for the above-mentioned bulk situations, or more precisely for the bulk T and O symmetry class crystals, we have

$$w_{\text{DMI}}^{\text{bulk}} = D\vec{M} \cdot \left(\vec{\nabla} \times \vec{M}\right) = D\left(M_x\left(\frac{\partial M_z}{\partial y} - \frac{\partial M_y}{\partial z}\right) + M_y\left(\frac{\partial M_x}{\partial z} - \frac{\partial M_z}{\partial x}\right) + M_z\left(\frac{\partial M_y}{\partial x} - \frac{\partial M_x}{\partial y}\right)\right), \tag{3.38a}$$

where the symbol w is commonly used for the energy density. Next, for the C_{nv} symmetry class materials and, importantly, for the 2D interfacial systems case (this case is fundamental for multilayered systems with dominating perpendicular anisotropy), the energy density equals (Bogdanov and Yablonskii 1989, Bogdanov and Hubert 1994, Bogdanov and Shestakov 1998, Butenko 2013)

$$w_{\text{DMI}}^{C_{nv}} = D\left(M_x\left(\frac{\partial M_z}{\partial x}\right) + M_y\left(\frac{\partial M_z}{\partial y}\right) + M_z\left(-\frac{\partial M_x}{\partial x} - \frac{\partial M_y}{\partial y}\right)\right), \tag{3.38b}$$

for the D_n class systems

$$w_{\text{DMI}}^{D_n} = D\left(M_x\left(\frac{\partial M_z}{\partial y}\right) + M_y\left(-\frac{\partial M_z}{\partial x}\right) + M_z\left(\frac{\partial M_y}{\partial x} - \frac{\partial M_x}{\partial y}\right)\right), \tag{3.38c}$$

for the D_{2d} class systems

$$w_{\text{DMI}}^{D_{2d}} = D\left(M_x\left(\frac{\partial M_z}{\partial y}\right) + M_y\left(\frac{\partial M_z}{\partial x}\right) + M_z\left(-\frac{\partial M_y}{\partial x} - \frac{\partial M_x}{\partial y}\right)\right), \tag{3.38d}$$

while for the following cases, more than one D constant is required. Thus, for the C_n class systems, we have

$$w_{\text{DMI}}^{C_n} = D\left(-M_x\left(\frac{\partial M_z}{\partial x}\right) - M_y\left(\frac{\partial M_z}{\partial y}\right) + M_z\left(\frac{\partial M_x}{\partial x} + \frac{\partial M_x}{\partial y}\right)\right) +$$

$$+ D'\left(-M_x\left(\frac{\partial M_z}{\partial y}\right) + M_y\left(\frac{\partial M_z}{\partial x}\right) + M_z\left(\frac{\partial M_y}{\partial x} + \frac{\partial M_x}{\partial y}\right)\right) +$$

$$D''\left(M_x\left(\frac{\partial M_y}{\partial z}\right) - M_y\left(\frac{\partial M_x}{\partial z}\right)\right), \tag{3.38e}$$

and, for the S_4 class system

$$w_{\text{DMI}}^{S_4} = D\left(-M_x\left(\frac{\partial M_z}{\partial x}\right) + M_y\left(\frac{\partial M_z}{\partial y}\right) + M_z\left(\frac{\partial M_x}{\partial x} - \frac{\partial M_y}{\partial y}\right)\right) +$$

$$+ D'\left(-M_x\left(\frac{\partial M_z}{\partial y}\right) - M_y\left(\frac{\partial M_z}{\partial y}\right) + M_z\left(\frac{\partial M_y}{\partial x} + \frac{\partial M_x}{\partial y}\right)\right). \tag{3.38f}$$

The ratio D'/D determines the orientation of a plane in which spiral structures are created, while the D'' constant determines the effect of spiral structure propagation along the magnetic easy axis (z axis) for a given coordinate system.

The symbols of T, O, C, D, and so on of the so-called symmetry groups were introduced by Arthur Moritz Schoenflies, a German mathematician.

To support and remind the reader of the meaning of the crystal class notion, we provide a more extended version for the just-used names. The full names of the appropriate classes where DMIs can be found are as follows:

- T class of symmetry group (tetroidal) in the cubic crystal system: it is the noncentrosymmetric point symmetry (the inverted elementary crystal cell, with respect to an arbitrary point, is not aligned with the original cell), and more specifically, it is the enantiomorphic (chiral) point symmetry (the inverted elementary crystal cell, with respect to an arbitrary point, cannot be covered with the original cell via any arbitrary rotation).
- O class of symmetry group (gyroidal) in the cubic crystal system: it is the enantiomorphic (chiral) point symmetry.
- C_n class of symmetry group ($n = 3, 4, 6$; n-fold rotation axis):
  - C_2 class (sphenoidal) belongs to the monoclinic crystal system: it is the enantiomorphic polar point symmetry (it is the enantiomorphic point symmetry, but equipped with the polar axis, the unique direction in space where physical properties are different at both ends of the axis).
  - C_4 class (tetragonal-pyramidal) belongs to the tetragonal crystal system; it is the enantiomorphic-polar point symmetry.
  - C_6 class (hexagonal-pyramidal) belongs to the hexagonal crystal system; it is the enantiomorphic-polar point symmetry.
- C_{nv} class of symmetry group ($n = 3, 4, 6$; n-fold rotation axis, n mirror planes containing the rotation axis):
  - C_{2v} class (rhombic-pyramidal) belongs to the orthorhombic crystal system: it is the polar point symmetry (it is the noncentrosymmetric point symmetry,

but equipped with the polar axis, the unique direction in space where physical properties are different at both ends of the axis).

- C_{4v} class (ditetragonal-pyramidal) belongs to the tetragonal crystal system; it is the polar point symmetry.
- C_{6v} class (dihexagonal-pyramidal) belongs to the hexagonal crystal system; it is the polar point symmetry.
- D_n class of symmetry group (n = 3, 4, 6; n-fold rotation axis, n 2-fold axes perpendicular to the n-fold rotation axis):
  - D_2 class (rhombic-disphenoidal) belongs to the orthorhombic crystal system; it is the enantiomorphic point symmetry.
  - D_4 class (tetragonal-trapezohedral) belongs to the tetragonal crystal system; it is the enantiomorphic point symmetry.
  - D_6 class (hexagonal-trapezohedral) belongs to the hexagonal crystal system; it is the enantiomorphic point symmetry.
- D_{2d} class (tetragonal-scalenohedral) of the symmetry group in the tetragonal crystal system (twofold rotation axis, the two twofold axes perpendicular to the twofold rotation axis, the two mirror planes passing between the twofold axes): it is the noncentrosymmetric point symmetry.
- $S_4 = C_{4i}$ class (tetragonal-disphenoidal) of symmetry group in the tetragonal crystal system (it has a fourfold so-called improper rotation axis, which consists of a rotation about one axis followed by an inversion): it is the noncentrosymmetric point symmetry.

The essential point of the above consideration is that the DMI can be met in multilayered 2D systems (nonmagnetic/ferromagnetic/nonmagnetic or metal-oxide/ferromagnetic/nonmagnetic) with a surface-broken symmetry, potentially applied in contemporary magnetoelectronic devices. This reveals new possibilities like the so-called Dzyaloshinskii–Moriya skyrmions, local objects of physical nature similar to vortex states which will be shortly described in the next section.

The above-derived expressions for energy densities can be adopted in the LLG-type micromagnetic equations in order to test the DMIs. Below, some representative cases are shown for permalloy-like materials. Since the resulting effective field is calculated as the functional derivative of the energy density, namely

$$\vec{H}_{\mathrm{DMI}} = -\frac{1}{\mu_0 M_s} \frac{\delta w_{\mathrm{DMI}}}{\delta \vec{M}}, \tag{3.39}$$

for the chosen symmetries, we have (cf. eqs. (3.38a)–(3.38d))

$$\vec{H}_{\mathrm{DMI}}^{\mathrm{bulk}} = -\frac{2D}{\mu_0 M_s} \left[\vec{i}\left(\frac{\partial M_z}{\partial y} - \frac{\partial M_y}{\partial z}\right) + \vec{j}\left(\frac{\partial M_x}{\partial z} - \frac{\partial M_z}{\partial x}\right) + \vec{k}\left(\frac{\partial M_y}{\partial x} - \frac{\partial M_x}{\partial y}\right) \right], \tag{3.40a}$$

$$\vec{H}_{\text{DMI}}^{\text{Cnv}} = -\frac{2D}{\mu_0 M_s} \left[\vec{i}\left(\frac{\partial M_z}{\partial x}\right) + \vec{j}\left(\frac{\partial M_z}{\partial y}\right) + \vec{k}\left(-\frac{\partial M_x}{\partial x} - \frac{\partial M_y}{\partial y}\right) \right],$$

(3.40b)

$$\vec{H}_{\text{DMI}}^{D_n} = -\frac{2D}{\mu_0 M_s} \left[\vec{i}\left(\frac{\partial M_z}{\partial y}\right) + \vec{j}\left(-\frac{\partial M_z}{\partial x}\right) + \vec{k}\left(\frac{\partial M_y}{\partial x} - \frac{\partial M_x}{\partial y}\right) \right],$$

(3.40c)

$$\vec{H}_{\text{DMI}}^{D_{2d}} = -\frac{2D}{\mu_0 M_s} \left[\vec{i}\left(\frac{\partial M_z}{\partial y}\right) + \vec{j}\left(\frac{\partial M_z}{\partial x}\right) + \vec{k}\left(-\frac{\partial M_y}{\partial x} - \frac{\partial M_x}{\partial y}\right) \right],$$

(3.40d)

$$-\frac{2D'}{\mu_0 M_s} \left(-\vec{i}\left(\frac{\partial M_z}{\partial y}\right) + \vec{j}\left(\frac{\partial M_z}{\partial x}\right) + \vec{k}\left(\frac{\partial M_y}{\partial x} + \frac{\partial M_x}{\partial y}\right) \right) - \frac{2D''}{\mu_0 M_s} \left(\vec{i}\left(\frac{\partial M_y}{\partial z}\right) - \vec{j}\left(\frac{\partial M_x}{\partial z}\right) \right)$$

(3.40e)

$$\vec{H}_{\text{DMI}}^{S} = -\frac{2D}{\mu_0 M_s} \left(-\vec{i}\left(\frac{\partial M_z}{\partial x}\right) + \vec{j}\left(\frac{\partial M_z}{\partial y}\right) + \vec{k}\left(\frac{\partial M_x}{\partial x} - \frac{\partial M_y}{\partial y}\right) \right) -$$

$$-\frac{2D'}{\mu_0 M_s} \left(-\vec{i}\left(\frac{\partial M_z}{\partial y}\right) - \vec{j}\left(\frac{\partial M_z}{\partial y}\right) + \vec{k}\left(\frac{\partial M_y}{\partial x} + \frac{\partial M_x}{\partial y}\right) \right),$$

(3.40f)

respectively. The results of simulations are depicted in Figs. 3.8–3.10 for two types of nanodisks and the above-mentioned four symmetries.

The results are asymmetrical in several aspects, being a consequence of the fundamental directional asymmetry of the vector product (comp. eq. (3.35), for example). From the hysteresis curve perspective for all DMI cases, the curves are biased, up- and left-shifted (Dhiman et al. 2022, Grochot et al. 2024). Also, the occurrence and dynamics of skyrmions are different for the ascending and descending parts of the hystereses.

At the end, let us mention that, in general, the following expression (Landau and Lifshitz 1980)

$$L_{ij}^{(k)} = M_i \frac{\partial M_j}{\partial k} - M_j \frac{\partial M_i}{\partial k},$$

(3.41)

containing magnetization gradients (spatial derivatives) creates the base for the aforementioned symmetry considerations. The adequate expression based on magnetization gradients defines the so-called skyrmion charge:

$$S = \frac{1}{4\pi} \iint_{\Sigma} \vec{M} \cdot \left(\frac{\partial \vec{M}}{\partial x} \times \frac{\partial \vec{M}}{\partial y} \right) dx dy,$$

(3.42)

the quantity of interest for the contemporary magnetic-based digital signal processing based on skyrmions being still under development.

After this short overview of micromagnetic methodology in general, and the special case of the DMI in low-dimensional systems, the next section will give an introduction to skyrmions, which have been examined in detail during the last years.

(a)

(b)

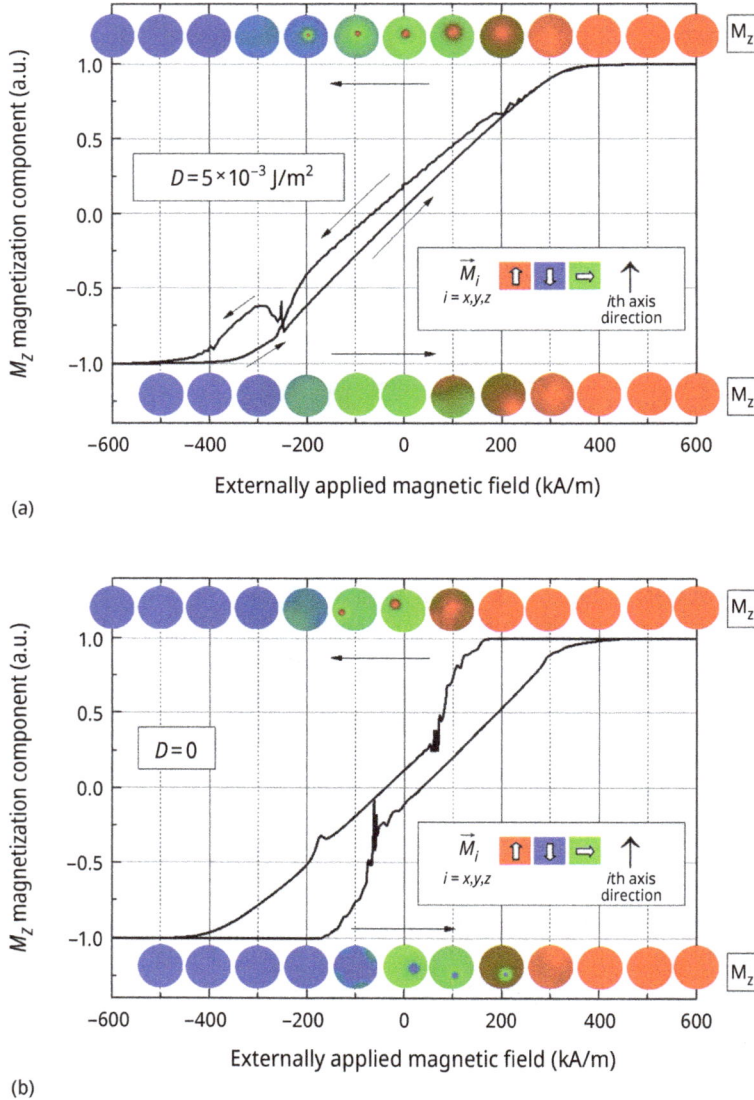

Fig. 3.8: Results of micromagnetic simulations for nanodisks of diameter 100 nm and height 2 nm. The material parameters – anisotropy constants $K1 = K2 = 0$, magnetic polarization at saturation $J_s = 0.48$ T, exchange constant $A = 8.78 \times 10^{-12}$ J/m, damping constant $\alpha = 0.01$, and DMI constant $D = 5 \times 10^{-3}$ J/m² – are consistent with a permalloy-like DMI material. The Dzyaloshinskii–Moriya interaction induces asymmetry in magnetization dynamics and the occurrence of a skyrmion state (a), while for the case $D = 0$, the hysteresis curve is symmetrical with the vortex-like states being visible both for descending and increasing parts of the hysteresis. The external field is directed along the z-axis and is perpendicular to the sample plane.

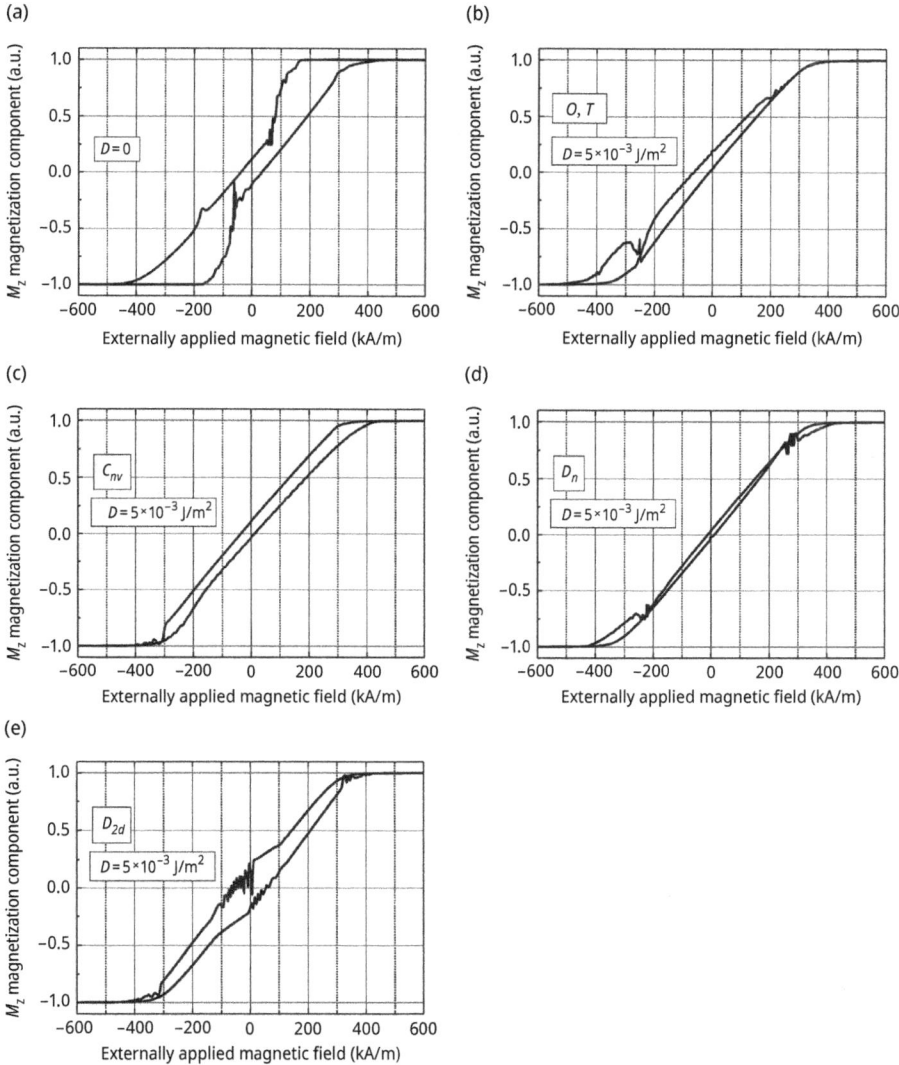

Fig. 3.9: The comparison of results of micromagnetic simulations for a nanodisk of diameter 100 nm and height 2 nm for the different symmetry classes of the permalloy DMI-like material: no DMI case (a), O or T (bulk material) class (b), C_{nv} (2D materials) class (c), D_n class (d), $D2_d$ class (e). The material parameters for simulations are as follows: the anisotropy constants $K1 = K2 = 0$, the magnetic polarization at saturation $J_s = 0.48$ T, the exchange constant $A = 8.78 \cdot 10{-12}$ J/m, damping constant $a = 0.01$, and the DMI constant $D = 5 \cdot 10{-3}$ J/m². The external field is directed along the z-axis and is perpendicular to the sample plane.

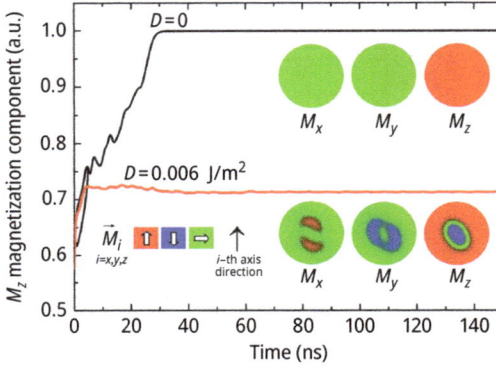

Fig. 3.10: Results of micromagnetic simulations for a nanodisk of diameter 100 nm and height 10 nm. The material parameters – anisotropy constants $K1 = K2 = 0$, magnetic polarization at saturation $J_s = 0.48$ T, exchange constant $A = 8.78 \cdot 10-12$ J/m, damping constant $\alpha = 0.01$, and DMI constant $D = 5 \cdot 10-3$ J/m^2 – are consistent with a permalloy-like DMI material. There is no external field provided – the magnetization vector evolves from its initial state and goes into its quasi-equilibrium state with M_z equal to about 0.7 (b). For the DMI-excluded case (a), the magnetization evolves into the saturation state with $M_z = 1$ (the perpendicular anisotropy magnetization state).

3.2 Skyrmions

Skyrmions were first proposed by Tony Skyrme as topologically stable soliton rotating field configurations to explain effects in the atomic nucleus (Skyrme 1962; Skyrme and Brown 1994). More important for the scope of this book, however, are the magnetic skyrmions that can be found with (Mühlbauer et al. 2009) or without external magnetic field (Jonietz et al. 2010; Heinze et al. 2011).

Theoretical explanations are, among others, based on chiral interactions stabilizing the intrinsic energy of a magnetic skyrmion, such as the aforementioned double-exchange mechanism, the Heisenberg exchange interaction (Heinze et al. 2011), or in most cases, the DMI (Heinze et al. 2011; Kiselev et al. 2011; Nagaosa and Tokura 2013). The latter, an antisymmetric exchange interaction between neighboring spins due to the spin–orbit coupling (Dzyaloshinskii 1958; Moriya 1960), results in a spin canting which leads to weak ferromagnetic behavior in an antiferromagnet, as described in the last section.

Such magnetic skyrmions can be formed in different ways: Similar to the above-described vortex cores, they also have a core with the magnetization oriented perpendicular to the plane, for example, pointing "down." The outer edge of the skyrmion, however, consists of spins pointing in the opposite direction, that is, "up" in the example here, opposite to the purely in-plane magnetic moments around vortex states. Now let's assume how the magnetic moments between both extrema – core and border – should be oriented if sudden changes in the magnetization orientation should be avoided.

One possibility is that directly around the core, the next "circle" of magnetic moments is not pointing exactly down, but each spin is pointing a little bit away from the core, too. In the next circle, the magnetic moments are pointing less down and further outside until they point only outside and are completely oriented in the plane. Going further from the core to the edge, this process goes on, with the magnetic moments in the next circle pointing slightly upward, those in the next circle a little bit more, until at the edge of the skyrmion, all magnetic moments point upward. This sort of skyrmion is called a Néel skyrmion or hedgehog skyrmion.

Starting from the same conditions – the core pointing down and the edge spins pointing up – there is another possibility to solve the problem of how the magnetic moments between these extrema should be oriented from an energetic perspective. Instead of canting to the outside, they can also be tilted along the circumference of the circle on which they are located around the core. In this way, in the middle between core and edge, all magnetic moments are again oriented in-plane, but not all pointing from core to edge (or vice versa); instead, they are now all pointing along the circumference, that is, to the next spin on the same circle. This type of skyrmion is called a Bloch skyrmion or spiral skyrmion.

Finally, a third type was found recently, combining both aforementioned systems. Assuming to look on top of a skyrmion, the "upper" and "lower" magnetic moments, for example, will point outward, while the "left" and "right" ones will point to the inside. Between them, an average orientation is taken. This mixed state is called an anti-skyrmion (Nayak et al. 2017).

Besides the principal interest in such systems in basic research, the technological possibilities offered by skyrmions in the area of spintronics are even larger than those given by vortices. Using spin currents or spin waves, magnetic skyrmions can principally be moved (Fert et al. 2013; Zhang et al. 2015b) and could thus be used in Racetrack applications (Fert et al. 2013; Zhang et al. 2014, 2015c; Zhou and Ezawa 2014).

This is why diverse research groups work on this emerging topic. A general task is the stability of skyrmions in different situations, which should be as high as possible to avoid data loss if skyrmions are used to store and transport data. Uzdin et al. (2018) have calculated energy surfaces and lifetimes of magnetic skyrmions, finding dependencies on the temperature, on the width of a track, and on nonmagnetic impurities.

Varentsova et al. (2018) found the stability of isolated skyrmions to be related to the skyrmion size as well as different material parameters. By adjusting these parameters, the energy barrier protecting the skyrmion from collapsing could be increased, while the shape of the skyrmion was modified from an arrow-like form to a shape similar to magnetic bubbles.

Guslienko (2018) reported on Néel skyrmions in ultrathin magnetic nanodots. He found that while Néel skyrmions could be the ground state for small dots, they could only be metastable in large films, with the border between both regimes depending on the out-of-plane anisotropy.

The stability of isolated magnetic skyrmions in sputtered nanowires and nanodots at room temperature was investigated by Juge et al. (2018). They found the skyrmion dimensions to be tunable by the lateral dimensions of the nanostructures and a small external magnetic field, as well as being influenced by pinning, while pinning and an out-of-plane magnetic field were important factors for the stability.

Another task is how to drive skyrmions. Tomasello et al. (2018) have suggested a combination of the spin-Hall effect with an anisotropy gradient. They showed that the latter mostly drives the skyrmions perpendicular to the gradient, combined with a slower motion parallel to it. The latter can be increased by a suitable spin-Hall torque, resulting in an acceleration of the skyrmions.

Driving skyrmions in an insulating magnetic nanowire was theoretically investigated by Psaroudaki and Loss (2018). They used time-dependent oscillating magnetic field gradients, which exerted a driving force on the intrinsic magnetic excitations of the skyrmions. The automatic coupling of the external field with the magnons, however, resulted in time-dependent dissipation of the skyrmions. Driving the magnon bath at resonance led to a helical movement of the skyrmions superposing the periodic bounded motion.

A magnetic field gradient was used by Zhang et al. (2018a) to move skyrmions. By applying a radial field gradient, they could rotate the skyrmions on a certain radius with a correlated speed. This led to a disintegration of the rotating ensemble, followed by the creation of a shell-like structure of discrete circular paths, which could be reversed by reversing the field direction.

Another study of the influence of a magnetic field gradient was performed by Liang et al. (2018). They investigated driving isolated skyrmions and anti-skyrmions in a frustrated magnetic nanostrip. While in the bulk system, a Hall-like motion was found, the lateral restrictions in the nanostrip completely suppressed this Hall motion, which resulted in acceleration along the direction of the field gradient.

Interestingly, at finite temperatures, Brownian motion of skyrmions and magnetic domain walls may occur, following different diffusion laws for different damping parameters. For very small skyrmions, this diffusion can be suppressed in the case of vanishing or very small damping coefficients (Miltat et al. 2018).

In strained chiral magnets, distorted skyrmions occur, which move anisotropically due to a gradient magnetic field and an electrical current. Their speed depends periodically on the directions of the external forces. Superposing the uniform motion, a harmonic oscillation was found with the frequency depending on the DMI (Chen et al. 2018b).

A strain in a ferromagnetic film can, on the other hand, be used to stabilize skyrmions. Shi and Wang (2018) found that a nonuniform strain with a cosine profile could be used to stabilize skyrmions in an FeGa film in the absence of a magnetic field. By adding a magnetic field or a spin-polarized current pulse, these metastable skyrmions could be partly transferred into a helical phase, resulting in a mixture of both phases. Skyrmions and helical phases could additionally be moved by a spin-polarized current.

As Li et al. (2018) have shown, a strain pulse can even be used to create a single stable skyrmion, while afterward, a strain is used to pin it in a nanowire while it is pushed through this wire by a current.

Besides the motion in a defined direction, skyrmions can also show precessional modes. Liu et al. (2018), for example, found the dimensions of the skyrmions in nanodisks to be correlated with the dimensions of the disks, and further identified several additional modes for increasing aspect ratios of the disks and correspondingly of the skyrmions. In elliptical nanodisks, they found new mixed modes, like different rotation and oscillation modes.

The preconditions for the formation of skyrmions should not be forgotten in this list. Behera et al. (2018) have studied how skyrmions are formed from a first single bubble domain in ferromagnets with perpendicular anisotropy and different shapes. They found that depending on the DMI and the uniaxial anisotropy, either no skyrmions at all could be formed, or no Néel skyrmions, or skyrmions could be created without these restrictions. In addition, they underlined the importance of the shape of the ferromagnetic nanoparticle for the final magnetic state.

Another parameter influencing the possible creation of different types of skyrmions is the distance to the surface. Zhang et al. (2018b) examined the depth dependence of skyrmion structures in Cu_2OSeO_3 by resonant elastic X-ray scattering. They found an exponential change from Néel- to Bloch-type skyrmions along several hundred nanometers from the surface into the bulk, which was attributed to small changes in the material properties along this distance, that is, an influence of the interface up to several hundred nanometers into the bulk material.

Important parameters of skyrmions are the overall size and the wall width. Wang et al. (2018) found that these properties depend on the exchange stiffness, the anisotropy, the strength of the DMI, and the external magnetic field.

An interesting effect occurs when a ferromagnetic layer with perpendicular anisotropy is covered with a superconducting layer. In this case, Vadimov et al. (2018) showed that even without DMI, stable skyrmions could be found for a broad range of geometrical and material parameters.

Another material-related effect was found by Hsu et al. (2018). They investigated an Fe double layer on an Ir(111) substrate, which they loaded with atomic hydrogen. Interestingly, adding the hydrogen enabled skyrmion formation in an external magnetic field, opposite to the pure Fe double layer. Theoretical calculations showed that hydrogenation modified both the Heisenberg exchange interaction and the DMI, and thus allows for optimization of materials for the occurrence and stability of skyrmions.

One of the most interesting properties of skyrmions is their ability to form skyrmion crystals, that is, to arrange themselves in a geometric form that can be described as a two-dimensional lattice. Yu et al. (2018), for example, have investigated the dynamical transition from a hexagonal-lattice skyrmion crystal at zero field to an amorphous state, triggered by an external magnetic field. At small fields, skyrmion microcrystals are formed, similar to colloidal crystallization, until a phase separation occurs between domains with skyrmion crystals and those with topological defects. In this way, the nucleation and annihilation dynamics of metastable skyrmions can be understood better.

El Hog et al. (2018) have shown that such a skyrmion crystal is created in a certain range of applied magnetic field and DMI. At low temperatures, this skyrmion crystal was found to be quite time-stable, while for a defined higher temperature, it shows a phase transition into a paramagnetic phase. The skyrmion dynamics below this phase transition were shown to follow a stretched exponential law (Diep et al. 2018).

Skyrmions are, similar to vortex states, often discussed for data storage and transport applications. Yang et al. (2018) found, for example, a twisted skyrmion state at the border between two antiparallel magnetic domains, stabilized by antiferromagnetic coupling at the border. Using this state, skyrmions with opposite polarity could be switched between both states using spin-polarized currents. Yang et al. thus suggested using a double-track racetrack memory working with skyrmions as well as skyrmion polarity-based logic gates.

An antiferromagnetic skyrmion was used to create a single-skyrmion transistor, consisting of a source, drain, barrier, and a skyrmion island. The barrier region was controlled by strain which further controlled the number of skyrmions that could pass the barrier, enabling even single-skyrmion transport from the skyrmion island to the drain (Zhao et al. 2018).

Skyrmions can also be used in the emerging field of neuromorphic computing (cf. Chapter 7.12). Here, mostly new software is developed, enabling faster calculation in such areas as pattern recognition or classification. However, besides neuromorphic software working on conventional complementary metal-oxide-semiconductor (CMOS) devices, several groups also investigate possibilities to create new hardware, aiming at reduced energy consumption and better compatibility of hardware and software. Especially the separation between calculation and data storage, which is typical for common computers, should be avoided. This means that new devices are necessary, which allow for direct mapping of synaptic and neuronal functions and, at the same time, for synaptic data storage. Magnetic skyrmions may be one possible solution for this task since they are quite stable and, on the other hand, need very low depinning current densities. Chen et al. (2018a) have published design ideas for corresponding skyrmion-based devices and architectures in which skyrmions can be moved. They found by device-to-system simulations that the energy consumption of such devices could be two orders of magnitude lower than in common CMOS implementations.

Instead of the aforementioned skyrmion lattice, Pinna et al. (2018) used a skyrmion gas to restructure a random signal into an identical, but uncorrelated, copy. In their model, skyrmion–skyrmion as well as skyrmion–edge interactions were taken into account. In this way, a low-energy, small-area device could be created which works similarly to an integrate-and-fire neuron and could thus be utilized in the area of neuromorphic computing.

3.3 Hopfions

Besides skyrmions, there are other topical solitons such as three-dimensional Hopfions (Hopf 1931). Magnetic skyrmions and hopfions, from a geometric perspective, are topologically stable, but their creation cannot be explained as a continuous and linear transformation from a fully homogeneous and static space of magnetization vectors. In this sense, hopfions are treated as topological defects. Importantly, the well-known Néel or Bloch domain walls should be treated as one-dimensional topological solitons. While magnetic hopfions have been investigated for more than a decade now, the number of experimental and theoretical studies is still relatively low compared to skyrmions and other nontrivial magnetic states. Here we give an overview of the recent state of research on magnetic hopfions and an outlook on their potential future application in data storage devices.

Firstly, a brief overview of theoretical findings in the field of hopfions is given. A hopfion, as a topologically nontrivial vector field configuration, can be described by mapping the coordinate space to the order parameter space, i.e., in the case of magnetic hopfions, to the magnetization field **M(r)**, with so-called topological charges (Hopf index, Hopf charge, or linking number) giving the mapping degree of the coordinate space to the unit sphere **M(r)**/|**M(r)**| (Guslienko 2023). The Hopf-index differs for different homotopy classes and is an integer in infinite samples. For toroidal hopfions, it is given by the product of the planar winding and the twisting of the magnetization configuration (Guslienko 2023). Alternatively, the Hopf index is calculated as a volume integral of the dot product $A \cdot B$ with the emergent field vector potential A and the emergent magnetic field B (Guslienko 2024):

$$H = \frac{1}{(4\pi)^2} \int dV \, AB$$

Toroidal magnetic hopfions generally have an emergent magnetic field, which leads to the topological Hall effect and skyrmion Hall effect of toroidal magnetic hopfions (Guslienko 2023). Electronic scattering was found to contain a skew-scattering component responsible for the Hall effect, although the average emergent magnetic field is $\langle B(r) \rangle = 0$, which is sometimes believed to prohibit the Hall effect in noncollinear magnetic structures (Pershoguba et al. 2021).

The emergent magnetic field of a magnetic hopfion also produces emergent magneto-multipoles with an emergent magnetic toroidal moment and an emergent magnetic octupole component, which enable a nonlinear hopfion Hall effect under an AC driving current and more generally define nonlinear hopfion dynamics (Liu et al. 2022).

On the other hand, the components of the emergent magnetic field define the gyrovector in a magnetic hopfion, whose axial component of an axially symmetric magnetic hopfion with Hopf index ± 1 is approaching zero in large cylindrical dots, while the in-plane angular component is unequal to zero for all dot sizes (Popadiuk et al. 2023). These values are important prerequisites for the calculation of the topological and the skyrmion Hall effect of toroidal hopfions.

Other authors reported different specific results of theoretical assumptions and calculations. In magnets with topological-chiral magnetic interaction, the orientation of a hopfion with respect to the underlying lattice was attributed to the spin-chiral interaction (Grytsiuk et al. 2020). The helicity and angular momentum of a magnetic hopfion were calculated based on a scalar product due to invariance requirements under the ten-parameter conformal group in three-dimensional Euclidean space $C10(3)$ (Fernandez-Corbaton and Vavilin 2023). Special solutions, such as hopfions coupled to magnetic fluxes or hopfions on a 3D lattice dimer model, were also investigated in theory (Samoilenka and Shnir 2018, Bednik 2019).

Based on new experimental methods to investigate 3D magnetic structures on the nanoscale, different topological magnetic structures were found (Donnelly and Scagnoli 2020), such as Bloch points (Donnelly et al. 2017) and merons (Ezawa 2011), besides the hopfions discussed here. Such three-dimensional topological solitons typically have a polar texture, which leads to specific functionalities such as their chirality, which suggests using them in nano-spintronics devices (Govinden et al. 2023).

For hopfions, the structure can be defined by the winding number n, which twists out of plane m times, giving the simplest expression of the Hopf index $H = n \cdot m$, as depicted exemplarily in Fig. 3.11 (Balakrishnan et al. 2023). Comparing Fig. 3.11a and 3.11b, both have the same winding number $n = 1$, but m differs, which can be seen by following one of the colored lines from an arbitrary starting point until this point is

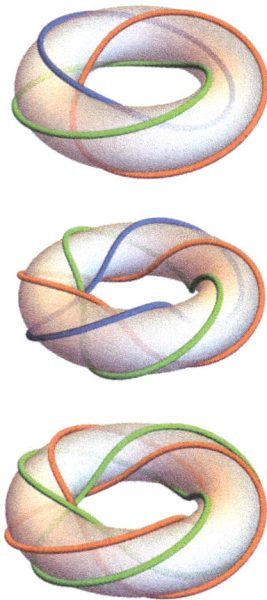

Fig. 3.11: Preimages on a torus for a hopfion vortex: (a) $n = 1$, $m = 1$, Hopf index $H = 1$; (b) $n = 1$, $m = 2$, $H = 2$; (c) $n = 2$, $m = 3$, $H = 6$ (reprinted from Balakrishnan et al. (2023), Copyright (2023), with permission from Elsevier).

reached again. A higher winding number, here $n = 2$ (Fig. 3.11c), means that an arbitrary starting point on one of the colored lines is only reached after following the line for 2 in-plane windings around the torus. Among the various possible torus knots, those with $n = 1$ or $m = 1$ are topologically equivalent to a circle and called unknots, while coprime values of n and m lead to nontrivial knots (Fig. 3.11c) (Balakrishnan et al. 2023).

Going further, many more different morphologies of magnetic hopfions can be calculated, as depicted in Fig. 3.12 for ferromagnets in which the micromagnetic exchange constant J is positive, while the other micromagnetic constants (here called B and C) can be varied (Rybakov et al. 2022). It should be mentioned that the hopfions with Hopf indices 1 and 2 show axial symmetry for $C = 6\,B$, while the other extrema $B = 0$ and $C = 0$ lead to differences from this axial symmetry even for the simplest hopfions (Rybakov et al. 2022). As a special case, fractional hopfions have partial linking and correspondingly nonintegral Hopf indices (Yu et al. 2023). For frustrated magnets, Rybakov et al. found only hopfions if an external magnetic field was taken into account, while for ferromagnets hopfions could occur for a broad range of combinations of the three micromagnetic constants and the nearest-neighbor distance in a lattice (Rybakov et al. 2022).

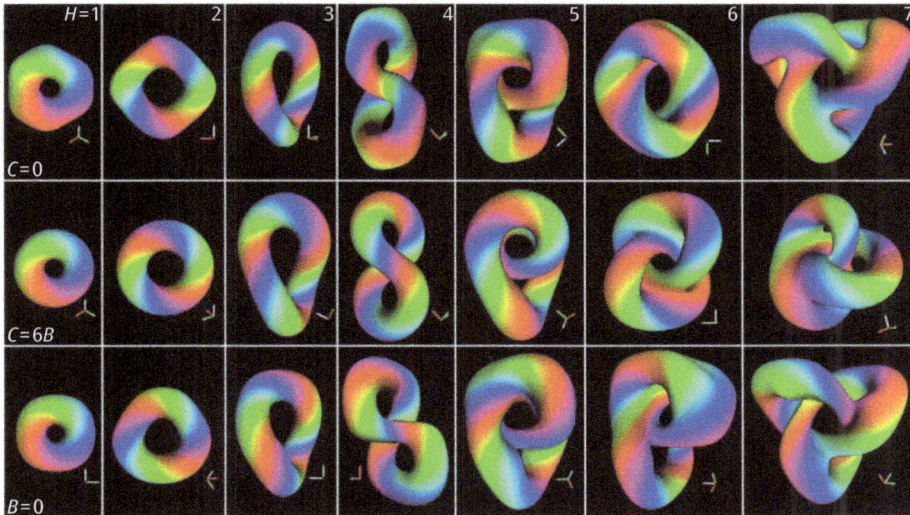

Fig. 3.12: Morphology of magnetic hopfions. Shown magnetization fields are energy minimizers of corresponding homotopy classes defined by the Hopf index H. Each row illustrates isosurfaces with $\Theta = \pi/2$ ($n_z = 0$). Note that the complexity of the hopfion shape and size increases with the Hopf index. For $C = 6B$, the hopfion with $H = 6$ is similar to the linked tori state, while for $C = 0$ and $B = 0$, the hopfion with $H = 7$ is similar to a trefoil knot (reprinted from Rybakov et al. (2022), originally published under a CC-BY license).

A further differentiation is possible between Bloch-type and Néel-type hopfions, which are topologically equivalent, i.e., their Hopf index is identical, but their cross-sections differ between a Bloch-type skyrmion and a Néel-type skyrmion (Li et al. 2022). It is possible to transform a Bloch hopfion into a monopole-antimonopole pair or a toron (Liu et al. 2018, Raftrey and Fischer 2021) as well as to transform a Néel hopfion into a Néel toron (Li et al. 2022). In bulk helimagnets, two types of topologically equivalent hopfions with $H = 1$ exist, but with all magnetization vectors reversed (Metlov 2023). Anti-hopfions with a Hopf index of −1 were described by Leonov and found to transform into torons in bulk helimagnets if they are contracted (Leonov 2023). A stable hopfion in a nanocylinder of a chiral magnet and its transformation from a target skyrmion due to a perpendicular magnetic anisotropy were shown by Sutcliffe (2018). Embedding a hopfion into a helix or conic background results in a heliknoton, which is stable in a broad range of magnetic fields and cubic chiral magnet film thicknesses (Kuchkin et al. 2023, Voinescu et al. 2020). Castillo-Sepúlveda et al. (2021) showed that in toroidal nanoparticles, hopfions were created by an Oersted magnetic field for large toroidal structures, while other magnetic configurations were found for other geometries.

Besides simulations of these structures, it is necessary to observe hopfions also experimentally with suitable techniques. Kent et al. (2021), for example, produced hopfions in magnetic multilayer systems and made them visible by X-ray photoelectron emission microscopy (X-PEEM) and magnetic transmission X-ray microscopy (MTXM), enabling differentiation between a hopfion, a magnetic toron, and a target skyrmion. Gao et al. (2024) used Lorentz transmission electron microscope images to study the transformation between hopfion and toron in nanostripes and stepped nanostructures, as well as the morphology change of hopfions under geometrical constrictions. Over-focus Lorentz images were also used by Zheng et al. (2023) to make hopfion rings in a cubic chiral magnet visible, as depicted in Fig. 3.13.

The stability of magnetic hopfions must be understood to enable their use in information storage. It depends on the geometrical and material parameters of the hopfions' host magnets as well as external magnetic fields. Tai and Smalyukh (2018) found hopfions with different Hopf indices in noncentrosymmetric solid magnetic nanostructures with a perpendicular magnetic anisotropy and showed that the combination of surface anisotropy and Dzyaloshinskii–Moriya interaction (DMI) stabilized the hopfions. For a toroidal nanoring with DMI, Corona et al. (2023) found the shape anisotropy to nucleate hopfions, while planar rings necessitated a perpendicular magnetic anisotropy. For a discrete lattice with competing exchange interactions, Lobanov and Uzdin (2023) calculated the magnetic hopfion lifetime and found that larger hopfions were more stable, while the stability of smaller hopfions depended more strongly on lattice parameters and was only possible at low temperatures. Similarly, Sallermann et al. (2023) found the energy barrier for the decay of a hopfion in a bulk magnet to depend on the size of the hopfion compared to the lattice constant.

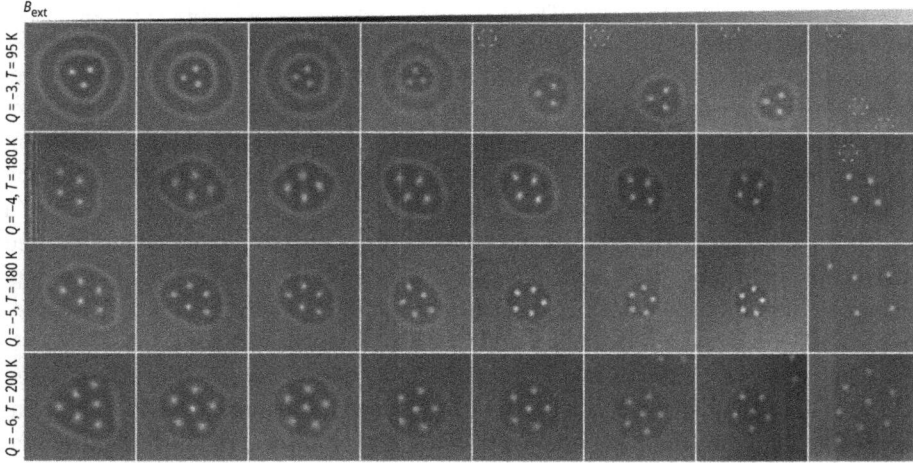

Fig. 3.13: Evolution of hopfion rings surrounding skyrmion strings, shown as a function of the external magnetic field from left to right in the form of experimental over-focus Lorentz images. The topological charge Q (i.e., the Hopf index) and the temperature at which each row was recorded are given on the left side. The value of the perpendicular field increases from left to right between around 150 mT and around 450 mT. The image sizes are around 450×450 nm^2. The dashed circles in the images for $Q = -3$ and $Q = -4$ mark spots of low contrast, which result from the nucleation of chiral bobbers or dipole strings during the collapse of the hopfion ring (reprinted from Zheng et al. (2023), originally published under a CC-BY license).

In chiral magnets, Khodzhaev and Turgut (2022) used bulk and interface DMI to stabilize Bloch and Néel hopfions, respectively. The special case of the collision of two hopfions was investigated by Arrayás and Trueba (2017), showing theoretically that the respective angular momentum of the hopfions changed the interference when the hopfions collided. Tai et al. (2022) showed the stability of hopfions in chiral magnets with perpendicular boundary conditions for a defined range of not-too-large magnetic fields and anisotropy, while outside this region hopfions were metastable, or magnetic torons were formed.

It should be mentioned that hopfions also occur in other material classes, such as chiral liquid crystals, where the self-assembly of hopfions in periodic arrays and linear chains and other interesting effects were found (Tai et al. 2022, Ackerman et al. 2015) which can stimulate research on magnetic hopfions. Here, however, only magnetic hopfions are discussed.

Besides long-term stability, ideally at temperatures near room temperature, it is necessary to move potential information-storing magnetic solitons in a defined way to make them usable for data processing devices. Some research groups have thus investigated the possibility of driving hopfions by an external electric field.

Sobucki et al. (2023), e.g., investigated the motion of hopfions in a ferromagnetic stripe due to a nonuniform electric field using simulations and found that the velocity of the hopfions moving in an electric field gradient was constant. Liu et al. (2023) used an AC driving current which worked on the emergent multipoles via the spin transfer torque and found that a current applied along the symmetry axis of the hopfion resulted in nonlinear dynamics, while an AC current perpendicular to the hopfion symmetry axis led to the nonlinear hopfion Hall effect, both correlated with translation and rotation of the hopfion. The spin transfer torque was also the source of hopfion dynamics in another work by Liu et al. (2020) where the current-driven dynamics of a magnetic hopfion in a frustrated magnet were investigated. Here, the authors showed analytically and numerically the entangled translation, rotation, and dilation of hopfions with Hopf index $H = 1$ and suggested investigating hopfions with higher Hopf indices to find more exotic dynamics.

Wang et al. (2019) found that hopfions were driven by an electric current based on a combination of spin-transfer torque (STT) and spin-Hall torque (SHT). They mentioned hopfions' vanishing gyrovector, which excluded undesired Hall effects (opposite to the previously mentioned nonzero gyrovector found by Popadiuk et al. 2023). Interestingly, they found a difference between Néel-type hopfions, which moved with the current direction due to STT and SHT, and Bloch-type hopfions, which moved along the current direction by STT or transverse to the current direction due to SHT [46]. Masell and Everschor-Sitte (2021) described the movement of hopfions due to STTs along the applied current, while frustrated magnets allowed motion and rotation of an STT-driven hopfion with $H = 1$, accompanied by inflation or deflation of the hopfion for different current directions. For the special case of fractional hopfions, Yu et al. (2023) varied the pulsed current strength and current direction, leading to an extension or compression of a fractional hopfion bundle.

Spin wave spectra could be used to distinguish between hopfions and other 3D solitons, such as torons or target skyrmions, as Raftrey and Fischer (2021) showed. They found characteristic patterns in real space and corresponding characteristic amplitudes in frequency space, which enabled using them as fingerprints for the respective 3D topological spin textures.

Bo et al. (2021) calculated the spin excitation spectrum as well as the correlated spin-wave modes of a magnetic hopfion. They found fewer resonance peaks for hopfions than for skyrmion tubes, which they attributed to the suppressed vertical spin-wave modes of hopfions due to internal topological defects. Besides, they showed the possibility that breathing and rotation modes could hybridize in the case of an excitation in the z-orientation. The authors also suggested these individual spin-wave modes as a possibility to detect magnetic hopfions by microwave magnetic field excitation.

Similarly, Sobucki et al. (2022) found four groups of modes differing in the rotation sense, the position along the radial direction, and regarding oscillations in the vertical cross-section, which they also suggested as a fingerprint for hopfion identification.

In a micromagnetic simulation of spin wave scattering by magnetic hopfions, Saji et al. (2023) found spin waves propagating along the symmetry axis of the hopfions to be deflected by the magnetic texture, leading to a convergent or divergent lens effect for different spin wave propagation directions.

The spin wave dispersion of a magnonic crystal constructed from linearly arranged hopfions at regular distances was investigated by Medlej et al. (2024). They showed a modification of the spin wave dispersion by the hopfions, leading to allowed and forbidden bands, as depicted in Fig. 3.14. The authors thus suggested using a hopfion lattice as a potential magnonic crystal with a well-defined band structure.

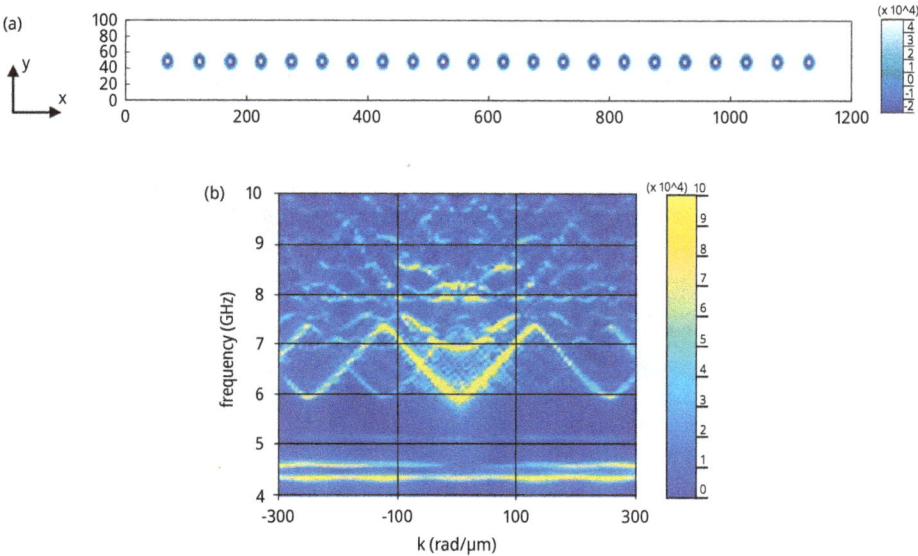

Fig. 3.14: (a) 22 magnetic hopfions along the (*XY*) plane; (b) dispersion relations for the propagation of spin waves (SWs) in a magnonic waveguide are examined in the absence of an external magnetic field. The color scale represents the intensities of the spin waves (reprinted from Medlej et al. (2024), Copyright (2024), with permission from Elsevier). (original figure was rearranged)

Finally, possibilities are reviewed to use hopfions in spintronics applications for data storage and processing. Hopfions, as a three-dimensional spin texture, belong to the natural candidates for data storage in three-dimensional magnetic structures.

One of the new data storage techniques which have been investigated in detail during the last decade is the racetrack technology, first proposed by Parkin et al. (2008), in which domain walls were moved in a controlled way through magnetic nanowires to store, read, and write information. Domain walls in such nanowires could be moved by spin-polarized currents (Fernandez-Roldan and Chubykalo-Fesenko 2022), where the chirality of the Oersted field associated with the current could either speed up the domain wall movement or reduce it significantly (Schöbitz et al. 2019). The next step in the design of racetrack memories was using skyrmions due to several advantages compared to domain walls, such as their topological stability and the very low spin-polarized current density necessary to move them (Kang et al. 2016). Skyrmion racetrack memories, on the other hand, still have disadvantages, such as the Magnus force in ferromagnetic materials causing a deflected skyrmion motion (Jin et al. 2016), or the topological Hall effect pushing skyrmions to the edges of the racetrack (Ladak et al. 2022). This problem could be solved by using hopfions instead of skyrmions (Wang et al. 2019, Ladak et al. 2022, Arava and Phatak 2023).

Such a hopfion-based racetrack memory was suggested by Göbel et al. (2020). They showed that the hopfions' emergent magnetic field resulted in a topological Hall signature, which was independent of anomalous and conventional Hall effects, vanished globally, and could be switched by electric currents or magnetic fields. In such a hopfion-based racetrack memory, data would be written in the form of a hopfion or its absence at a defined position. Since the hopfions would move along the racetrack without transverse deflection, as mentioned before, they could be more easily transported and measured than skyrmions. The calculated signal for a hopfion moving along the racetrack (Fig. 3.15a) is given in Fig. 3.15b (Göbel et al. 2020).

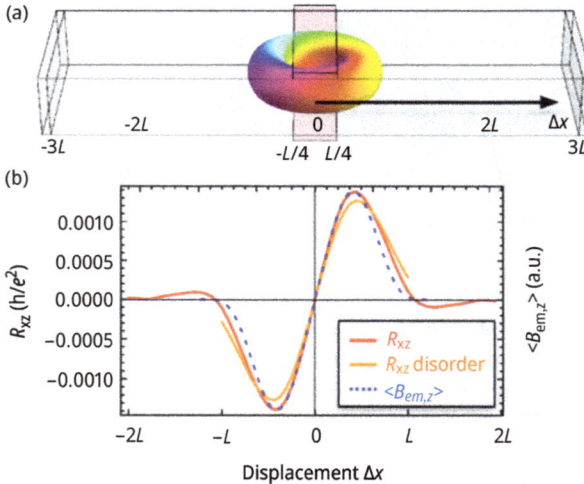

Fig. 3.15: Local detection of hopfions in a nanostripe: (a) the considered geometry with the displacement Δx of the hopfion with respect to the detecting leads (light red), as indicated. A current flowing between the gray contacts (along x) leads to a Hall voltage between the two light red contacts (along $\pm z$); (b) antisymmetric Hall resistance signal, calculated for a perfect geometry (red) and taking into account disorder (orange). The average emergent field between the two red leads is shown as a blue-dashed curve (reprinted from Göbel et al. (2020), originally published under a CC-BY license).

Another approach to using hopfions for data processing was suggested by Zhang et al. (2023), who proposed a hopfion-based neural network for a possible neuromorphic device. Neuromorphic computing is a brain-inspired way to build a computer that should work like neuronal networks and overcome the "von Neumann bottleneck" of recent computers, where data processing and storage are faster than data transport (Emma 1997). Neuromorphic hardware includes diverse spintronic elements, such as magnetic tunnel junctions, which could be used to enable "learning" and "forgetting," but could also be based on skyrmions, which could be driven along defined nano-tracks or be used in different other ways for neuromorphic hardware (Blachowicz and Ehrmann 2020). To move forward from skyrmions to hopfions in neuromorphic computing, Zhang et al. investigated the polarization-dependent hopfion motion and the frequency-dependence of magnon scattering at hopfions (Zhang et al. 2023). The correlation between spin-wave polarization and the Hall angle of the hopfion enabled programming the hopfion's 3D path. Neuromorphic computing could thus use the full 3D movement of a hopfion to realize meta-learning and the nonlinear magnon-hopfion scattering to encode neuron connections (Zhang et al. 2023).

A more general approach was given by Ghosh and Grytsiuk (2020), who discussed "or-bitronics," i.e., the interaction between electronic orbitals and the electron spin or other degrees of freedom. They explained that noncollinear spin systems, such as hopfions, might show a broad range of orbital magnetism due to scalar spin chirality and thus have generally great potential for spintronic applications and brain-inspired computing.

Generally, while many requirements for their potential use in data storage and processing have thus already been investigated, only a few research groups have shown concepts for the use of hopfions in racetrack memories, in neuromorphic com-puting, or other data processing approaches yet. We hope that this overview will stim-ulate further theoretical and experimental research on hopfions as a promising base for future data storage and processing devices.

3.4 Magnetization dynamics for temperatures above 0 K

The present topic extends the description of magnetization dynamics with the use of LLG-type equations beyond the standard 0 K case. This issue is very important for practical reasons, since knowledge about it is fundamental for understanding the functionality of real magnetoelectronic devices working in a wide range of tempera-tures and, of course, for understanding the magnetization vector evolution expressed by the adequate formulas. The whole set of related topics is sometimes named heat-assisted dynamics.

The goal of this section is thus to introduce a quantitative description for $T > 0$ K, or simply, to extend the formalism met in previous equations. Our derivations will give us information about the statistically averaged equilibrium values of the magne-tization, and first of all, will introduce a modified version of the LLG equation, where new thermal random fields will occur. Another important point is that in the proper evolution equation, the length of the magnetization vector is not conserved and undergoes relaxation mechanisms never found for $T = 0$.

Let us start our considerations with a short reminder of the local magnetic moment formula projected onto the effective local magnetic field direction inside the magnetic material. It equals

$$p_{zs} = g_s m_s \mu_{\mathrm{B}}, \tag{3.43}$$

where g_s is the spin-electron g-factor and m_s is the magnetic spin quantum number. Since the ferromagnetism of metals depends in principle on the electron spin, for simplicity we assume $g_s \approx 2$ and $m_s = 1/2$. Before presenting the magnetic moment equation of evolution, we make our formalism more practical by extending the expression to work for the whole set of electrons in a magnetic material instead of using a single particle formalism.

In the case of a many-electron set and for temperatures above absolute zero, there are many possibilities for the magnetic moment orientations and values of the total energy U of such a system. The energy is a function of the temperature T. This fact can be adopted by an operational formalism using a probability distribution function. The function, the so-called Boltzmann distribution, can be written as

$$f = e^{-(U/kT)}. \tag{3.44}$$

This distribution function measures the competition between thermal energy kT, with k being the Boltzmann constant, and the characteristic energy of the underlying physical effect U, or more precisely, it measures the probability of occurrence of a given energy state in the presence of a given temperature. The characteristic state energy U in magnetic materials results from interactions between the magnetic moments and the aforementioned effective magnetic fields, represented by the magnetic induction B, and can be calculated for the single magnetic moment case, locally, from the elementary formula

$$U = -\vec{p}_m \cdot \vec{B}. \tag{3.45}$$

Since the magnetic moment is quantized, and $p_m = \mu_{\mathrm{B}} g \sqrt{J(J+1)}$, its length can be calculated with the use of the total angular momentum quantum number J. The above energy interaction term is thus equal to

$$U = -\mu_{\mathrm{B}} g \sqrt{J(J+1)} B \cos\theta, \tag{3.46}$$

with θ being the angle between the magnetic field vector and the magnetic moment vector. Equivalently, it can be written more simply by the introduction of the magnetic quantum number m_J, which counts the possible space orientations of the magnetic moment vector referenced to the direction of the local effective magnetic field vector. The energy can be expressed in this way as

$$U = -\mu_\text{B} g B m_J. \tag{3.47}$$

Next, the population of magnetic moments, strictly speaking the number of magnetic moments per volume (concentration), possessing a given orientation, can be weighted by the probability distribution function in the following way:

$$n(m_J) = A f(m_J) = A e^{-(U/kT)} = A e^{\left((\mu_\text{B} g_s B m_J)/kT\right)}, \tag{3.48}$$

with A being an unknown factor to be revealed later. Thus, the total number of magnetic moments contributing to a given value of net magnetization at a given temperature, calculated per volume, equals

$$n = \sum_{m_J=-J}^{m_J=+J} A \exp\left((\mu_\text{B} g B m_J)/kT\right) = A \sum_{m_J=-J}^{m_J=+J} \exp\left((\mu_\text{B} g B m_J)/kT\right). \tag{3.49}$$

The concentration n is the well-known quantity derived from the basic atomic material parameters. Thanks to this, the constant A can be easily calculated from

$$A = \frac{n}{\sum_{m_J=-J}^{m_J=+J} \exp\left((\mu_\text{B} g B m_J)/kT\right)}. \tag{3.50}$$

This is why the population of magnetic moments of the given m_J can be precisely acquired from

$$n(m_J) = n \frac{\exp\left((\mu_\text{B} g B m_J)/kT\right)}{\sum_{m_J=-J}^{m_J=+J} \exp\left((\mu_\text{B} g B m_J)/kT\right)}. \tag{3.51}$$

Thus, the magnetization resulting from a specific population of moments equals

$$M(m_J) = n(m_J) p_{mz} = n(m_J) \mu_\text{B} g \sqrt{J(J+1)} \cos\theta = n(m_J) \mu_\text{B} g m_J, \tag{3.52}$$

and the total magnetization, superimposed from all the possible cases, can be calculated as the weighted average of different orientations represented by different m_J, namely

$$M = n \mu_\text{B} g \frac{\sum_{m_J=-J}^{m_J=+J} m_J \exp\left((\mu_\text{B} g B m_J)/kT\right)}{\sum_{m_J=-J}^{m_J=+J} \exp\left((\mu_\text{B} g B m_J)/kt\right)}. \tag{3.53}$$

The total magnetization, then calculated with the use of the above approach, has the physical meaning of the weighted average value of differently oriented magnetizations for a given temperature $T > 0$.

The next step in derivation will lead us to the expression of the total magnetization, averaged in the above meaning, which is commonly known as the Brillouin function. It is based on some transformations of eq. (3.53).

To begin, let us make the following substitution:

$$x = \frac{g\mu_B B}{kT}, \tag{3.54}$$

and let us use the mathematical transformation

$$M = n\mu_B g \frac{\sum_{m_J=-J}^{m_J=+J} m_J e^{m_J x}}{\sum_{m_J=-J}^{m_J=+J} e^{m_J x}} = n\mu_B g \frac{d}{dx}\left(\ln\left(\sum_{m_J=-J}^{m_J=+J} e^{m_J x}\right)\right). \tag{3.55}$$

In this way, the m_J parameter can be extracted out of the summation in the numerator, and the resulting series can be expressed using hyperbolic functions

$$\sum_{m_J=-J}^{m_J=+J} e^{m_J x} = \frac{\sinh(((2J+1)/2)x)}{\sinh(x/2)}. \tag{3.56}$$

Using the formula for the first derivative of $\sinh(x)$, we obtain the final form of the total averaged magnetization at equilibrium

$$M = ng\mu_B \left[\frac{2J+1}{2} ctgh\left(\frac{2J+1}{2}\frac{g\mu_B B}{kT}\right) - \frac{1}{2} ctgh\left(\frac{1}{2}\frac{g\mu_B B}{kT}\right)\right], \tag{3.57}$$

or in the more commonly known form as the Brillouin function $B_J(x)$ (the function in the square brackets below):

$$M = ngJ\mu_B \left[\frac{2J+1}{2J} ctgh\left(\frac{2J+1}{2J}\frac{g\mu_B B}{kT}J\right) - \frac{1}{2J} ctgh\left(\frac{1}{2J}\frac{g\mu_B B}{kT}J\right)\right] = ngJ\mu_B B_J(x). \tag{3.58}$$

The important point is, as mentioned earlier, that the most significant contribution to strong ferromagnetism comes from the exchange coupling of spin electrons in *3d* shells. In such a case, the elementary magnetic moment equals

$$p_m = \mu_B g\sqrt{J(J+1)} \cong \mu_B g_s \sqrt{\frac{1}{2}\left(\frac{1}{2}+1\right)} = 2\mu_B \sqrt{\frac{1}{2}\left(\frac{1}{2}+1\right)}, \tag{3.59}$$

where, for simplicity, we have assumed $g = g_s = 2$. This is why the general formula for magnetization is reduced to

$$M = n\mu_B \left[2ctgh\left(\frac{2\mu_B B}{kT}\right) - ctgh\left(\frac{\mu_B B}{kT}\right)\right], \tag{3.60}$$

but because of

$$2ctgh(2x) = 2\frac{\cosh(2x)}{\sinh(2x)} = 2\frac{\cosh^2(x) + \sinh^2 x}{2\sinh(x)\cosh(x)} = 2\frac{\cosh^2(x)}{2\sinh(x)\cosh(x)} + 2\frac{\sinh^2 x}{2\sinh(x)\cosh(x)}$$

$$= ctgh(x) + tgh(x)$$

$$(3.61)$$

the formula for M can be simplified to

$$M = n\mu_B \left[tgh\left(\frac{\mu_B B}{kT}\right) \right]. \tag{3.62}$$

This relatively simple formula expresses the magnetization of a given sample as a function of temperature.

Now, we will derive complementary formulas for magnetization, providing the opportunity to introduce temperature limits for the validity of the above approach, the limit known as the Curie temperature, when the magnetic material loses its ferromagnetic properties.

The effective magnetic field inside a material, the magnetic induction B in eq. (3.62), is always the superposition of the externally applied magnetic field and some internal fields of quantum origin. Thus,

$$B = \mu_0 (H_{ext} + H_{int}) = \mu_0 (H_{ext} + \alpha_{int} M) \tag{3.63}$$

with α_{int} being the so-called internal field constant, a linear parameter connecting the internal field intensity and the actual magnetization. This linear condition means the level of temperature-introduced disorder is quite high, and we are away from complete saturation – this is the reason to use the Boltzmann probability distribution function. The important point is, however, that the external field (the magnetic induction) is of the order of 1 T, while the internal field induction is 10^3 times larger (cf. considerations from Chapter 1), so that using the $H_{ext} \ll H_{int}$ condition we obtain

$$B = \mu_0 (H_{ext} + H_{int}) = \mu_0 (H_{ext} + \alpha_{int} M) \cong \mu_0 \alpha_{int} M, \tag{3.64}$$

and from the above results

$$M = n\mu_B tgh\left(\frac{\mu_B \mu_0 \alpha_{int} M}{kT}\right), \tag{3.65a}$$

$$M = n\mu_B tgh(y), \tag{3.65b}$$

where, for clarity, we have introduced the new variable $y = \mu_B \mu_0 \alpha_{int} M / (kT)$. From this newly introduced variable, we can calculate another relation, a linear one, between the magnetization M and the variable y, namely

$$M = \frac{kT}{\mu_B \mu_0 \alpha_{int}} y. \tag{3.66}$$

Both these equations, eqs. (3.65b) and (3.66), can be processed simultaneously using, for example, a graphical approach or adequate computer-based convergent algorithms to find the magnetization at a given temperature T. The question arises now from where the α_{int} parameter can be taken to make our calculations fully operative. This issue is related, as was already mentioned, to the notion of the Curie temperature T_C. Namely, for high enough temperatures, when $T \to T_C$, the magnetization approaches zero, $M \to 0$, and in eq. (3.65a) $tgh(y) \to y$, thus we have in this case

$$M = n\mu_B y = n\mu_B \frac{\mu_B \mu_0 \alpha_{int} M}{kT_C}.$$ (3.67)

After simplification of M on both sides of the above, we finally have

$$1 = n\frac{\mu_B^2 \mu_0 \alpha_{int}}{kT_C}$$ (3.68)

or

$$\alpha_{int} = \frac{kT_C}{n\mu_0 \mu_B^2}.$$ (3.69)

This is a clear message now: the internal field constant is a direct function of the experimentally accessible Curie temperature T_C. Putting back the just obtained result into eq. (3.66), we get

$$M(y) = n\mu_B \frac{T}{T_C} y,$$ (3.70)

from one side, and again the counterpart eq. (3.65b) from the second side. Solving this system of equations, graphically or numerically, it could be good to know the possible range of the y variable. Estimating the saturation value, assuming some typical values of the Curie temperature (10^3 K), the concentration (10^{28} m^{-3}), and the magnetization at saturation (10^5 A/m), we obtain

$$y_s = \frac{1}{n\mu_B} M_s \frac{T_c}{T} \approx \frac{1}{10^{28}\text{m}^{-3}10^{-23}\text{Am}^2} 10^5 \text{A/m} \frac{10^3\text{K}}{T} = \frac{10^3\text{K}}{T}.$$ (3.71)

Thus, for a rational range of temperatures from 1 K up to 10^3 K, the variable y falls in the range of $\langle 10^0 \div 10^3 \rangle$.

To summarize our considerations, we solve the following set of two equations:

$$\begin{cases} M(y) = n\mu_B tgh(y) \\ M(y) = n\mu_B \frac{T}{T_C} y \end{cases}, \quad y \in \langle 1; 10^3 \rangle$$ (3.72)

or the self-consistent form of the equation

$$tgh(y) = \frac{T}{T_C}y, \tag{3.73}$$

with the use of the two material parameters: T_C and n, and with the temperature T treated as the free parameter. The obtained value of y can then be back-substituted into one of the $M(y)$ to get the final result.

For the most commonly known element of Fe, the read-out values (Fig. 3.16) of the crossed variable y_{cross} are 7.50 and 3.47 for temperatures of 140 and 300 K, respectively. Subsequently, the calculated magnetizations at these temperatures are equal to $7.93 \cdot 10^5$ and $7.86 \cdot 10^5$ A/m, respectively, assuming the concentration $n_{Fe} = 8.49 \cdot 10^{28}$ m^{-3}. Additionally, the calculated values of the magnetization for Fe, Co, and Ni for temperatures of 140, 300, and 400 K are given in Tab. 3.1.

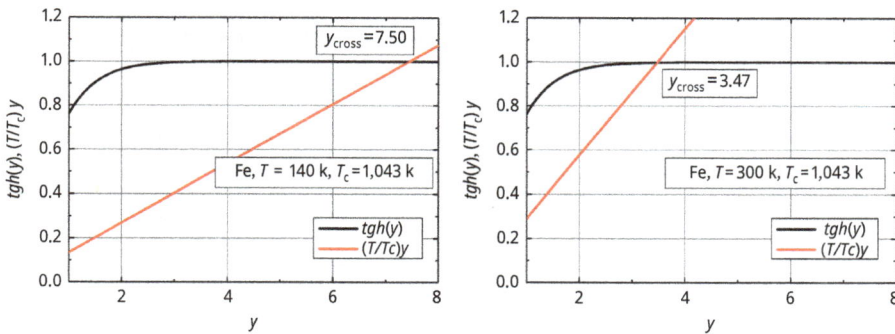

Fig. 3.16: Self-consistent solving of equations leading to the determination of the magnetization at temperatures of 140 K (left panel) and 300 K (right panel) for the element Fe.

Tab. 3.1: The concentrations, Curie temperatures, and equilibrium magnetizations at 140, 300, and 400 K for the ferromagnetic elements Fe, Co, and Ni, respectively.

Element	n (m^{-3})	T_c (K)	M (140 K) (A/m)	M (300 K) (A/m)	M (400 K) (A/m)
Fe	$8.49 \cdot 10^{28}$	1,043	$7.87 \cdot 10^5$	$7.86 \cdot 10^5$	$7.79 \cdot 10^5$
Co	$9.10 \cdot 10^{28}$	1,388	$8.43 \cdot 10^5$	$8.44 \cdot 10^5$	$8.41 \cdot 10^5$
Ni	$9.14 \cdot 10^{28}$	627	$8.48 \cdot 10^5$	$8.19 \cdot 10^5$	$7.46 \cdot 10^5$

The results, derived from eqs. (3.67)–(3.70), are dependent on the linear factor $T \cdot y$. Hence, on the other hand, the $tgh(y)$ term is significantly nonlinear. A detailed analysis for the above materials, for a wider range of temperatures, using the approach presented here, is depicted in Fig. 3.17. It is visible that for temperatures below 400 K, the magnetization of Fe and Co is, to a good approximation, constant. The drop of magnetization for Ni is more significant, which results from the lowest value of the equivalent Curie temperature. The calculated values are also presented in Tab. 3.2.

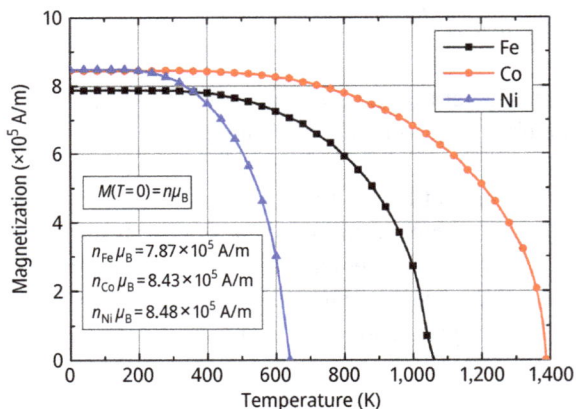

Fig. 3.17: The temperature dependence of magnetization for Fe, Co, and Ni. The values of magnetization in the bottom-left frame are valid for $T = 0$.

Tab. 3.2: The equilibrium magnetizations of Fe, Co, and Ni for the (0 K, T_c) range of temperatures.

T (K)	M_{Fe} (10^5 A/m)	M_{Co} (10^5 A/m)	M_{Ni} (10^5 A/m)	T (K)	M_{Fe} (10^5 A/m)	M_{Co} (10^5 A/m)
0	7.874	8.433	8.476	720	6.577	8.006
20	7.874	8.433	8.476	740	6.424	7.958
40	7.872	8.433	8.473	760	6.311	7.896
60	7.872	8.432	8.476	780	6.124	7.820
80	7.875	8.433	8.478	800	5.919	7.777
100	7.874	8.433	8.476	820	5.757	7.723
120	7.872	8.436	8.467	840	5.517	7.605
140	7.874	8.430	8.478	860	5.324	7.524
160	7.875	8.429	8.478	880	5.049	7.432
180	7.868	8.432	8.467	900	4.756	7.382
200	7.866	8.433	8.435	920	4.445	7.267
220	7.872	8.435	8.416	940	4.116	7.139
240	7.863	8.429	8.370	960	3.696	7.058
260	7.871	8.436	8.330	980	3.255	6.967
280	7.863	8.438	8.251	1,000	2.718	6.805
300	7.859	8.439	8.192	1,020	2.002	6.693
320	7.851	8.438	8.089	1,040	0.707	6.572
340	7.828	8.429	7.951	1,060	0	6.376
360	7.827	8.421	7.835	1,080		6.234
380	7.803	8.427	7.654	1,100		6.082
400	7.791	8.409	7.462	1,120		5.920
420	7.768	8.421	7.267	1,140		5.749
440	7.739	8.395	7.018	1,160		5.498
460	7.674	8.385	6.716	1,180		5.306
480	7.646	8.370	6.424	1,200		5.104
500	7.587	8.354	6.083	1,220		4.818

Tab. 3.2 (continued)

T (K)	M_{Fe} (10^5 A/m)	M_{Co} (10^5 A/m)	M_{Ni} (10^5 A/m)	T (K)	M_{Fe} (10^5 A/m)	M_{Co} (10^5 A/m)
520	7.537	8.341	5.623	1,240		4.596
540	7.501	8.334	5.183	1,260		4.287
560	7.398	8.302	4.618	1,280		3.966
580	7.356	8.282	3.920	1,300		3.554
600	7.247	8.239	3.001	1,320		3.208
620	7.161	8.212	1.509	1,340		2.687
640	7.054	8.205	0	1,360		2.066
660	6.975	8.141		1,380		1.090
680	6.879	8.098		1,420		0
700	6.711	8.081				

The obtained magnetizations are the equilibrium steady-state values, and in order to avoid misunderstanding, we will now use the $M_{eq}(T)$ symbol in considerations about the Landau–Lifshitz–Boltzmann (LLB) equation, instead of the M one, to mark this meaning more clearly.

Thus, there are the following three issues to be managed for this special physical problem:

1) The dependence of the magnetization on the temperature (which was solved in this chapter),
2) The thermal fluctuations of the effective magnetic field surrounding the given magnetic moment in the medium or just the existence of thermal noise,
3) The new mechanism of longitudinal relaxation of the magnetization.

The thermal dependence of physical properties means the length of the locally counted magnetization vector is no longer constant, which is a consequence of energy outflow and inflow for the given physical system – strictly speaking, the total energy density consists of the four well-known components: the Zeeman (external), demagnetizing, exchange, and magnetocrystalline energy density.

In order to make the magnetization vector adaptable for quantitative analysis and to introduce the temperature factor, there is a need to immerse a physical system in an external thermal bath to exchange energy in a nondeterministic way (Fig. 3.18), while the total energy of the system and the bath is constant. In this way, the system evolves, influenced by external energy fluctuations, and relaxes into the local energy equilibrium state at a given temperature. This approach enables the reformulation of the LLG equation and the introduction of the new LLB equation (Garanin 1997; Garanin et al. 2004; Evans et al. 2012; McDaniel 2012).

To provide the reader with some deeper understanding, apart from the strict mathematical formalism, what will be given below are some intuitive considerations.

Our main goal is to introduce a description of the new longitudinal (parallel to the magnetization vector) thermal damping or relaxation mechanism for the magneti-

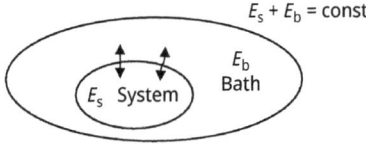

Fig. 3.18: General idea of the possibility of variations in the physical system's energy. The magnetization vector length is not conserved during its evolution.

zation vector. Until now in the book, only the transverse mechanism (perpendicular to the magnetization vector) of damping was used, and it was represented by the $-\left[\vec{M}\times\left(\vec{M}\times\vec{H}\right)\right]$ term in the LLG equation at $T = 0$ K. Using simple algebraic derivations, we see this term consists of two parts:

$$-\left[\vec{M}\times\left(\vec{M}\times\vec{H}\right)\right] = -\vec{M}\left(\vec{M}\cdot\vec{H}\right) + \vec{H}M^2, \tag{3.74}$$

which, presented graphically, gives us the hint (Fig. 3.19) that the longitudinal modifications of the magnetization can be realized by a scalar product of the $(\vec{M}\cdot\vec{H})$ type. This is the first point.

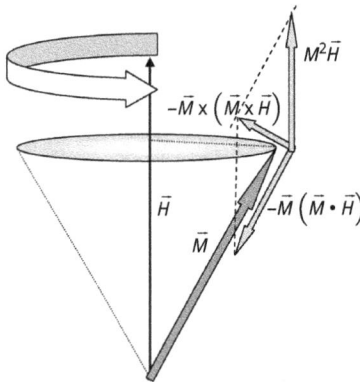

Fig. 3.19: The components of the transverse damping term in the LLG equation. The change of the magnetization vector is described with the use of the scalar product of the magnetization and the external field vectors.

The second issue is how to introduce the equivalent temperature-modified vector $\vec{M}$ (T) into the LLB equation. For the case of $T = 0$, we have in the LLG equation the following repeated term:

$$\frac{\vec{M}(0)}{M_s}. \tag{3.75}$$

For the $T > 0$ K equation, there is a need to introduce equilibrium and nonequilibrium magnetizations for the given temperature. The equivalent term can be introduced as follows:

$$\frac{M_{eq}(T)}{M^2}\vec{M}, \tag{3.76}$$

with $M_{eq}(T)$ being the equilibrium value of the magnetization obtained from the Brillouin function derived in the previous paragraph. Thus, the newly introduced equation of motion can be written in the following form (McDaniel 2012):

$$\frac{d\vec{M}}{dt} = -\gamma\left[\vec{M}\times\left(\vec{H}_{eff}+\vec{H}^{th,0}\right)\right] + \alpha_1\gamma\frac{M_{eq}(T)}{M^2}\left[\vec{M}\cdot\left(\vec{H}_{eff}+\vec{H}^{th,1}\right)\right]\vec{M} - \alpha_2\gamma\frac{M_{eq}(T)}{M^2}$$
$$\left[\vec{M}\times\left(M\times\left(\vec{H}_{eff}+\vec{H}^{th,2}\right)\right)\right], \tag{3.77}$$

where the meanings of the longitudinal damping coefficient α_1 and the transverse damping coefficient α_2 are as follows:

$$\alpha_1 = \lambda\cdot\frac{2T}{3T_C}\cdot\frac{2q}{\sinh(2q)}\cdot\frac{M(0)}{M_{eq}(T)}\bigg|_{H=0} \tag{3.78}$$

and

$$\alpha_2 = \lambda\cdot\left[\frac{\tanh(q)}{q} - \frac{T}{3T_C}\right]\cdot\frac{M(0)}{M_{eq}(T)}\bigg|_{H=0}, \tag{3.79}$$

respectively, with

$$q = \frac{3T_C}{T}\cdot\frac{1}{2(S+1)}\cdot\frac{M_{eq}(T)}{M(0)}\bigg|_{H=0}. \tag{3.80}$$

The sense of the symbols in the above expressions is as follows: λ is the damping parameter of the classical LL equation valid for a single magnetic moment, T_C is the Curie temperature, $M(0)$ is the magnetization (obviously at equilibrium) at $T = 0$, $M_{eq}(T)$ is the equilibrium magnetization at the elevated temperature T derived from the Brillouin function, S is the total spin of a given magnetic material, which can be derived from the $3d$ shell analysis, and where $\vec{H}^{th,0}$, $\vec{H}^{th,1}$, $\vec{H}^{th,2}$ are the white noised Langevin fields which fulfill the conditions

$$\langle\vec{H}^{th,i}\rangle = 0, i = 0,1,2 \tag{3.81}$$

and

$$\langle\vec{H}_\mu^{th,i}(t)\vec{H}_\nu^{th,i}(t')\rangle = kT\frac{2}{\gamma V\alpha_i M_{sat}(0)}\delta_{ij}\delta_{\mu\nu}\delta(t-t'), \quad i,j=1,2 \quad \mu,\nu=x,y,z, \tag{3.82}$$

where V is the local, microscopic volume around a given point in the magnetic sample. We can assume that the $\vec{H}^{th,0}$ field in the undamped term of the LLB equation is not correlated with the damped terms. This fact and the $\langle\vec{H}^{th,0}\rangle = 0$ property allow us to skip this term in the equation.

Considerations of the limiting cases for the longitudinal damping coefficient α_1 and the transverse damping coefficient α_2 give us some additional insight into the nature of effects, namely,

$$\lim_{T\to 0} q \to \infty \Rightarrow \frac{2q}{\sinh(2q)} \to 0, \frac{T}{T_C} \to 0, \frac{M_{eq}(T)}{M(0)} \to 1, \alpha_1 \to 0, \tag{3.83a}$$

and next

$$\lim_{T\to 0} q \to \infty \Rightarrow \frac{\tanh(q)}{q} \to 0, \frac{T}{T_C} \to 0, \alpha_2 \to 0. \tag{3.83b}$$

Thus, at $T = 0$, the damping coefficients within the system-bath model vanish (Fig. 3.20).

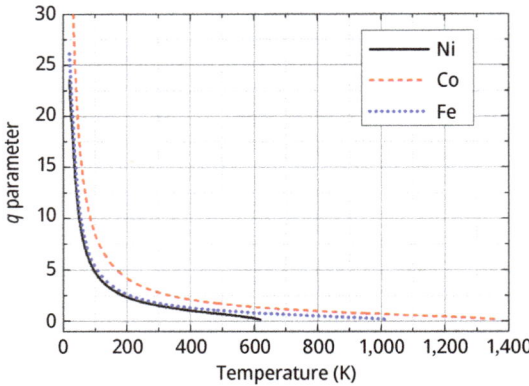

Fig. 3.20: The temperature dependence of the q-factor (eq. (3.38)). At the limiting case of the Curie temperature, the factor approaches 0.

It is worth estimating the values of the above parameters as well as the values of thermal field amplitudes, assuming typical values of $\lambda \approx 10^{-2}$. For the well-known model of the $3d$ shell origin of ferromagnetism, the material spin number was set to $S = 2\cdot1/2 = 1$, $S = 3\cdot1/2 = 3/2$, $S = 4\cdot1/2 = 2$ for nickel, cobalt, and iron, respectively. Numerical results are presented in Fig. 3.21. The values of damping coefficients are singular at $T = T_C$, and by assumption can be treated as equal for $T > T_C$.

To complete the content of this chapter, some general comments related to the formalism that enabled the transition from LLG to LLB equations are given. The approach is based on the probability density f conservation equation given in the following form:

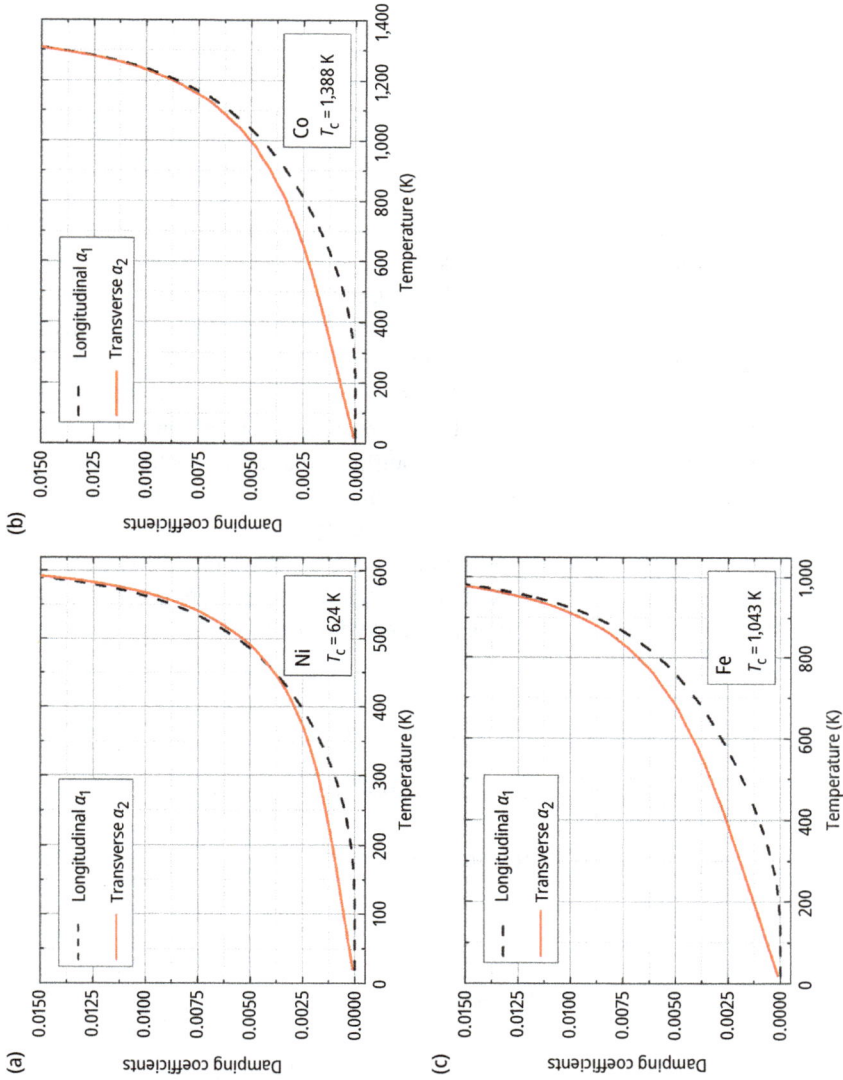

Fig. 3.21: Temperature-dependent longitudinal and transverse damping coefficients in the LLB equation for Ni (a), Co (b), and Fe materials (c).

$$\frac{\partial f}{\partial t} + \frac{\partial}{\partial \vec{n}} \vec{J} = 0, \tag{3.84}$$

where the probability current $\vec{J}$ equal to

$$\vec{J} = \left\{ \gamma \left(\vec{n} \times \vec{H} \right) - \gamma \lambda \left[\vec{n} \times \left(\vec{n} \times \vec{H} \right) \right] + \frac{\gamma \lambda}{\mu_0} T \left[\vec{n} \times \left(\vec{n} \times \frac{\partial}{\partial \vec{n}} \right) \right] \right\} f \tag{3.85}$$

flows along the direction of the unit vector $\vec{n}$. This is the Fokker–Planck equation. The physical meaning of the above will now be explained.

In this type of fundamental approach, the magnetization dynamics is considered on a unit radius sphere spanned in all possible directions of the unit magnetization vector – the radius of the sphere is normalized and equals 1. On the other hand, the sphere can be represented by the probability density distribution, equivalent to that sphere, in the sense that the value of the probability density is constant in time for $T = 0$, which means there is no probability inflow and outflow, thus $\partial f/\partial t = 0$. Hence, the overall surface integral of that probability density obviously equals 1. For the case $T > 0$, there is a probability flow, and the generalized probability current $\vec{J}$ is expressed by the vector perpendicular to the mentioned probability sphere – the $\vec{n}$ normal vector – and to the effective magnetic field vector $\vec{H}$. Knowing this, it can be written for a given point of the sphere that the current vector can be spanned with the use of the following three vectors:

$$\left(\vec{n} \times \vec{H} \right), \left[\vec{n} \times \left(\vec{n} \times \vec{H} \right) \right], \left[\vec{n} \times \left(\vec{n} \times \frac{\partial}{\partial \vec{n}} \right) \right], \tag{3.86}$$

which give us two possibilities for the two equivalent pairs of vectors:

$$\left(\vec{n} \times \vec{H} \right), \left[\vec{n} \times \left(\vec{n} \times \vec{H} \right) \right] \tag{3.87}$$

or

$$\left(\vec{n} \times \vec{H} \right), \left[\vec{n} \times \left(\vec{n} \times \frac{\partial}{\partial \vec{n}} \right) \right]. \tag{3.88}$$

It can be said that both pairs determine a local plane, being tangential to the sphere at the point where the flow takes place, and hence the probability current flow vector $\vec{J}$ is perpendicular to that plane (Fig. 3.22).

Typical probability functions of the Boltzmann type usually have the form $f = \exp$ (−energy/kT). The solution of the Fokker–Planck equation is then

$$f(\vec{n}) \sim \exp \left[-\mu_0 \vec{H} \cdot \vec{s}/kT \right], \quad \mu_0 \vec{H} = \vec{B}, \tag{3.89}$$

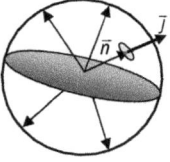

Fig. 3.22: The theoretical scene for thermal dynamics of the magnetization. The unit radius sphere represents possible directions of the magnetization orientation, while the given direction is described by the probability current flow $\vec{J}$ at the surface in that direction $\vec{n}$.

where $\vec{s}$ is the unit vector representing the spin, which can be found in the magnetic moment vector $\vec{p}_m = \mu_0 \vec{s}$. Next, starting from the following integral, calculating the average value of the spin

$$\vec{m} = \langle \vec{s} \rangle = \int \vec{n} f(\vec{n}) d^3 \vec{r} \tag{3.90}$$

and using the above Fokker–Planck equation, we come after some derivations to the LLB equation (Garanin 1997)

$$\frac{\partial m}{\partial t} = \gamma \left(\vec{m} \times \vec{H} \right) - \Gamma_{\parallel} \left(1 - \frac{\vec{m}\vec{m}_0}{m^2} \right) \vec{m} - \Gamma_{\perp} \frac{[\vec{m} \times (\vec{m} \times \vec{m}_0)]}{m^2} \tag{3.91}$$

with characteristic parallel and transverse damping terms and the corresponding efficiencies of damping, $\Gamma_{\parallel}$ and $\Gamma_{\perp}$, respectively. The details of all derivations can be found in the literature. It should be emphasized that there are several subtypes of the LLB, and this topic is still under intense investigation. Let us mention, for example, an approach similar to the above one, but treating a temperature process as an impulse in the time domain and using the notion of scattering rates in a similar way to the transport equation used in semiconductors (Mayergoyz et al. 2012).

After this overview, the next chapter will explain how exotic magnetic states and properties of thin-film or nanostructured samples can be detected.

4 Sensing magnetization dynamics

Based on the Faraday effect, two approaches, such as the magneto-optical Kerr effect (MOKE) and the diffracted magneto-optical Kerr effect (DMOKE) extension, are well-established experimental techniques used in magnetoelectronic laboratories today. In DMOKE experiments, the incident laser beam probes a set of periodically distributed micro-objects, while the light intensity and its state of polarization, measured in the obtained diffraction orders, provide information about magnetization states during their evolution. This method is effective in situations when the distance between single objects on the surface sample is of the order of the applied wavelength. Typically, the light is in the visible range; for example, its wavelength is equal to 532 nm, which is a typical green laser wavelength. From a practical perspective, DMOKE can be used for testing bit-patterned media (BPM) or in technological situations where a set of magnetic nanodevices or nanostructures has to be investigated during production processes. Also, from the optical science perspective, the method collects correlative information about nano-objects probed by the focused laser beam. In other words, it can be used to test the uniformity of magnetization dynamics for a given set of individual elements creating a given system. Hence, the MOKE signal is measured at the zeroth diffractive order (Fig. 4.1).

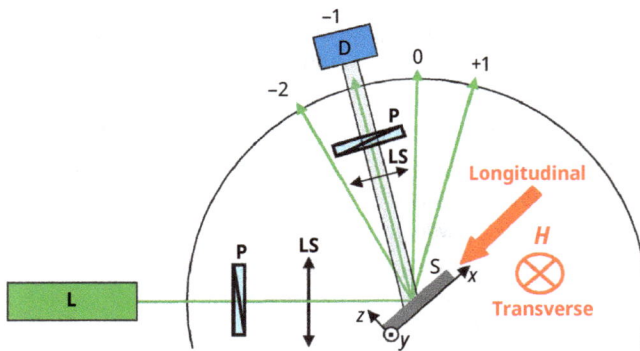

Fig. 4.1: Sample plane is defined as the x–y plane of the DMOKE (MOKE) interaction, while the z-axis is perpendicular to the sample (S) surface. Description: laser (L), lenses (LS), polarizers (P), and detector (D). The externally applied magnetic field direction is shown for the two possible experimental configurations: longitudinal and transverse, respectively. The third, commonly used MOKE configuration is the case when the field vector is perpendicular to the sample surface (parallel to z-axis) and is called polar configuration.

Let us start now with a quantitative analysis of the underlying effects. In Fig. 4.2, a selected part (four nano-objects) of a larger periodical oval-like objects set is presented. The structure possesses repeating spatial periods, d_x and d_y, measured in hori-

https://doi.org/10.1515/9783111383736-004

zontal and vertical directions of the sample, respectively. These periods are needed for the definition of the so-called reciprocal lattice vector:

$$\vec{g} = \frac{2\pi}{d_x}\vec{i} + \frac{2\pi}{d_y}\vec{j}, \tag{4.1}$$

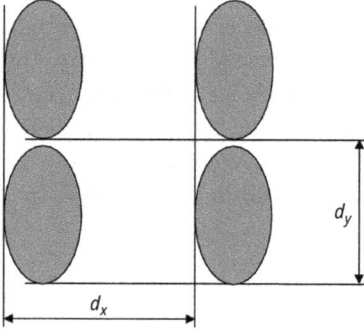

Fig. 4.2: Uniformly distributed magnetic ellipses of constant thickness forming a two-dimensional system of magnetic nano-objects.

being used in calculating the light intensities obtained at different diffractive orders. In the specific experimental situations, the Kerr signal depends on the $\vec{g} \cdot \vec{r}$ scalar product or, more explicitly, on the $e^{in\vec{g} \cdot \vec{r}}$ factor with n being the diffractive order number and $\vec{r}$ being the position vector at the sample surface. Hence, for $n = 0$, we deal with the classical MOKE case, which is used on flat surfaces, but can also be transferred to surfaces of a certain roughness (Blachowicz et al. 2017).

The measured signal, after reflection from a sample, consists of two parts: non-magnetic and magnetic ones. The magnetic component of the signal depends on an integral, the so-called form factor, being a function of the magnetization vector component oriented in the plane of interaction $M(x, y)$, that is the sample plane $(x–y)$, which is the case for the transverse or longitudinal MOKE configurations. To remind the reader of the formal definitions of these configurations, we assume that the incident and reflected beams are allocated in the $(x–z)$ plane, and the external magnetic-field vector is parallel to the y-axis (transverse MOKE, $M(x, y) = M_y$) or the vector is parallel to the x-axis (longitudinal MOKE, $M(x, y) = M_x$). The sample plane is the $(x–y)$ one. In this case, the magnetic form factor equals (Grimsditch et al. 2002)

$$f_n^m = \iint M(x,y)e^{in\vec{g} \cdot \vec{r}}dx \, dy, \tag{4.2}$$

where the integration is carried out over a single object surface, while the eventual system periodicity is included by the use of the $\vec{g} \cdot \vec{r}$ product. The result of the integration is a complex number. It consists of real and imaginary parts, and the obtained "complex" intensity is proportional to

$$I^m \sim \mathrm{Re}\left[f_n^m\right] + A_n \mathrm{Im}\left[f_n^m\right], \tag{4.3}$$

with the A_n factor being dependent on Fresnel's r_{pp}^m/r_{pp} ratio, that is, the ratio of the magnetic and the nonmagnetic Fresnel's coefficients. The (pp) subscript represents the so-called transverse magnetic (TM) polarized light for both incoming and outgoing rays (the electric field vector of the incident beam lies in the plane of incidence, the plane defined by the following vectors: the wave vector of the beam propagation direction and the vector normal to the sample surface) and the transverse MOKE geometry. The (pp) convention for marking the polarization state is another choice of notation against the TM one. Hence, the case of another MOKE geometry and the respective calculated intensity is similar to the one presented here. The issue is, however, that, in general, the measured optical signal intensity contains both the nonmagnetic and the magnetic contributions of the subsequent electric fields, namely

$$E_{\mathrm{out}} = E_{\mathrm{in}}\left(r_{pp}f_n + r_{pp}^m f_n^m\right). \tag{4.4}$$

From the above expression, and after some very simple derivations, we get the proportionality of the magnetic form factor f_n^m to the ratio r_{pp}^m/r_{pp}. Thus, remembering that all the aforementioned quantities are complex numbers, and what is important, especially for the complex electric field amplitudes, for both the incident and the reflected beams, the reflected intensity is proportional to $E_{\mathrm{out}}E_{\mathrm{out}}^*$, the real number measurable by a photodetector. Hence, the intensity is proportional to

$$I^m \sim \mathrm{Re}\left[f_n^m\right] - \frac{\mathrm{Im}\left[r_{pp}^m/r_{pp}\right]}{\mathrm{Re}\left[r_{pp}^m/r_{pp}\right]} \mathrm{Im}\left[f_n^m\right]. \tag{4.5}$$

From this, the meaning of the factor A_n results, namely

$$A_n = -\frac{\mathrm{Im}\left[r_{pp}^m/r_{pp}\right]}{\mathrm{Re}\left[r_{pp}^m/r_{pp}\right]}, \tag{4.6}$$

while the nonmagnetic form factor f_n is only a shape-dependent quantity, which is obviously deducible from the simple integral:

$$f_n = \iint e^{in\vec{g}\cdot\vec{r}} dx dy. \tag{4.7}$$

The reader can search the literature for specific values of the A_n factor. For example, for a square system of 800 nm permalloy (Py) circular disks of 60 nm thickness, spaced equally with a period of 1,600 nm, the best agreement between experiment and theory was achieved for $A_n = -0.05\,n$. Also, in many practical situations $A_n = -(0 \div 0.3)n$ was found (Vavassori et al. 2003; Vavassori et al. 2007), where n is the diffraction order to be measured.

To provide a specific example of the DMOKE analysis, the square objects lattice case will now be presented. The single cell for this type is depicted in Fig. 4.3.

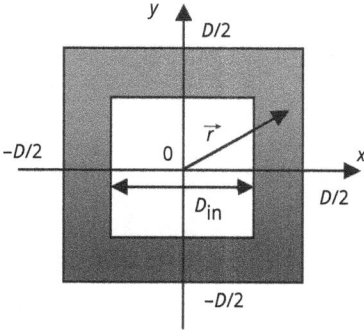

Fig. 4.3: A square single magnetic cell of external dimensions $D \times D$. The inner square cut has the size $D_{in} \times D_{in}$. The vector $\vec{r}$ points to an arbitrary position where the magnetization vector can be accounted for in order to calculate DMOKE amplitudes.

Assuming the squares are equally spaced in both the x and y directions, thus $d_x = d_y = d$, we obtain subsequent reciprocal vectors, phases, and factors, namely

$$\vec{g} = \frac{2\pi}{d}\vec{i} + \frac{2\pi}{d}\vec{j}, \tag{4.8a}$$

$$e^{in\vec{g}\cdot\vec{r}} = e^{in\left(\frac{2\pi}{d}x + \frac{2\pi}{d}y\right)} = e^{in\frac{2\pi}{d}(x+y)} = e^{in\frac{2\pi}{d}x} \cdot e^{in\frac{2\pi}{d}y}, \tag{4.8b}$$

$$f_n^m = \iint M(x,y)e^{in\frac{2\pi}{d}(x+y)}dxdy = \iint M(x,y)\cos\left(n\frac{2\pi}{d}(x+y)\right)dxdy + \\ + i\iint M(x,y)\sin\left(n\frac{2\pi}{d}(x+y)\right)dxdy, \tag{4.8c}$$

where the integration is carried out, as mentioned earlier, only at the single object area. This is why we obtain the relatively simple expressions representing the DMOKE signals in an arbitrary nth order:

$$\mathrm{Re}\left[f_n^m\right] = \iint M(x,y)\cos\left(n\frac{2\pi}{d}(x+y)\right)dxdy \tag{4.9a}$$

and

$$\mathrm{Im}\left[f_n^m\right] = \iint M(x,y)\sin\left(n\frac{2\pi}{d}(x+y)\right)dxdy. \tag{4.9b}$$

For $n = 0$, which is the common MOKE case, we obtain

$$\mathrm{Re}\left[f_n^m\right] = \iint M(x,y)dxdy \tag{4.10a}$$

and

$$\operatorname{Im}\left[f_n^m\right] = 0, \tag{4.10b}$$

respectively. Next, assuming the exemplary case of the transverse MOKE configuration (the case of the external magnetic field vector directed perpendicularly to the plane of incidence), $M(x,y) = M_y$, we have

$$\operatorname{Re}\left[f_n^m\right] = \iint M_y(x,y)\cos\left(n\frac{2\pi}{d}(x+y)\right)dxdy \tag{4.11a}$$

and

$$\operatorname{Im}\left[f_n^m\right] = \iint M_y(x,y)\sin\left(n\frac{2\pi}{d}(x+y)\right)dxdy. \tag{4.11b}$$

For the longitudinal configuration $M(x,y) = M_x(x,y)$ we get

$$\operatorname{Re}\left[f_n^m\right] = \iint M_x(x,y)\cos\left(n\frac{2\pi}{d}(x+y)\right)dxdy \tag{4.12a}$$

and

$$\operatorname{Im}\left[f_n^m\right] = \iint M_x(x,y)\sin\left(n\frac{2\pi}{d}(x+y)\right)dxdy, \tag{4.12b}$$

respectively. In practical situations, for a discretized set of data obtained from a micromagnetic calculation and the finite element nodes, integrals should be replaced with the appropriate sums. It is easy to imagine that the following expressions should fulfill this requirement for the respective transverse and longitudinal configurations $S_{Re}^{(y)}, S_{Im}^{(y)}, S_{Re}^{(x)}, S_{Im}^{(x)}$:

$$S_{Re}^{(y)} = \operatorname{Re}\left[f_n^m\right] \approx \sum_x\sum_y M_y(x,y)\cos\left(n\frac{2\pi}{d}(x+y)\right), \tag{4.13a}$$

$$S_{Im}^{(y)} = \operatorname{Im}\left[f_n^m\right] \approx \sum_x\sum_y M_y(x,y)\sin\left(n\frac{2\pi}{d}(x+y)\right), \tag{4.13b}$$

$$S_{Re}^{(x)} = \operatorname{Re}\left[f_n^m\right] \approx \sum_x\sum_y M_x(x,y)\cos\left(n\frac{2\pi}{d}(x+y)\right), \tag{4.13c}$$

$$S_{Im}^{(x)} = \operatorname{Im}\left[f_n^m\right] \approx \sum_x\sum_y M_x(x,y)\sin\left(n\frac{2\pi}{d}(x+y)\right). \tag{4.13d}$$

The only practical difference between the transverse and longitudinal cases results from the method of the DMOKE signal detection. The transverse case produces changes only in the reflected beam intensity, while for the longitudinal case, the state of polarization for the reflected beam is modified by changes in the magnetization and requires specific detection. Thus, for the transverse configuration, a single photodiode is the normal choice, while for the longitudinal case, two photodiodes (A, B), sensing intensities of both (*s*) and (*p*) polarization states in the outgoing beam, are used, and the DMOKE signal is proportional to the ratio $(A - B)/(A + B)$. Such polarization analysis usually requires adequate beam splitters, for example, a Wollaston prism.

In practice, DMOKE collects a signal from a given set of nano-objects. The micro-optics can cover regions of about 500–1,000 nm, and then the average information from the objects can be disturbed by statistically independent fluctuations of the distances between them, which is expressed by the quantity Δd – theoretically speaking, it is the standard deviation. The measure of this uncertainty can easily be calculated from the first derivatives of eqs. (4.13a)–(4.13d) and thus can be expressed as changes of the signals in the subsequent orders $\Delta S_{\mathrm{Re}}^{(y)}, \Delta S_{\mathrm{Im}}^{(y)}, \Delta S_{\mathrm{Re}}^{(x)}, \Delta S_{\mathrm{Im}}^{(x)}$:

$$\Delta S_{\mathrm{Re}}^{(y)} = \sum_x \sum_y \left[\frac{\partial M_y}{\partial d} \cos\left(n\frac{2\pi}{d}(x+y)\right) + M_y \sin\left(n\frac{2\pi}{d}(x+y)\right) \frac{2\pi(x+y)}{d^2} n \right] \Delta d, \quad (4.14a)$$

$$\Delta S_{\mathrm{Im}}^{(y)} = \sum_x \sum_y \left[\frac{\partial M_y}{\partial d} \sin\left(n\frac{2\pi}{d}(x+y)\right) - M_y \cos\left(n\frac{2\pi}{d}(x+y)\right) \frac{2\pi(x+y)}{d^2} n \right] \Delta d, \quad (4.14b)$$

$$\Delta S_{\mathrm{Re}}^{(x)} = \sum_x \sum_y \left[\frac{\partial M_x}{\partial d} \cos\left(n\frac{2\pi}{d}(x+y)\right) + M_x \sin\left(n\frac{2\pi}{d}(x+y)\right) \frac{2\pi(x+y)}{d^2} n \right] \Delta d, \quad (4.14c)$$

$$\Delta S_{\mathrm{Im}}^{(x)} = \sum_x \sum_y \left[\frac{\partial M_x}{\partial d} \sin\left(n\frac{2\pi}{d}(x+y)\right) - M_x \cos\left(n\frac{2\pi}{d}(x+y)\right) \frac{2\pi(x+y)}{d^2} n \right] \Delta d. \quad (4.14d)$$

At a first approximation, the partial derivatives $\partial M_x/\partial d, \partial M_y/\partial d$ can be omitted – in case that each single nano-object is of the same size – and the measure of spatial distribution uncertainty can be simplified to

$$\Delta S_{\mathrm{Re}}^{(y)} = \Delta d \frac{2\pi}{d^2} n \sum_x \sum_y \left[M_y \sin\left(n\frac{2\pi}{d}(x+y)\right)(x+y) \right], \quad (4.15a)$$

$$\Delta S_{\mathrm{Im}}^{(y)} = -\Delta d \frac{2\pi}{d^2} n \sum_x \sum_y \left[M_y \cos\left(n\frac{2\pi}{d}(x+y)\right)(x+y) \right], \quad (4.15b)$$

$$\Delta S_{\mathrm{Re}}^{(x)} = \Delta d \frac{2\pi}{d^2} n \sum_x \sum_y \left[M_x \sin\left(n\frac{2\pi}{d}(x+y)\right)(x+y) \right], \quad (4.15c)$$

$$\Delta S_{\mathrm{Im}}^{(x)} = -\Delta d \frac{2\pi}{d^2} n \sum_x \sum_y \left[M_x \cos\left(n\frac{2\pi}{d}(x+y)\right)(x+y) \right]. \quad (4.15d)$$

Sensing the random distribution of the distances between single nano-objects increases with the DMOKE order n and is inversely proportional to the second power of the average distance d between them. The total uncertainty is derived from the principle of propagation of individual uncertainties (Fig. 4.4), namely

$$\Delta S_{\mathrm{tot}} = \sqrt{(\Delta S_{\mathrm{Re}})^2 + A_n^2 (\Delta S_{\mathrm{Im}})^2} \quad (4.16)$$

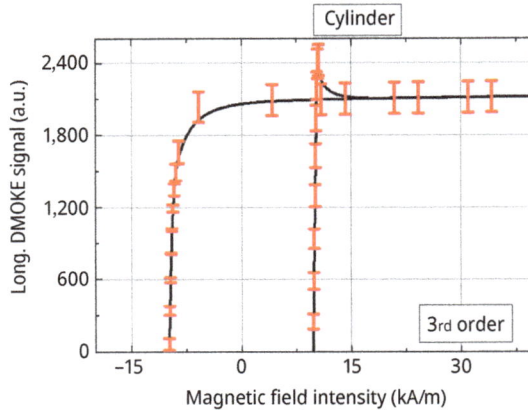

Fig. 4.4: The level of uncertainties (upper part of the hysteresis loop) depicted for the case of the longitudinal configuration of the DMOKE signal for the third diffraction order. The sample has a cylindrical shape with a diameter of 100 nm; the assumed uncertainty of distance distortion equals 10 nm, while the cylinders are spaced 600 nm apart.

Now, the next figures show some representative samples and respective DMOKE signals. The marked points (P_1, P_2, . . .) visible in the figures indicate the special moments in magnetization evolution characterized by a rapid change in its magnetic state.

The important point is that spatial uncertainties are never detectable in the zeroth diffraction order. The reader should carefully compare the relative amplitudes of signals, scales, and uncertainties for the subsequent cases presented here. Some regularities and tendencies are easy to notice (Figs. 4.5–4.21). More detailed simulations of several special geometries can be found in the literature (e.g., Ehrmann and Blachowicz 2017b, 2018; Blachowicz and Ehrmann 2016a, 2018).

At the end of this chapter, to give the reader some flavor of correlated effects in BPM, the results of simulations are presented in graphical form. The physical point of the effect is such that a super-cell, consisting of 64 nanoelements, adopts magnetic properties from its elementary basic constituents. The representative magnetic states are shown in duplicate, with and without magnetization vector fields. The four states visible in Figs. 4.22 and 4.23 are remanent ones, simulated at zero external magnetic field and can be applied as bits to store information. The size of the super-cell equals 1,040 nm × 1,040 nm and is thus roughly twice as large as the typical wavelength of light in the visible range so that magnetization dynamics are sensitive to the DMOKE approach. The interested reader is invited to read the respective works (e.g., Blachowicz et al. 2017).

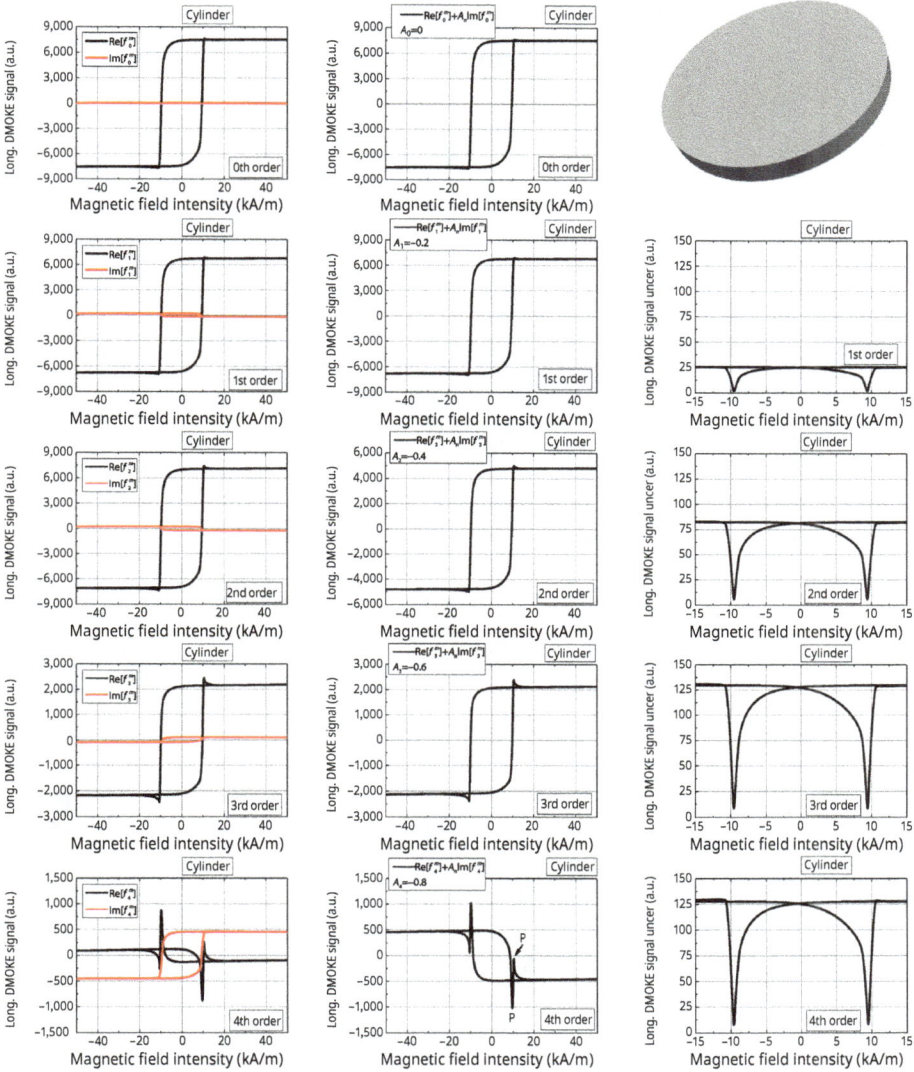

Fig. 4.5: DMOKE signals for the zeroth to the fourth order simulated in longitudinal configuration. Graphs depict the real and the imaginary parts (left column), their sum (middle column), and the uncertainty of the signals (right column) for a set of ferromagnetic cylinders spaced by 600 nm. The diameter of a single cylinder equals 100 nm, its height equals 10 nm, and the assumed uncertainty of the distance between the cylinders equals 10 nm. The samples are made from permalloy (Py). The externally applied field was swept in the range of ±300 kA/m. The arbitrarily assumed formula for the A_n factor is $A_n = -0.2n$.

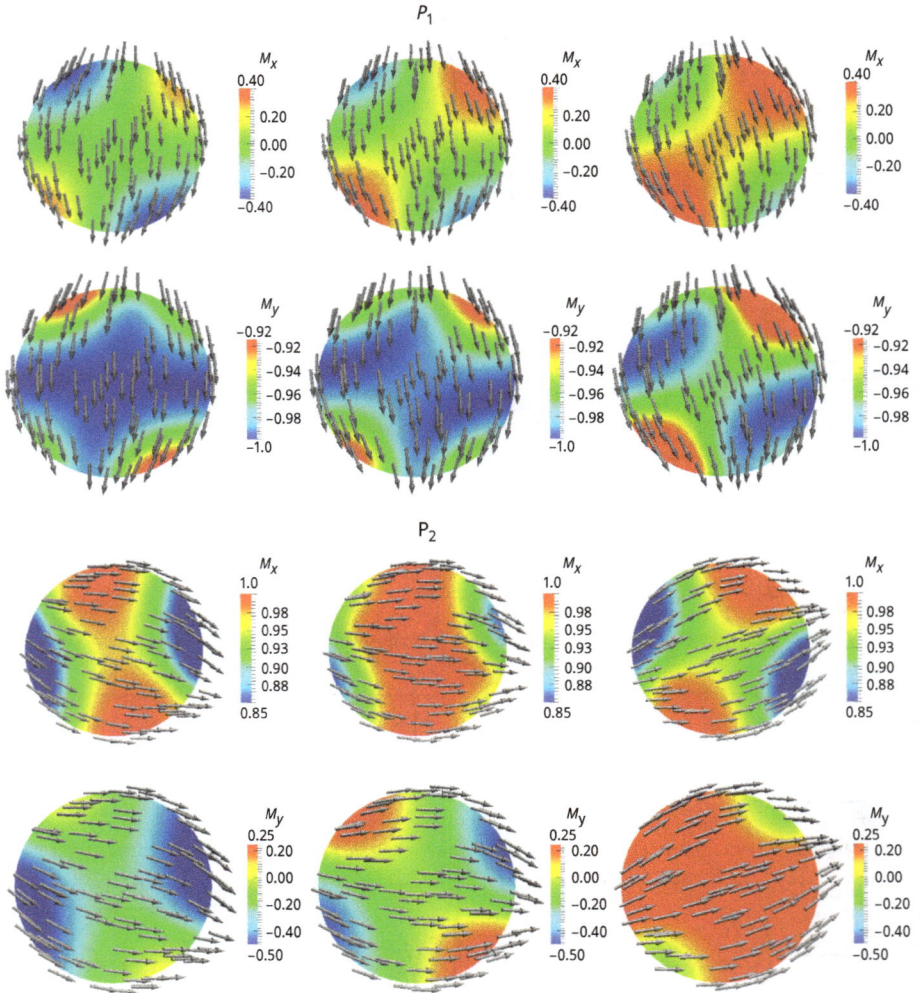

Fig. 4.6: The evolution of the magnetization is seen as the rapid change in the magnetization. The middle column is equivalent to the points P_1 and P_2 marked in Fig. 4.5, while the left and right panels are equivalent to earlier and later moments (spaced in time by approx. 1 ns), respectively.

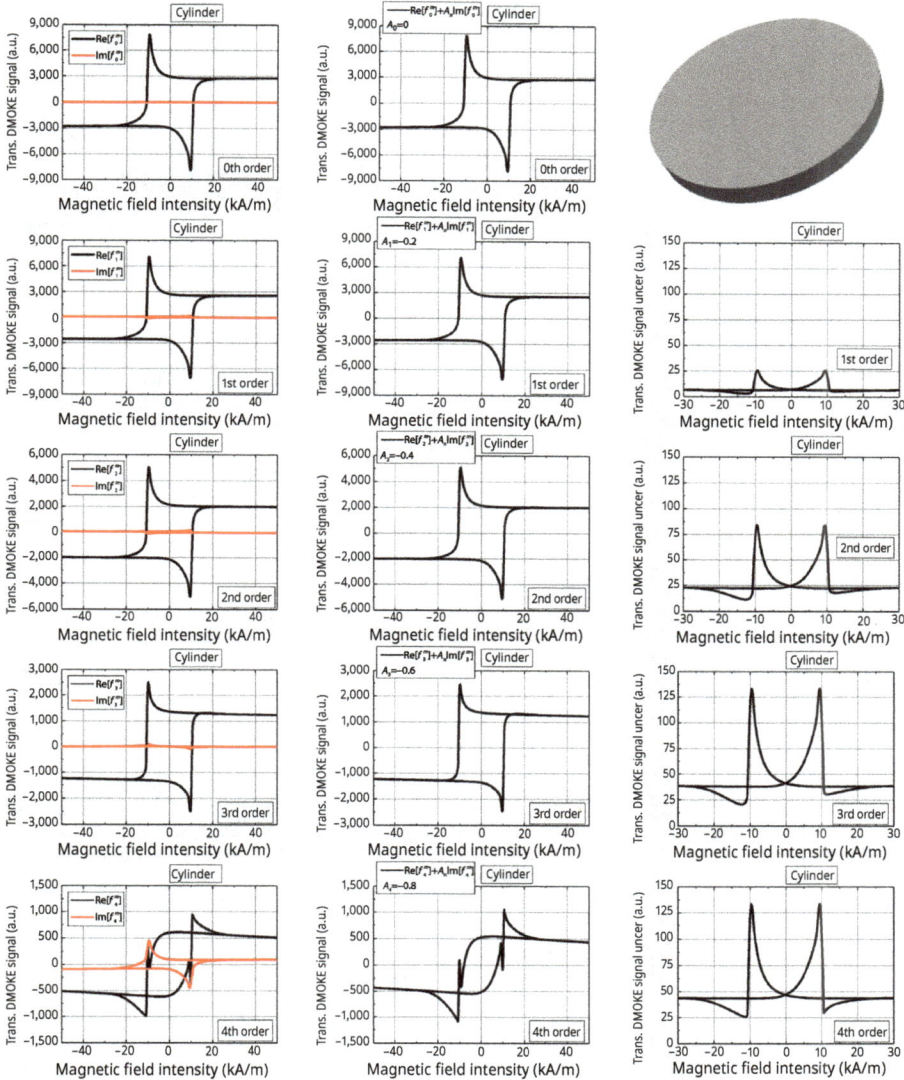

Fig. 4.7: DMOKE signals for the zeroth to the fourth order, simulated in transverse configuration. Graphs depict the real and the imaginary parts (left column), their sum (middle column), and the uncertainty of the signals (right column) for a set of ferromagnetic cylinders spaced by 600 nm. The diameter of a single cylinder equals 100 nm, its height equals 10 nm, and the assumed uncertainty of the distance between the cylinders equals 10 nm. The samples are made from permalloy (Py). The externally applied field was swept in the range of ±300 kA/m. The arbitrarily assumed formula for the A_n factor is $A_n = -0.2n$.

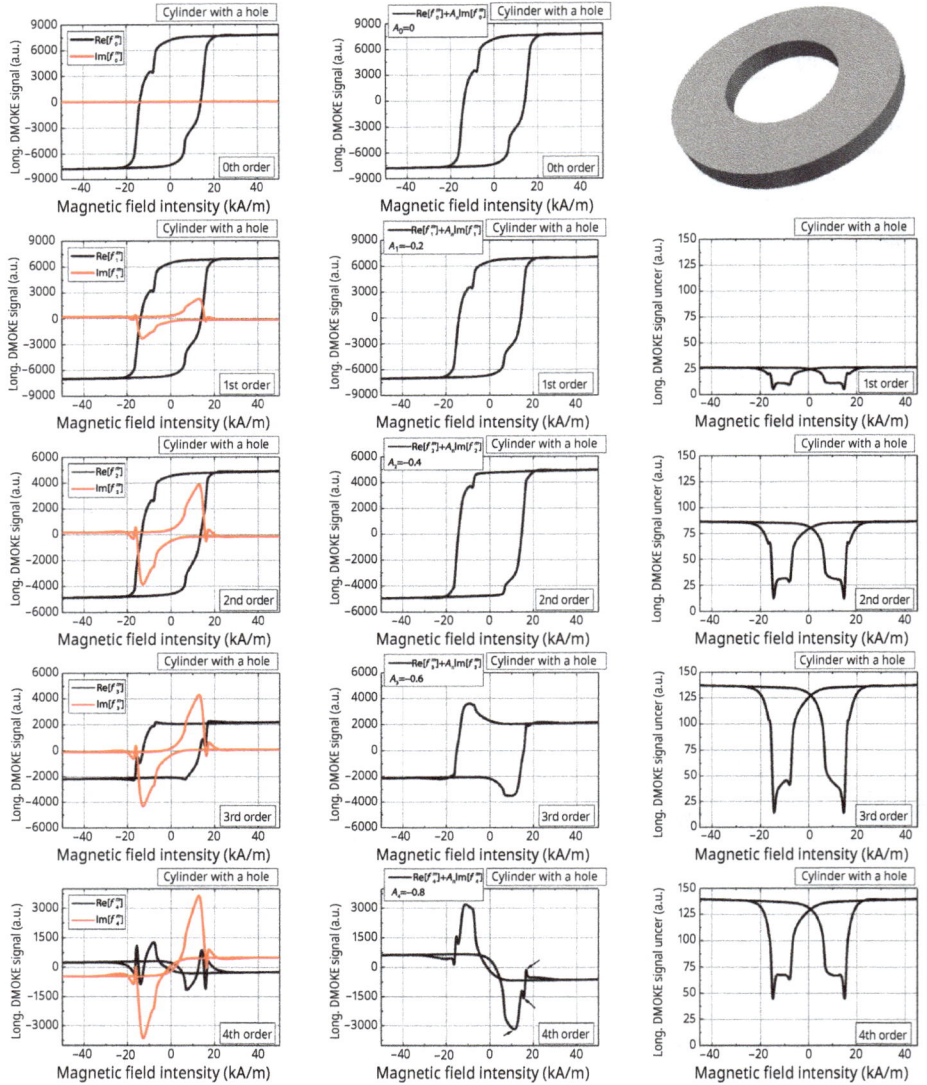

Fig. 4.8: DMOKE signals for the zeroth to the fourth order, simulated in longitudinal configuration. Graphs depict the real and the imaginary parts (left column), their sum (middle column), and the uncertainty of the signals (right column) for a set of ferromagnetic cylinders with a hole spaced by 600 nm. The diameter of a single cylinder equals 100 nm, its height equals 10 nm, the internal hole diameter equals 50 nm, and the assumed uncertainty of distance between the cylinders equals 10 nm. The samples are made from permalloy (Py). The externally applied field was swept in the range of ±300 kA/m. The arbitrarily assumed formula for the A_n factor is $A_n = -0.2n$.

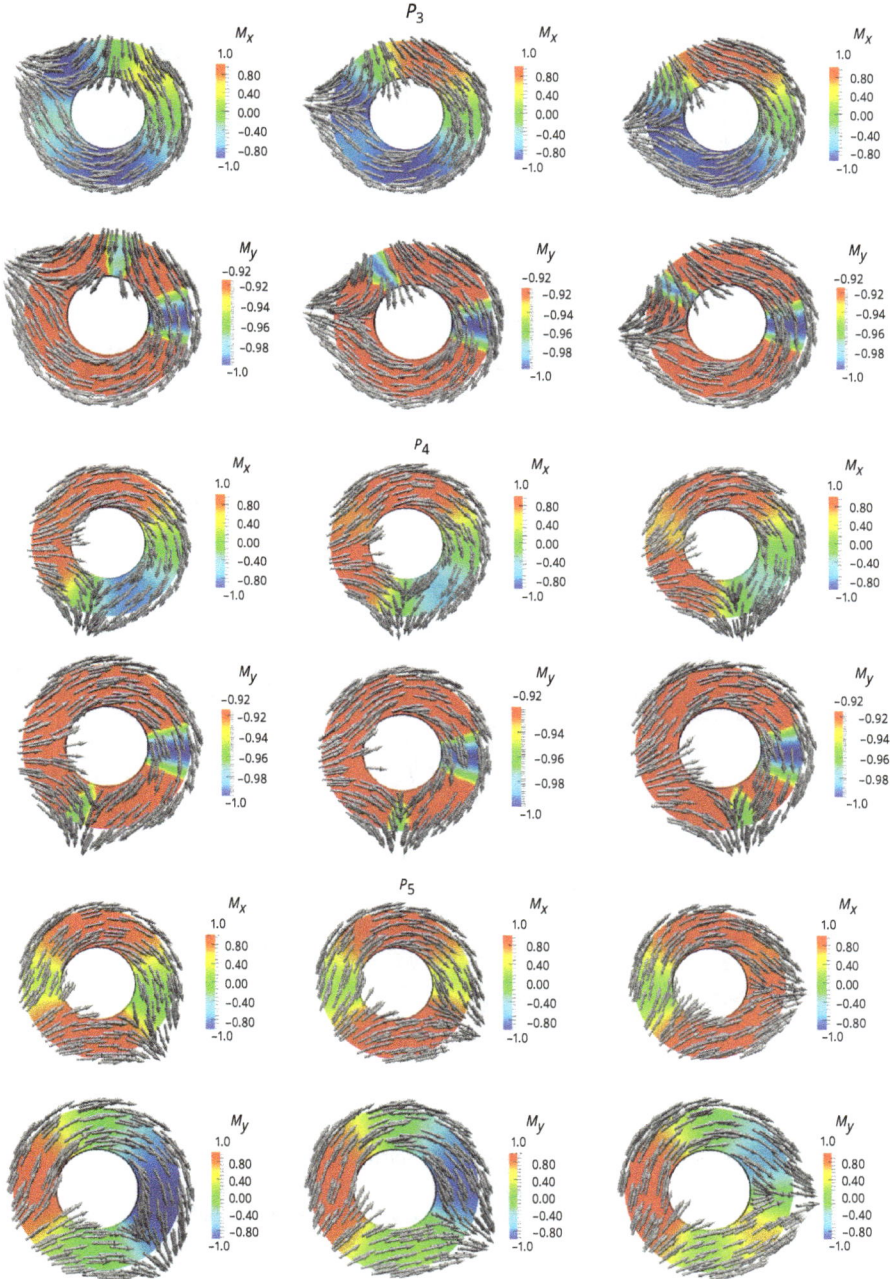

Fig. 4.9: The evolution of the magnetization is seen as a rapid change in the magnetization. The middle column is equivalent to the points P_3, P_4, and P_5 marked in Fig. 4.8, while the left and right panels are equivalent to earlier and later moments (spaced in time by approx. 1 ns), respectively.

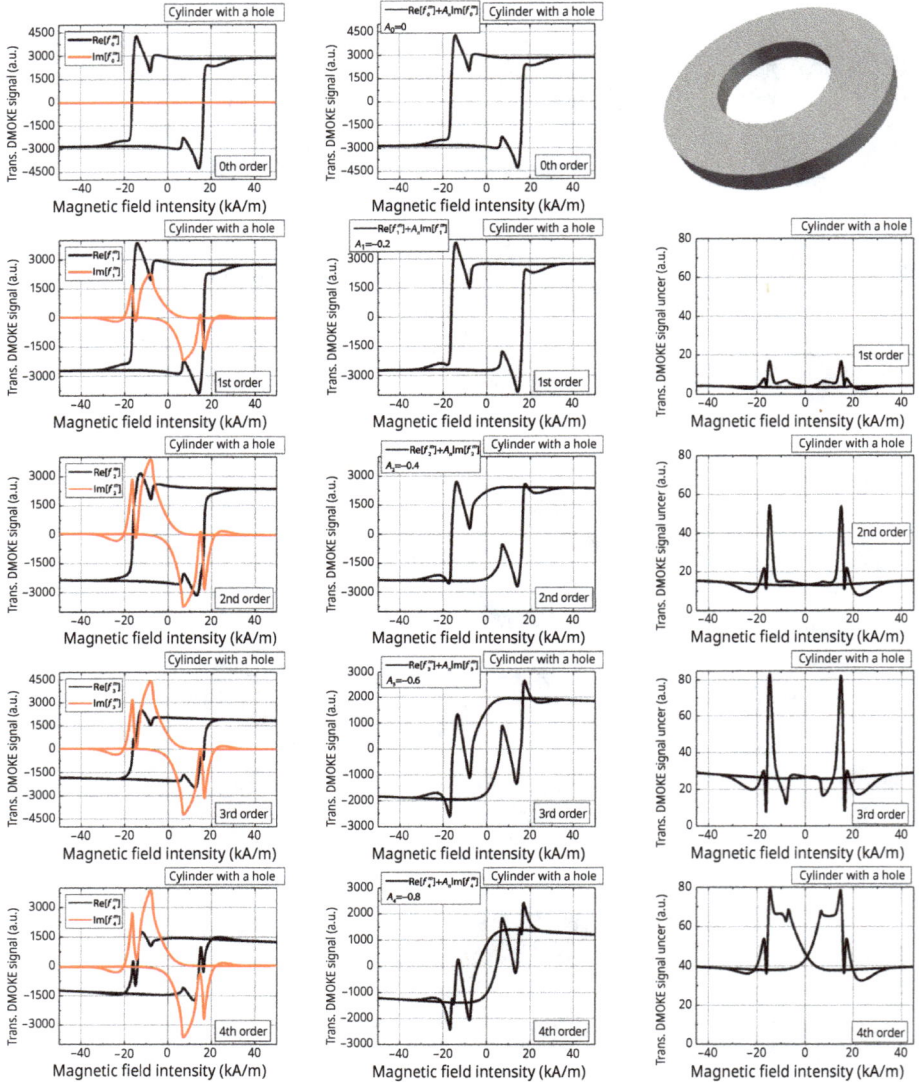

Fig. 4.10: DMOKE signals for the zeroth to the fourth order, simulated in transverse configuration. Graphs depict the real and the imaginary parts (left column), their sum (middle column), and the uncertainty of the signals (right column) for a set of ferromagnetic cylinders with a hole spaced by 600 nm. The diameter of a single cylinder equals 100 nm, its height equals 10 nm, the internal hole diameter equals 50 nm, and the assumed uncertainty of distance between the cylinders equals 10 nm. The samples are made from permalloy (Py). The externally applied field was swept in the range of ±300 kA/m. The arbitrarily assumed formula for the A_n factor is $A_n = -0.2n$.

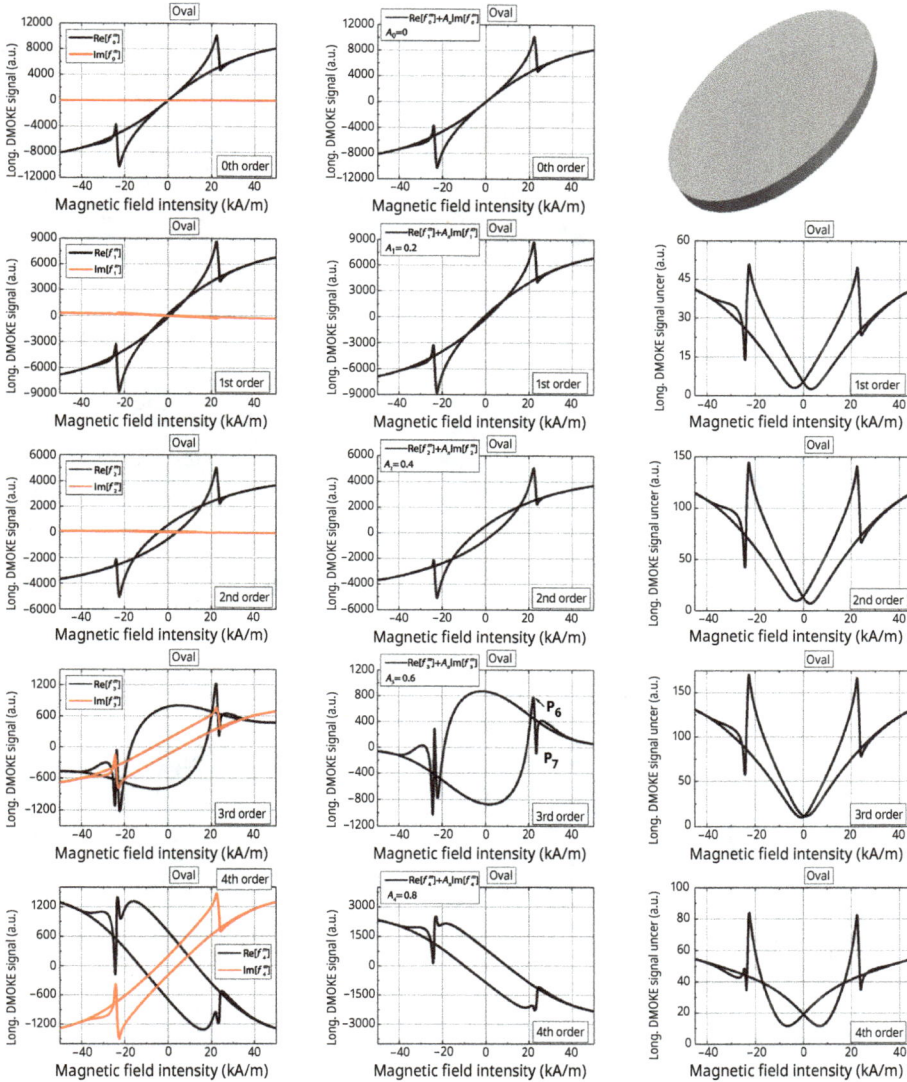

Fig. 4.11: DMOKE signals for the zeroth to the fourth order, simulated in longitudinal configuration. Graphs depict the real and the imaginary parts (left column), their sum (middle column), and the uncertainty of the signals (right column) for a set of ferromagnetic ovals spaced by 600 nm. The length of the longer oval's axis equals 100 nm, its height equals 10 nm, and the vertical axis of symmetry equals 50 nm. The assumed uncertainty of the distance between the ovals equals 10 nm. The samples are made from permalloy (Py). The externally applied field was swept in the range of ±300 kA/m. The arbitrarily assumed formula for the A_n factor is $A_n = -0.2n$.

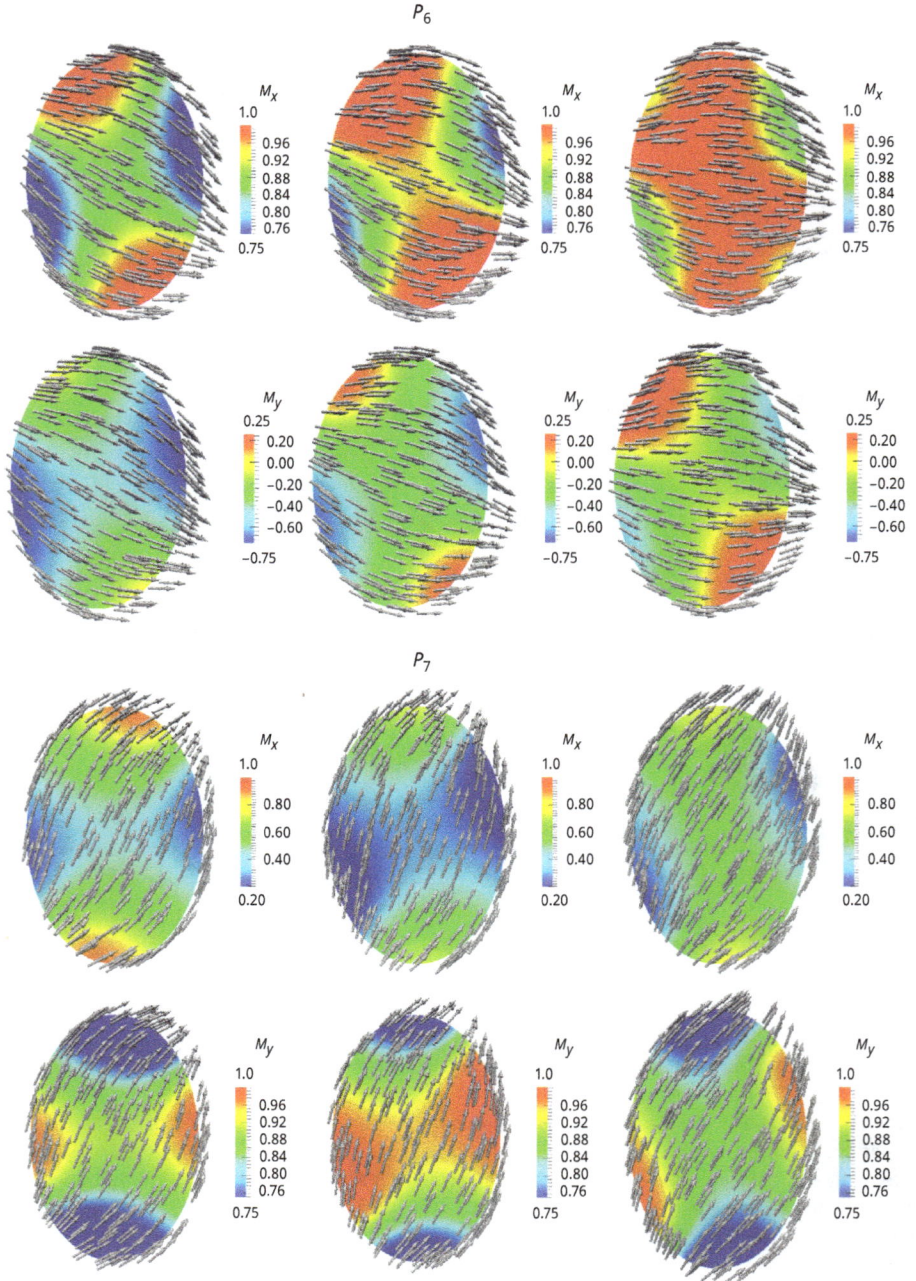

Fig. 4.12: The evolution of the magnetization is seen as the rapid change in the magnetization. The middle column is equivalent to the points P_6 and P_7 marked in Fig. 4.11, while the left and right panels are equivalent to earlier and later moments (spaced in time by approx. 1 ns), respectively.

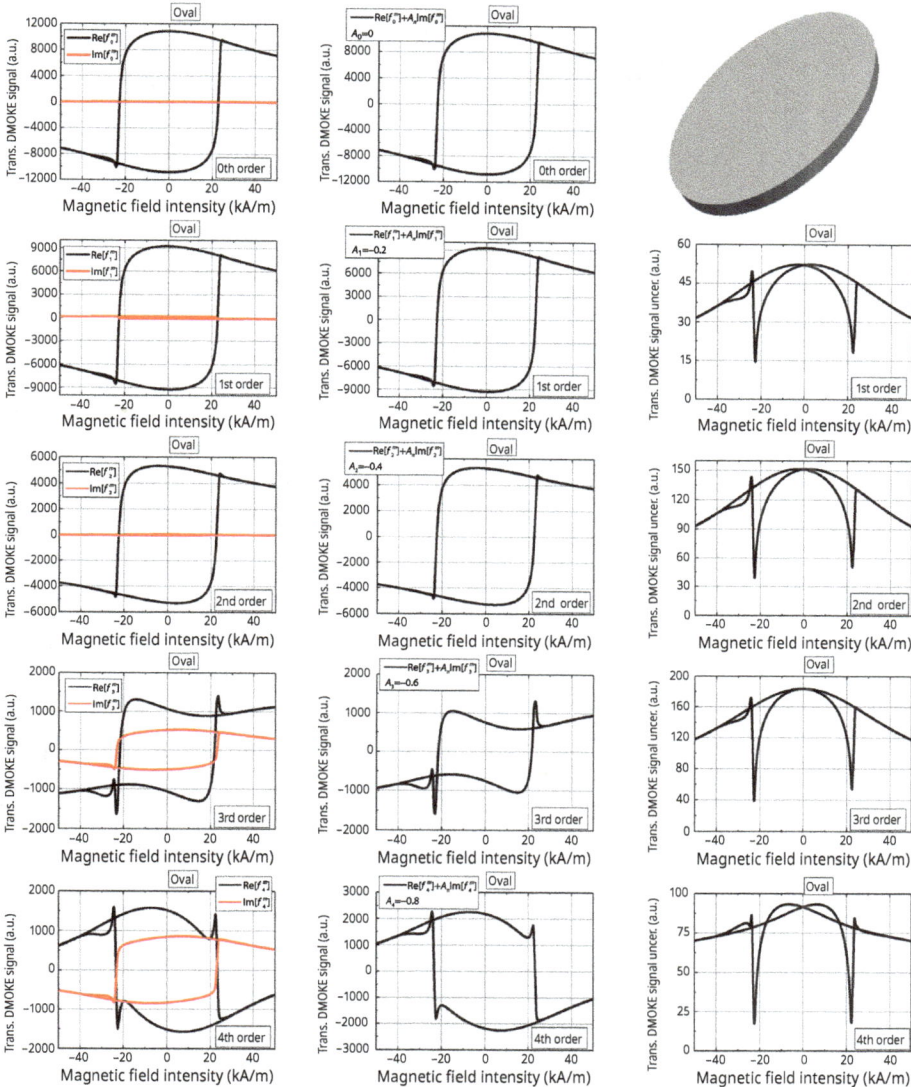

Fig. 4.13: DMOKE signals for the zeroth to the fourth order simulated in transverse configuration. Graphs depict the real and the imaginary parts (left column), their sum (middle column), and the uncertainty of the signals (right column) for a set of ferromagnetic ovals spaced by 600 nm. The length of the longer oval's axis equals 100 nm, its height equals 10 nm, and the vertical axis of symmetry equals 50 nm. The assumed uncertainty of the distance between the ovals equals 10 nm. The samples are made from permalloy (Py). The externally applied field was swept in the range of ±300 kA/m. The arbitrarily assumed formula for the A_n factor is $A_n = -0.2n$.

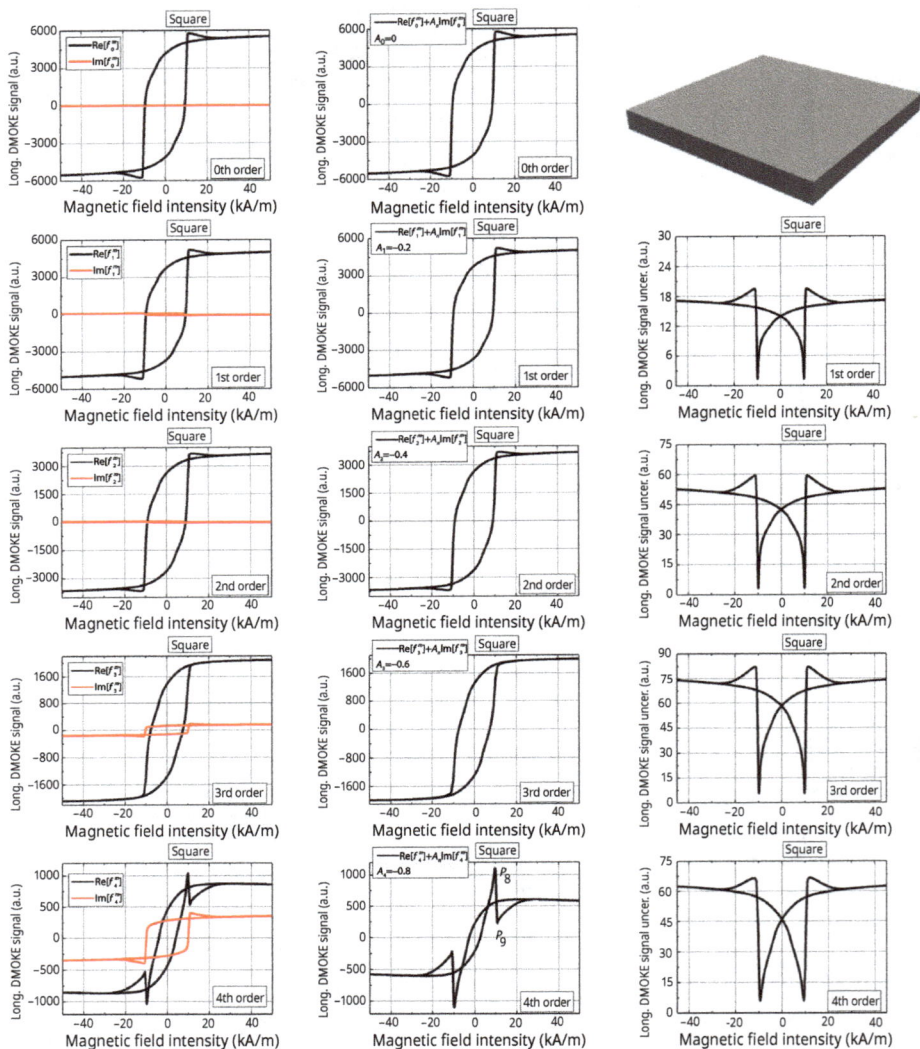

Fig. 4.14: DMOKE signals for the zeroth to the fourth order simulated in longitudinal configuration. Graphs depict the real and the imaginary parts (left column), their sum (middle column), and the uncertainty of the signals (right column) for a set of ferromagnetic squares spaced by 600 nm. The length of the square's side equals 100 nm, and its height equals 10 nm. The assumed uncertainty of the distance between the squares equals 10 nm. The samples are made from permalloy (Py). The externally applied field was swept in the range of ±300 kA/m. The arbitrarily assumed formula for the A_n factor is $A_n = -0.2n$.

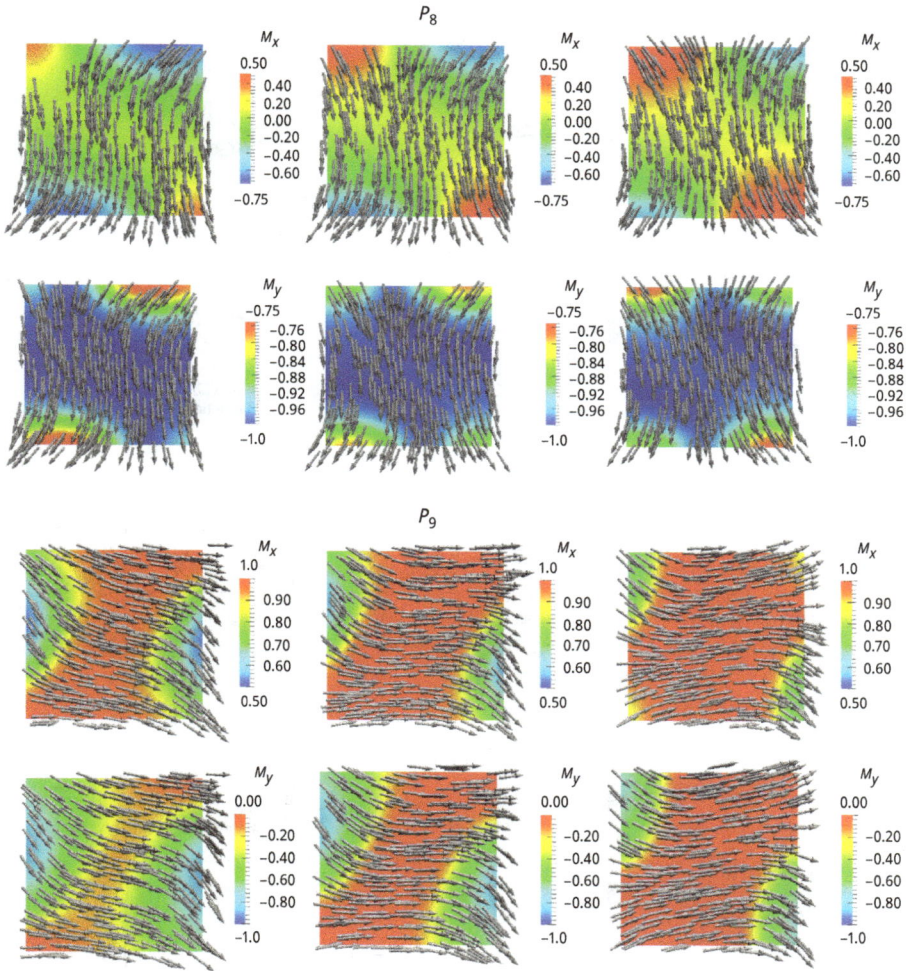

Fig. 4.15: The evolution of the magnetization is seen as the rapid change in the magnetization. The middle column is equivalent to the points P_8 and P_9 marked in Fig. 4.14, while the left and right panels are equivalent to earlier and later moments (spaced in time by approx. 1 ns), respectively.

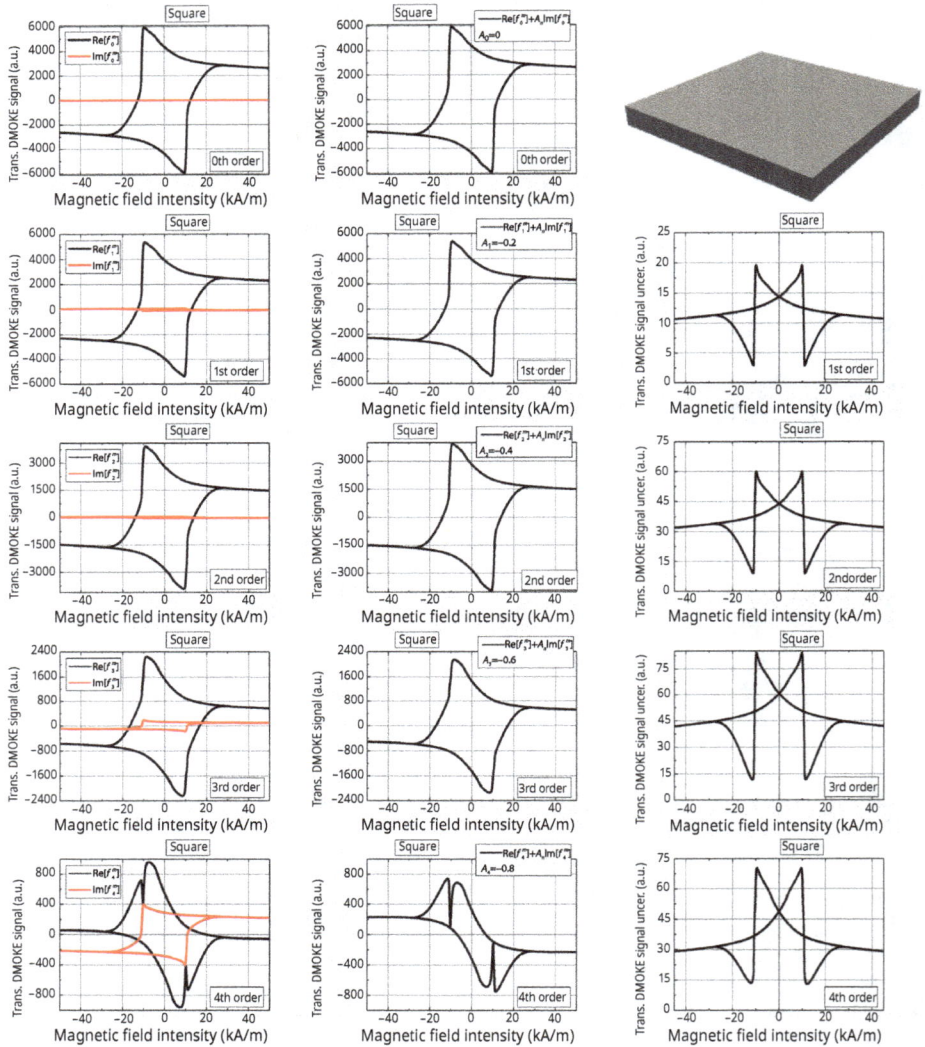

Fig. 4.16: DMOKE signals for the zeroth to the fourth order, simulated in transverse configuration. Graphs depict the real and the imaginary parts (left column), their sum (middle column), and the uncertainty of the signals (right column) for a set of ferromagnetic squares spaced by 600 nm. The length of the square's side equals 100 nm and its height equals 10 nm. The assumed uncertainty of the distance between the squares equals 10 nm. The samples are made from permalloy (Py). The externally applied field was swept in the range of ±300 kA/m. The arbitrarily assumed formula for the A_n factor is $A_n = -0.2n$.

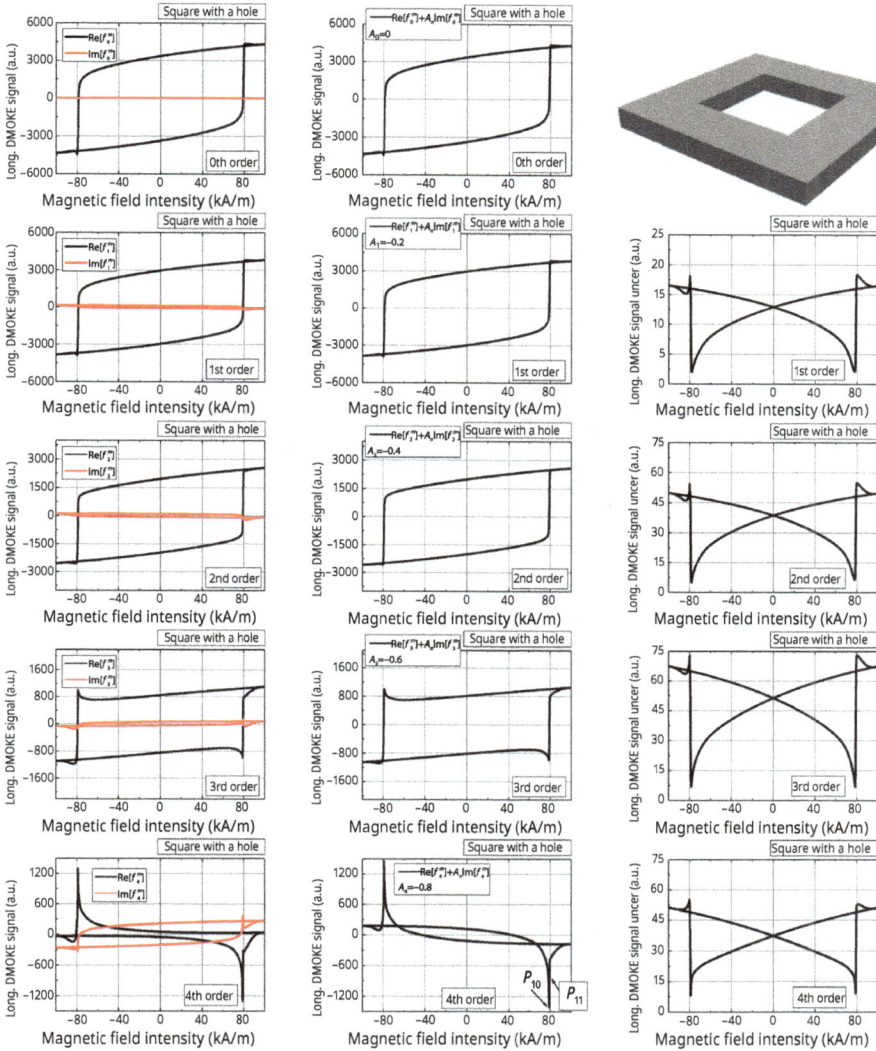

Fig. 4.17: DMOKE signals for the zeroth to the fourth order simulated in longitudinal configuration. Graphs depict the real and the imaginary parts (left column), their sum (middle column), and the uncertainty of the signals (right column) for a set of ferromagnetic square frames (squares with a hole), spaced by 600 nm. The length of the square's side equals 100 nm, the internal length equals 50 nm, and the height is equal to 10 nm. The assumed uncertainty of the distance between the cylinders equals 10 nm. The samples are made from permalloy (Py). The externally applied field was swept in the range of ±300 kA/m. The arbitrarily assumed formula for the A_n factor is $A_n = -0.2n$.

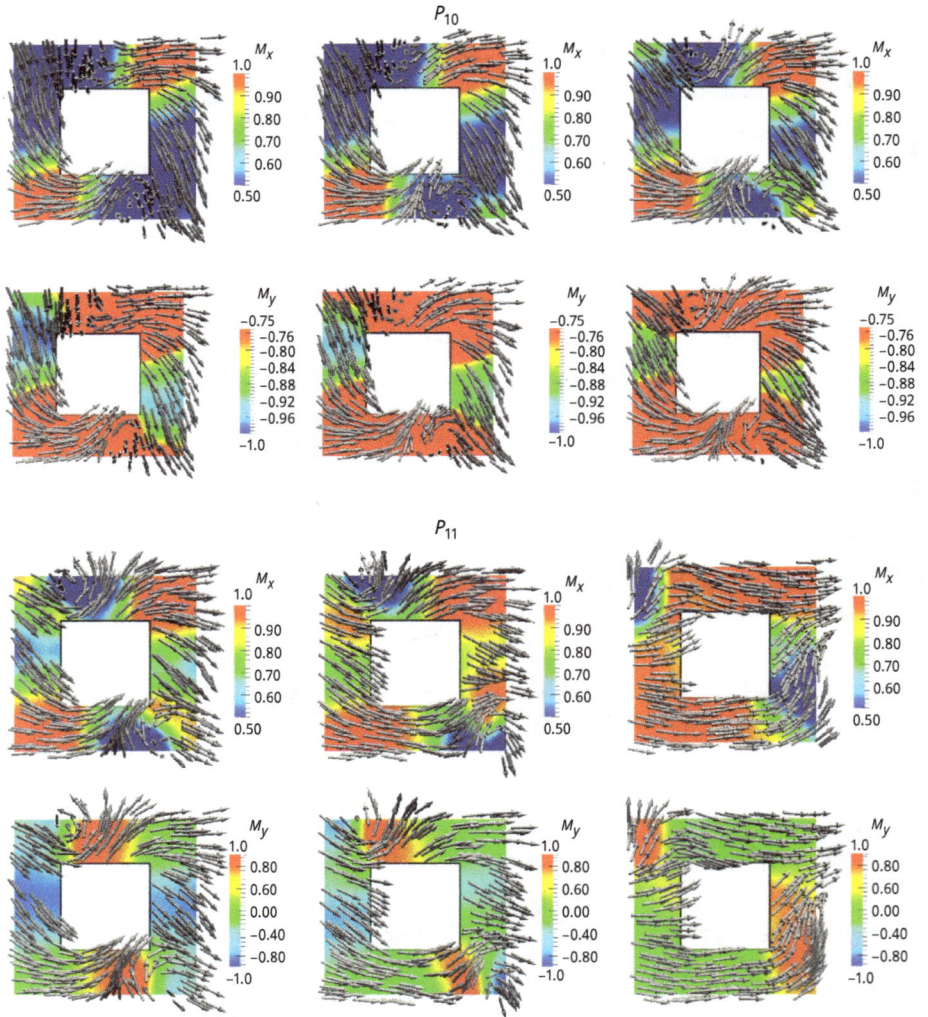

Fig. 4.18: The evolution of the magnetization is seen as the rapid change in the magnetization. The middle column is equivalent to the points P_{10} and P_{11} marked in Fig. 4.17, while the left and right panels are equivalent to earlier and later moments (spaced in time by approximately 1 ns), respectively.

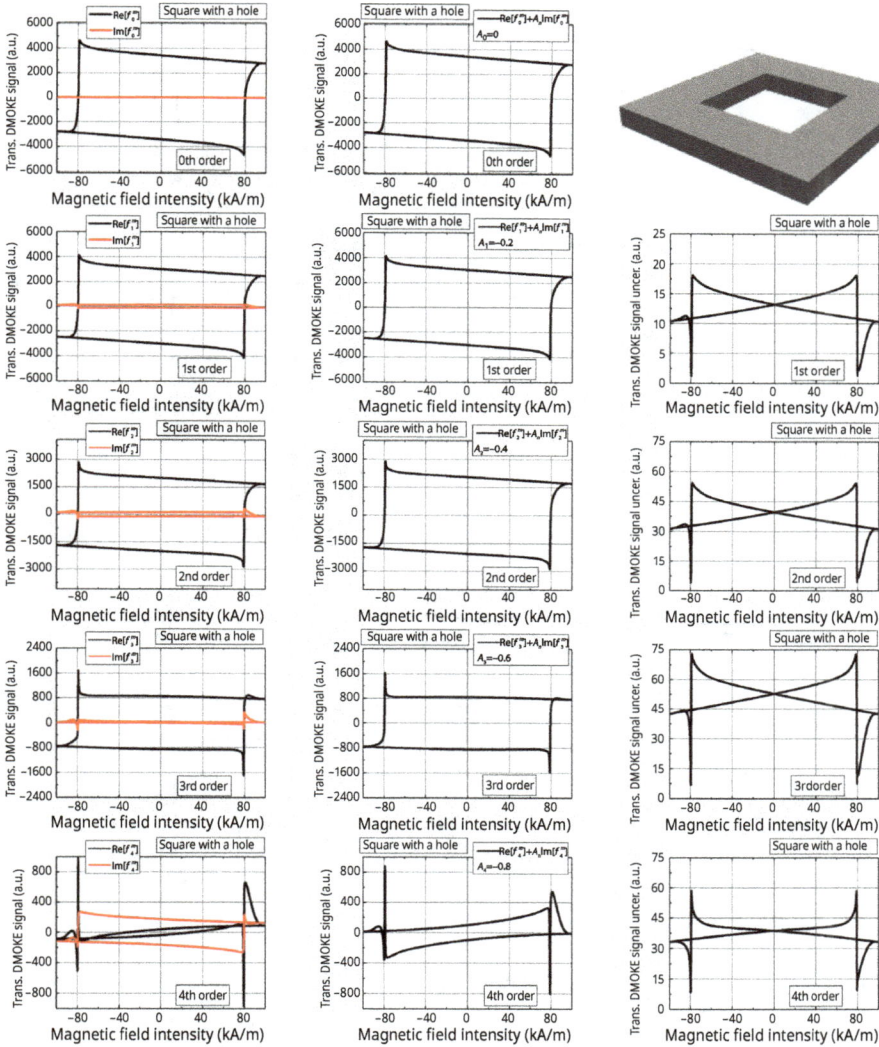

Fig. 4.19: DMOKE signals for the zeroth to the fourth order simulated in transverse configuration. Graphs depict the real and the imaginary parts (left column), their sum (middle column), and the uncertainty of the signals (right column) for a set of ferromagnetic square frames (squares with a hole), spaced by 600 nm. The length of the square's side equals 100 nm, the internal length equals 50 nm, and the height is equal to 10 nm. The assumed uncertainty of the distance between the cylinders equals 10 nm. The samples are made from permalloy (Py). The externally applied field was swept in the range of ±300 kA/m. The arbitrarily assumed formula for the A_n factor is $A_n = -0.2n$.

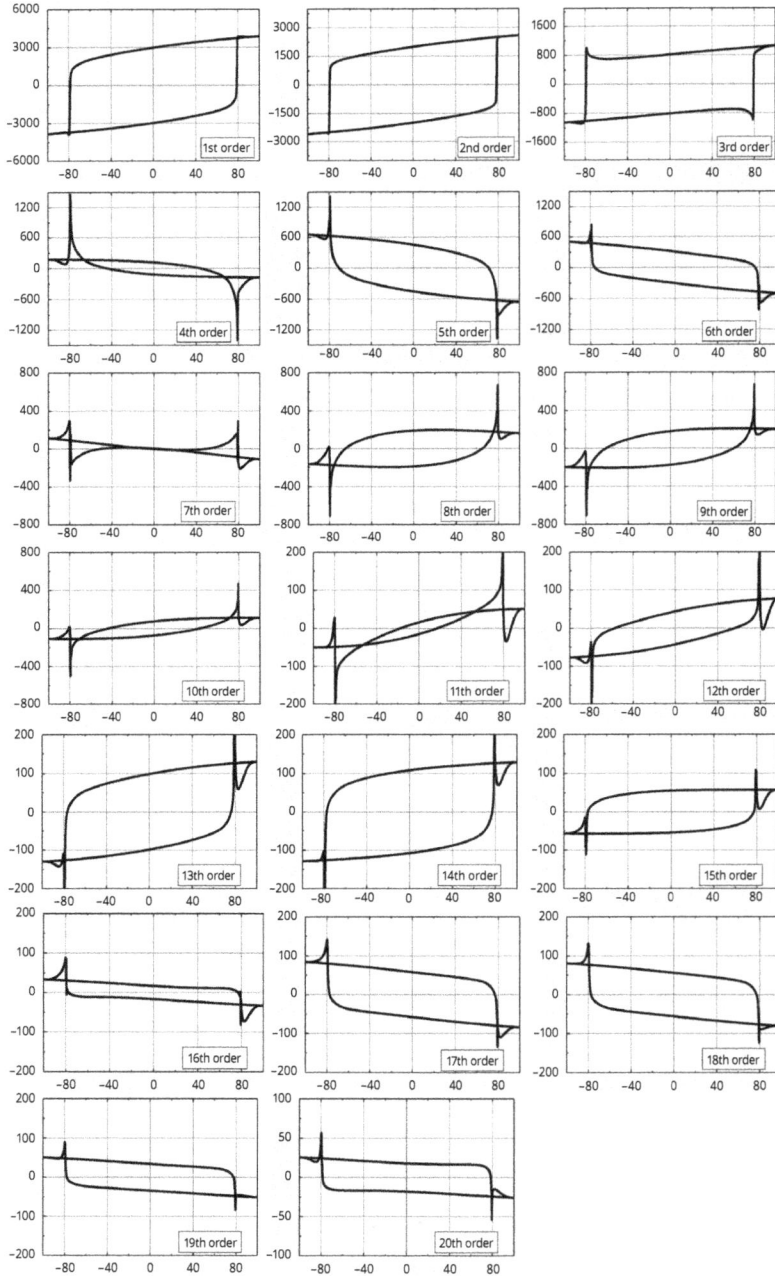

Fig. 4.20: Longitudinal configuration DMOKE signals for the first 20 diffractive orders obtained for the set of ferromagnetic squared frames (squares with a hole), spaced by 600 nm. The other material parameters are the same as for Figs. 4.17–4.19.

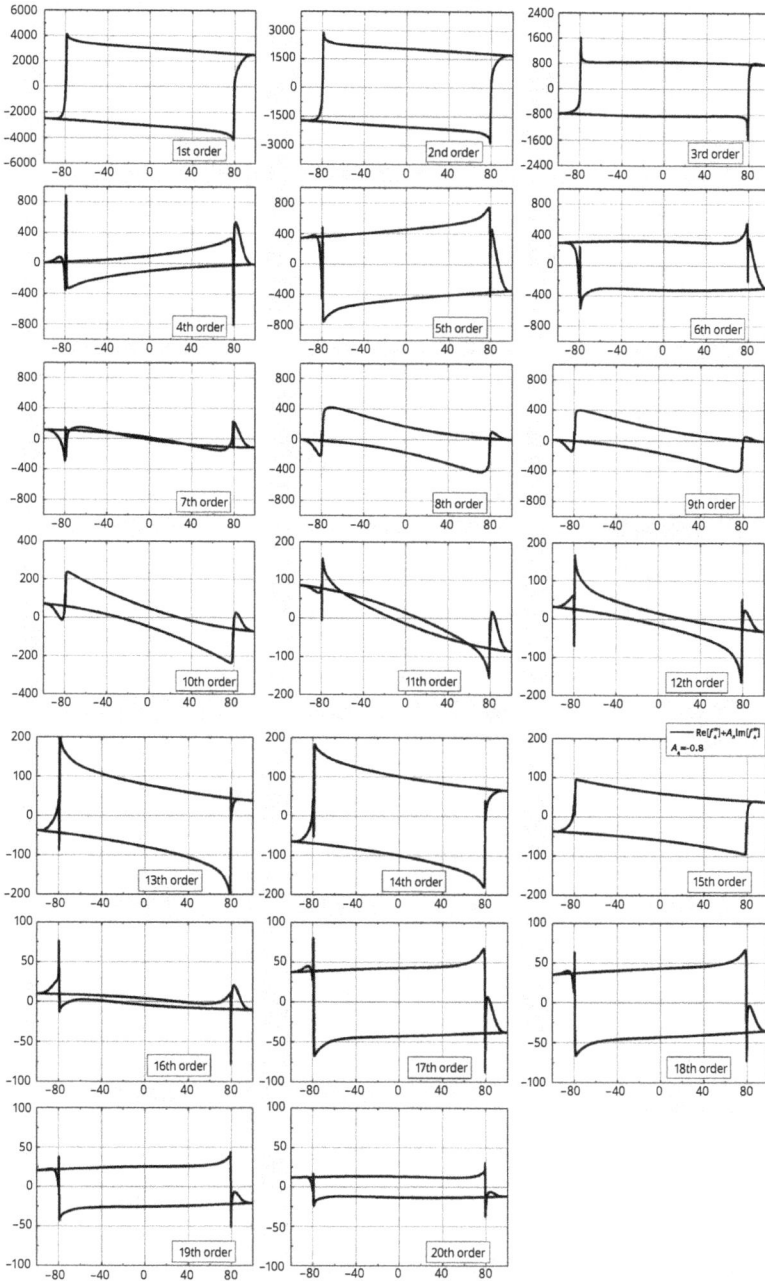

Fig. 4.21: Transverse configuration DMOKE signals for the first 20 diffractive orders obtained for the set of ferromagnetic squared frames (squares with a hole), spaced by 600 nm. The other material parameters are the same as for Figs. 4.17–4.19.

State 1

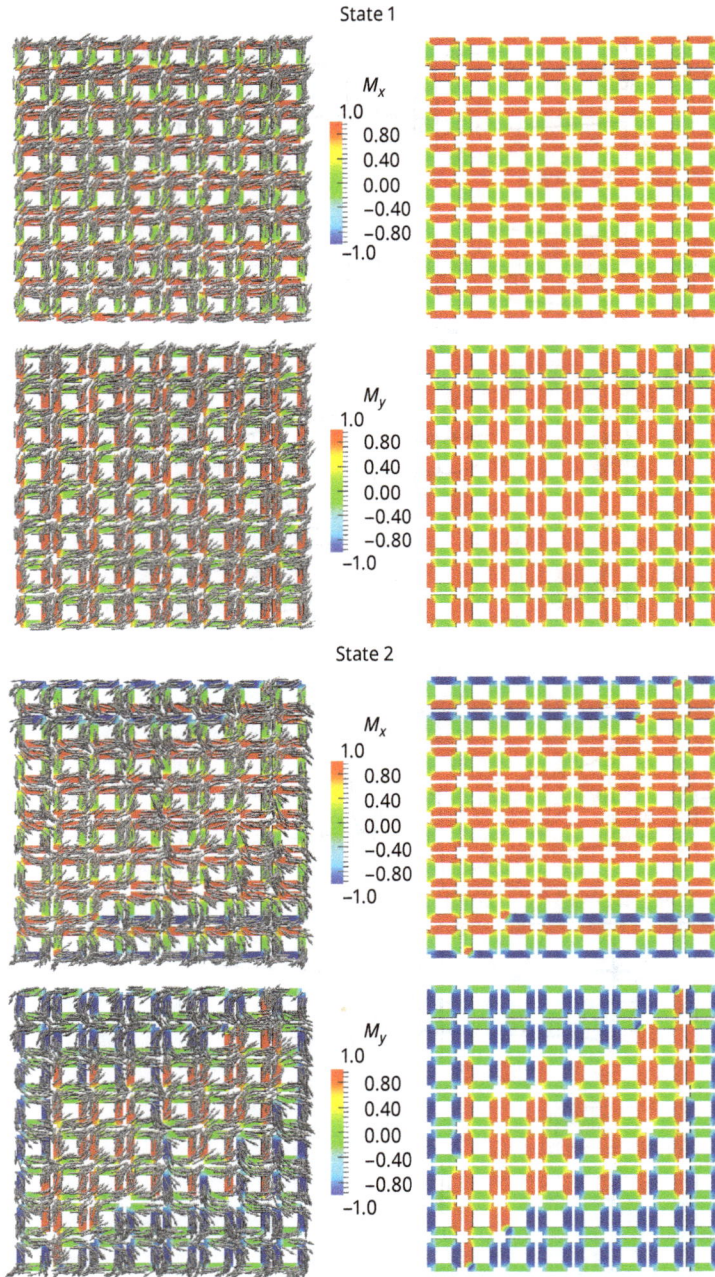

State 2

Fig. 4.22: The four remanent magnetic states in the 8 × 8 elements super-cell made from Fe. The total size equals 1,040 nm × 1,040 nm. The thickness of each single element is equal to 5 nm. The numbered states are explained in Fig. 4.23, presenting the corresponding hysteresis curve.

State 3

M_x

State 4

M_x

M_y

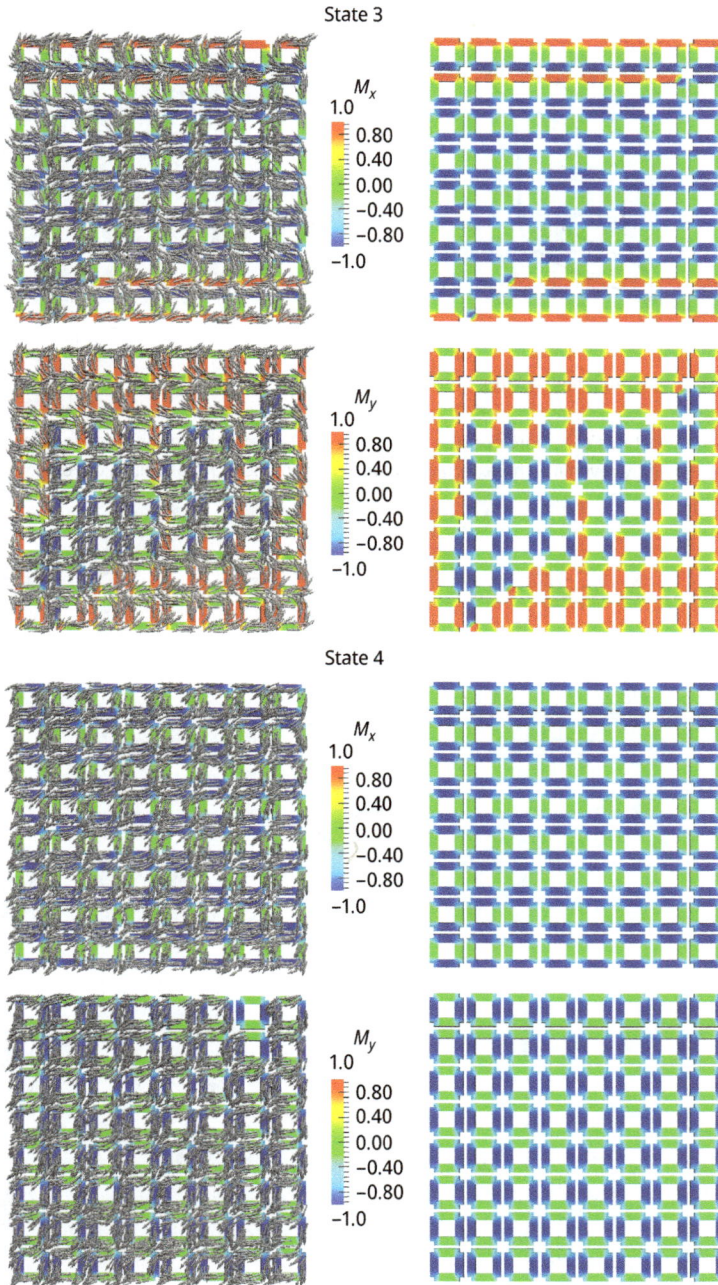

Fig. 4.22 (continued)

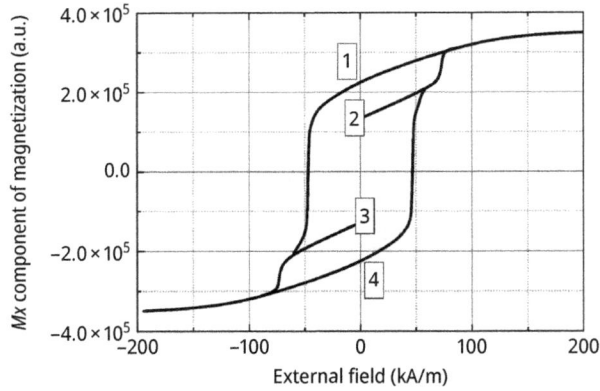

Fig. 4.23: Four stable magnetic states at remanence obtained in the 8 × 8 elements super-cell made from Fe. States 2 and 3 were reached by reversing the external field to zero and backward at values of +63 and −63 kA/m, respectively.

5 Contact effects in layered structures

Layered structures, especially stacks of thin films, can have large interfaces in comparison to bulk and edge effects. Especially when layers with different magnetic or conductive properties are in contact, new effects may occur. This chapter gives an overview of some of these effects that have to be taken into account when new spintronics devices are designed – or, on the other hand, are advantageous or even necessary to reach special technological goals, such as the exchange bias (EB) that is used in diverse devices in which hard and soft magnetic materials must be coupled, that is, materials with larger and smaller coercive fields. Other interfaces that are of special interest for spin injection and similar phenomena are described in Chapter 6.

5.1 Boundary conditions and magnetic interface effects

Some of the effects described in this chapter are based on well-known boundary conditions, derived from Maxwell's equations (Maxwell 1861), which have to be taken into account at boundaries between different materials.

One of these boundary conditions can be calculated using Gauss's law in a dielectric:

$$\nabla \cdot \vec{D} = \rho_{\text{free}} \tag{5.1}$$

with the displacement field $\vec{D} = \varepsilon_0 \vec{E} + \vec{P}\vec{D} = \varepsilon_0 \vec{E} + \vec{P}$. Here, $\vec{E}$ denotes the electric field, ε_0 the vacuum permittivity, and $\vec{P}$ the polarization density, that is, the density of electric dipole moments in a material. ρ_{free} means the free charges in the material. Gauss's law describes the situation that the electric field flow through a closed surface is proportional to the charge embedded in this surface – the charges are the sources of the electric field. It can also be related to electric field lines: Field lines start at positive electric charges and end at the negative ones, with the number of field lines being proportional to the total charge.

In addition, Gauss's law gives us a first boundary condition at an interface between two media 1 and 2. Integrating over it, we get the following:

$$\int \nabla \cdot \vec{D} dV = \int \vec{D} dS = \int \rho_{\text{free}} dV \tag{5.2}$$

with the surface area S of the closed volume over which the integration is performed, let us assume that the closed volume has the form of an upright cylinder, with the interface under examination in the middle between the top and bottom layers. Let us make the height h of this cylinder very small or, to be mathematically correct, infinitesimally small. The integral over the displacement field can then be written as the sum of the integrals over D for the bottom and the top layers, that is,

https://doi.org/10.1515/9783111383736-005

$$\int \vec{D} dS = \vec{D}_1 \, \overrightarrow{n_1} \delta A_1 + \vec{D}_2 \, \overrightarrow{n_2} \delta A_2 = \left(\vec{D}_2 - \vec{D}_1 \right) \vec{n} \, \delta A \tag{5.3}$$

since the normal vectors $\overrightarrow{n_1}$ and $\overrightarrow{n_2}$ of both sides of the interface are oriented antiparallel, with the definition of the surface charge density on the interface $\rho_{surface} = \rho_{free}.h$, using the infinitesimally small height h of the cylinder again, we get

$$\int \rho_{free} dV = \int \rho_{surface} dS = \rho_{surface} \delta A. \tag{5.4}$$

Combining the last two equations results in the boundary condition.

$$\left(\overrightarrow{D_2} - \overrightarrow{D_1} \right) \vec{n} = \rho_{surface}. \tag{5.5}$$

This means that the normal component of the displacement field $\vec{D}$ shows a jump at the interface between two media, as soon as there is any surface charge density on the interface.

On the other hand, Gauss's law for magnetism can be written as

$$\nabla \cdot \vec{B} = 0 \tag{5.6}$$

with the magnetic field $\vec{B}$. This means that no magnetic monopoles exist, opposite to the law for the electric field. It can also be interpreted as the fact that the overall magnetic flux through a closed surface is zero. "Closed surface" means a surface without boundaries, completely enclosing a certain volume, such as a ball or a torus.

From this equation, the next boundary condition can be calculated:

$$\left(\vec{B}_2 - \vec{B}_1 \right) \vec{n} = 0. \tag{5.7}$$

The normal component of the magnetic field, opposite to the normal component of the displacement field, is thus always continuous. It should be mentioned that in the case of magnetic monopoles being found, which would change Gauss's law on magnetism by exchanging the 0 with a source term for the magnetic field, this equation would have to be altered accordingly.

After deriving the normal components of the electric and the magnetic fields at an interface, the tangential components can be calculated from the other two Maxwell equations. This leads to the following equations for the tangential components of the electric and the magnetic fields:

$$\vec{n} \times \left(\overrightarrow{E_2} - \overrightarrow{E_1} \right) = 0, \tag{5.8}$$

$$\vec{n} \times \left(\overrightarrow{H_2} - \overrightarrow{H_1} \right) = \vec{j}. \tag{5.9}$$

Here, $\vec{H}$ denotes the magnetizing field $\vec{H} = \vec{B}/\mu_0 - \vec{M}$, and $\vec{j}$ is the conduction surface current density. This means that the tangential component of $\vec{H}$ is not necessarily continuous, while the flux of $\vec{B}$ is continuous.

Although these boundary conditions look very simple, they are not sufficient to estimate all possible interface effects occurring in diverse material combinations. Several interface effects have been discovered during the last decades, often only becoming visible after thin film production technologies have been developed further and further so that sufficiently flat, smooth interfaces could be produced. An overview of such effects is given in the next sections as well as in the next chapter.

5.2 Ferromagnetic–antiferromagnetic junction: EB and 90° coupling

By coupling a ferromagnet (FM) and an antiferromagnet (AFM), an interesting anisotropy can arise at the interface – the so-called EB. Opposite to shape anisotropy, magnetocrystalline anisotropy, and so on, the EB is a unidirectional anisotropy, which means that not one axis, but only one direction of such an axis is an easy direction for the magnetization, while the opposite direction is harder than expected from the other anisotropies. In a first approximation, this can be compared with the influence of an external magnetic field on a ferromagnetic sample, while in several FM/AFM bilayer systems, the impact of the EB cannot be described so easily.

An EB can generally occur in systems in which the Néel temperature of the AFM (its ordering temperature) is lower than the Curie temperature of the FM. To create an EB, the system must be cooled in an external magnetic field (or with the FM magnetized) through the Néel temperature. This results in a shift of the hysteresis loop, often to negative fields for a positive cooling field (Fig. 5.1), while the opposite effect is also possible, depending on the material system, layer heights, and so on. This shift is called the EB. At the same time, the hysteresis loop usually becomes broader. Sometimes, an additional asymmetry of the loop can be recognized.

The EB was first observed by Meiklejohn and Bean (1956) for ferromagnetic cobalt particles with an oxidized CoO surface. During the last decades, several material systems were found showing the same effect. In most cases, thin film systems are investigated since, in this way, the interface and the crystal orientations of the layers can be controlled in the best way. Besides, most applications are based on thin film technology. Some typical properties of EB systems are described now, while it should be taken into account that there are always possible exceptions.

Starting from low temperatures, the EB usually decreases with increasing temperature until it vanishes at the so-called blocking temperature. The blocking temperature can be much lower than the Néel temperature, especially for thin-film samples, while it is usually similar to the Néel temperature for systems with thicker monocrys-

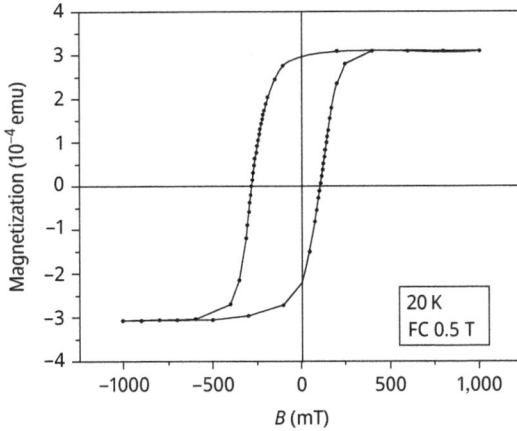

Fig. 5.1: Example hysteresis loop, measured on the EB system Co/CoO(111) after field cooling at 0.5 T to a temperature of 20 K.

talline antiferromagnetic structures (Nogués and Schuller 1999). Additionally, the blocking temperature depends strongly on the disorder in the AFM and the interface.

In several systems, the EB is approximately antiproportional to the layer thickness of the FM, suggesting that the EB is an interface effect (Nogués and Schuller 1999). Deviations are especially found for very thin and very thick layers. Investigations of the EB dependence on the AFM thickness give ambiguous results: On the one hand, the EB can be independent of the AFM thickness; on the other hand, it can increase with increasing AFM layer thickness until a certain thickness is reached and afterward either stay constant or decrease again. This behavior changes with different material systems and preparation conditions (Nogués and Schuller 1999).

The EB can also depend on the cooling field. This dependence differs from one system to another. Sometimes large cooling fields result in a reduction of the EB (Moran and Schuller 1996) or even a change in sign (Nogués et al. 1996, 2000); in other material systems, the EB increases with increasing cooling fields until it stays constant for high cooling fields (Ambrose and Chien 1998).

The interface roughness is of special importance in EB systems. If a monocrystalline antiferromagnetic substrate, for example, FeF_2, with a very flat surface is coupled to an FM, the system shows only a small EB. Roughening the surface before the FM is applied leads to a significant increase in the EB. In most thin film systems, such as Fe/FeF_2, the EB is large for smooth interlayers and decreases with increasing interface roughness (Nogués et al. 1999); it increases again after the first decrease with higher roughness in Fe/MnF_2 thin film systems (Leighton et al. 1999). For NiO as the AFM, no dependence of the EB on the AFM roughness is visible, while the EB can be increased by increasing the roughness slope (Hwang et al. 1998).

The EB depends not only on the interface roughness but also on the crystalline quality of the AFM. Systems from Fe and twinned FeF_2 films consisting of two sublattices that are rotated with respect to each other, for example, show a higher EB than systems prepared from monocrystalline AFMs (Nogués et al. 1999). The same effect can be found in thin CoO layers and CoO substrates combined with Co as FM (Moran et al. 1995). In polycrystalline films, the EB usually grows with reduced crystallite size, while the opposite behavior is also possible (Nogués and Schuller 1999). The size of the crystallites can be varied by modifying the growth temperature, growth rate, or pressure during preparation. These growth parameters influence other material properties, too, which explains the different experimental results.

If an antiferromagnetic crystal is cut along different orientations, the surface spins can be compensated or uncompensated, that is, oriented antiparallel or parallel. Opposite to the intuitive approach, an EB can also be found for compensated surfaces; partly, the effect is even larger here (Nogués and Schuller 1999). Besides, the magnetic moments are not always parallel to the surface. In FeF_2, for example, the spins are oriented parallel to the (110) surface and perpendicular to the (001) surface, and in the (101) orientation, they have an intermediate angle to the surface. The maximum EB is found for the (110) orientation, while it vanishes in the (001) orientation and is approximately half the maximum value in the (101) orientation (Nogués and Schuller 1999). Generally, an EB can be found in a broad variety of material systems and geometries, such as core-shell structures (Schneider et al. 2006), nanostructures (Blachowicz et al. 2023) or layered systems with various anisotropies (Ehrmann and Blachowicz 2011, Blachowicz and Ehrmann 2021) and can even influence systems above the Néel temperature (Tillmanns et al. 2009).

Besides these dependencies of the EB on different factors, some special features can typically be found in EB systems. The most obvious one is the significantly increased coercivity compared to a pure ferromagnetic film of identical thickness and growth parameters. This effect is of special importance below the blocking temperature and is strongly temperature dependent. In some systems, however, a maximum of the coercivity occurs around the blocking temperature, for example, in Fe/FeF_2 (110).

If several hysteresis curves are measured successively without field cooling again, a decrease in the EB is visible with an increasing number of magnetization reversal processes. This so-called training effect is of special importance in polycrystalline films, while it is much smaller in single crystals or epitaxial layers (Nogués and Schuller 1999).

Several theoretical models were developed to explain these – often contradictory – results. While they can explain the EB effect in general, the theoretically calculated EB values often differ from the experimentally found ones by some orders of magnitude. Such models first need to give a reason for the energy difference between the opposite magnetization directions.

In the first intuitive models, either perfectly compensated or uncompensated surfaces were used as the base of the calculations. Besides, the AFM is assumed to have an infinitely high anisotropy, that is, the AFM spins can only be oriented parallel or antiparallel (Ising model). For the compensated surface, in which the AFM spins have alternating orientations, the coupling between AFM and FM is alternately advantageous and disadvantageous, on average resulting in a vanishing net magnetization. For the uncompensated surface, in which all AFM spins have the same orientation, however, the coupling between AFM and FM is advantageous for one FM orientation and disadvantageous for the opposite one, resulting in a net interface energy and thus an EB field. While this intuitive model can explain the principle of the EB, the calculated values are between two and three orders of magnitude higher than the measured ones. On the other hand, perfectly flat surfaces, as assumed here, are not realistic. By introducing random surface roughness, however, the uncompensated surface becomes compensated, which lets the EB vanish. Apparently, this intuitive model is not suited to explain the EB effect in detail.

In the random-field model introduced by Malozemoff (1987, 1988, 1988a), a compensated spin structure with surface roughness is assumed. Opposite to the simple intuitive model, now domain walls perpendicular to the interface are allowed, that is, the long-range order in the AFM is repealed. This means that now the domain wall energy has to be taken into account besides the interface energy. The surface roughness is integrated by adding defect spins, protruding from the AFM layer into the FM area. In this way, a preferred direction of the FM is implied locally, but not globally. This results in domain formation, depending on the orientation of the defect spins parallel to the cooling field direction. All of these domains have a small spin surplus due to the direction of the cooling field, resulting in a net magnetization that is visible as the EB.

Another model for the EB was presented by the discoverers of this effect (Meiklejohn and Bean 1957; Meiklejohn 1962). They measured the angle-dependent torque of the magnetic samples in addition to hysteresis loops, allowing for evaluation of the magnetic anisotropy. In the EB system Co/CoO, they found an angle dependence of the EB proportional to $\cos\theta$, with the angle θ between the easy direction of the anisotropy and the external magnetic field used for measuring. Such a torque at the interface necessitates an uncompensated surface. In their calculation, the EB fields are again some orders of magnitude higher than the measured ones.

While the aforementioned models were based on (anti)parallel AFM spins, the model of Mauri allows spins in different layers to be rotated against each other (Mauri et al. 1987). This means that domain walls parallel to the interface (here assumed to be uncompensated) are possible. Assuming a thick AFM with uniaxial (twofold) anisotropy and a thin FM, the model gives the same result as the one obtained by Meiklejohn and Bean for weak interface coupling and results in an EB, which depends only on the anisotropies and coupling constants in the AFM for strong interface coupling, giving similar results as the random-field model in this case.

The model of Schulthess and Butler is based on the micromagnetic simulation of a 3D Heisenberg model (1998, 1999). The calculation is limited to FM and AFM single crystals with a perfectly flat interface in a single-domain state. Interestingly, in this calculation, the ferromagnetic spins are oriented perpendicular to the easy axis of the AFM, resulting in the so-called 90° coupling, which is often found in EB systems. However, this effect can only explain the increased coercivities, not the EB. By introducing defects along the perfectly flat surface by exchanging ferromagnetic spins with antiferromagnetic ones, the preferred spin orientation of the magnetization in the FM differs from the orientation of the 90° coupling. For an experimentally found spin surplus of 1% (Takano et al. 1997, 1998), realistic values of the EB are calculated. While the model of Schulthess and Butler underlines the importance of interface defects, it cannot derive them theoretically.

Opposite to the aforementioned ones, the model of Stiles and McMichael (1999) is based on a polycrystalline AFM. Since the single grains differ in dimensions and orientations, the energies of the crystallites are first evaluated independently before they are averaged over the whole layer. The grains are assumed to be too small to be divided into different domains. The FM is modeled as a single-domain structure. The AFM spins can rotate freely (Heisenberg model), taking into account the 90° coupling. Similar to the model of Mauri, two extremal cases are found, depending on the interface coupling strength. For a weak interface coupling, the result is similar to the one of Mauri, while for a strong interface coupling, the 90° coupling becomes dominant.

Another approach to model the EB qualitatively and quantitatively consistent with the experimental findings is the domain-state model (Nowak et al. 2002; Keller et al. 2002). This model showed that integrating nonmagnetic defects in the AFM can increase the EB by a factor of 2–3 (Miltényi 2000), since in this way domain formation in the AFM due to local random fields is supported. Such a random-field Ising model is similar to diluted AFMs in an external magnetic field, which are known to build a domain state after cooling in an external magnetic field below a critical temperature.

This domain state depends in 3D systems on the dilution by nonmagnetic defects and the temperature. In the respective phase diagram, the antiferromagnetic and the paramagnetic phases are separated by the domain state, which is assumed to be a spin-glass state. Interestingly, by field cooling, the system always reaches the spin-glass state in which domains are built due to statistically distributed defects, making domain formation energetically favorable. Cooling further to the antiferromagnetic state, the domains are frozen in a metastable state. Reaching the antiferromagnetic ground state with long-range order is prevented by the domain walls, which are pinned at nonmagnetic defects. Thus, the thermoremanent magnetization relaxes on very long time scales after switching off the cooling field (Kleemann 1993; Leitão et al. 1988). Such diluted AFMs in external magnetic fields have been investigated in detail by different research groups (Kleemann 1993; Belanger and Young 1991).

These properties of diluted AFMs in an external magnetic field explain a remanent magnetization of the AFM without further assumptions. The remanent magnetization, on the other hand, can couple with the FM in an EB system and thus can result in a unidirectional anisotropy. This coupling is stronger than usual external magnetic fields, but only relevant for the antiferromagnetic spins at the interface.

Transferring the idea of diluted AFMs in an external magnetic field to EB systems, Monte-Carlo simulations were performed, first based on one layer of ferromagnetic spins coupled to nine layers of antiferromagnetic ones (Miltényi et al. 2000). The ferromagnetic spins can rotate freely in-plane (Heisenberg model). This model not only shows an EB but also a vertical shift of the hysteresis loop due to the magnetization of the interface and the AFM bulk, a feature that is often observed in hysteresis measurements of EB systems. The EB results here from the magnetization of the AFM interface layer after field cooling, which works like an additional effective field on the FM. Additionally, the training effect can be explained by irreversible effects in the interface layer of the AFM, and the experimentally observed dilution dependence of the AFM as well as the cooling field dependence can be modeled in this way.

During the last years, the domain-state model was investigated further with respect to special problems in EB systems, for example, the training effect in dependence on the AFM structure (Biternas et al. 2009, 2010, 2014), and was extended to more sophisticated systems, such as FM/FM/AFM spin valves (Yüksel 2018). It could also be shown that the domain state model is not suited for each material system (Zhou et al. 2015).

Recently, research has been performed on diverse EB systems, such as Mn-substituted $NiCr_2O_4$, which shows a tunable EB due to a change in the dominating sublattice by changing the temperature (Barman and Ravi 2018), or $BaCo_{0.5}Ni_{0.5}F_4$, in which a spin-glass phase could be excluded experimentally, and the EB was instead attributed to two antiferromagnetic phases (Xu et al. 2018). Additionally, unexpected new effects could be found in long-known systems such as Co/CoO, in which the EB was originally discovered (Ehrmann and Blachowicz 2017d).

The technological relevance of the EB is mostly related to pinning one ferromagnetic layer in spin valves (Nawrocki et al. 2018; Chung et al. 2018), GMR or TMR systems (cf. Chapter 7). Besides being used as hard disk drive read/write heads, such systems can be utilized as sensors, for example, to measure small magnetic fields (Luong et al. 2018; Luong et al. 2018a).

5.3 Superconductor–normal conductor junction: Andreev reflection

At the border between a normal conductor and a superconductor, another special effect can occur that is named after the Russian physicist Alexander F. Andreev (Andreev 1964).

If an electron from the normal conductor reaches the interface to a superconductor, it cannot be simply transmitted into the superconductor if its energy is lower than the superconducting band gap (Blonder et al. 1982; Octavio et al. 1983; Beenakker 2000). Nevertheless, it can cross this border if a charge of $2\,e^-$ is transferred. This is what happens during the Andreev reflection (de Jongg and Beenakker 1995).

As depicted in Fig. 5.2, the electron can cross the interface barrier by forming a Cooper pair (CP) in the superconductor, while a hole is reflected back into the normal conductor. The same effect can occur when holes and electrons are exchanged.

Fig. 5.2: Andreev reflection: an electron can cross the interface between a normal conductor (NC) and a superconductor (SC) by forming a CP in the semiconductor; a hole is then reflected back into the normal conductor.

In terms of spintronics, it is important to know that the reflected hole has opposite spin (and velocity) to the original electron, while the CP needs to consist of an up and a down spin electron, necessitating an electron of opposite spin to the original one. This means that the probability of such an Andreev reflection strongly depends on the spin orientations in both parts of the structure since cotunneling of both electrons for a complete CP is prohibited from a fully spin-polarized material.

It should be mentioned that besides the "normal" Andreev effect, cotunneling from separate, independent normal conductors into one superconductor is also possible. The effect of cotunneling thus has a nonlocal character (Falci et al. 2001).

Especially for FMs or other spin-polarized materials used as normal conductors, the Andreev reflection probability is reduced with increased spin polarization. This, on the other hand, provides a possibility to measure the spin polarization in a normal conductive material. For this, a superconducting tip is pressed onto the normal material's surface. Measuring the conductance between the tip and the sample gives a measure of the spin polarization at this point. In this way, for example, it is possible to investigate the influence of an external magnetic field on the spin polarization of a normal conductor (Soulen Jr. et al. 1998).

The importance of the Andreev effect in spintronics is reflected by the relatively large amount of recent research on this topic. The Heusler alloy Co2FeSi, for example, was examined by epitaxial contact Andreev reflection spectroscopy, a technique in which epitaxially grown superconducting NbN on the Heusler alloy was used to measure the spin polarization of Co_2FeSi. This material is of special interest since it has the highest Curie temperature of approximately 1,100 K and the highest spontaneous magnetic moment among the Heusler alloys. Measurements showed spin polarization of approximately 54% and revealed that the technique of epitaxial contact Andreev reflection may also be interesting for producing superconductive spintronics devices (Shigeta et al. 2018).

Theoretical investigations of FM–superconductor junctions in the presence of the so-called Rashba and Dresselhaus interface spin–orbit coupling (special spin–orbit coupling effects typical for systems with uniaxial and cubic crystals without inversion symmetry, respectively) have not only shown that the Andreev reflection can be controlled by manipulating the orientation of the magnetization, but also a giant in- and out-of-plane magnetoanisotropy was predicted. For a highly spin-polarized FM, the magnetoanisotropic Andreev reflection was calculated to depend only on the spin–orbit fields. Thus, Andreev reflection spectroscopy can also be used to probe spin–orbit fields along the junction between an FM and a superconductor (Högl et al. 2015).

In a double tunnel junction consisting of FM/semiconductor/superconductor, strong Rashba spin–orbit coupling can occur in the semiconducting middle layer. In the half-metal limit, an unconventional equal-spin Andreev reflection can occur, resulting in the in-gap conductance becoming finite as long as the voltage is different from zero, predicting the creation of long-range triplet states in the half-metal (Wu and Meng 2016).

Equal-spin Andreev reflection, intermediated by spin-flip processes, and common Andreev reflection were also found simultaneously in junctions prepared from InAs quantum wells and spin-singlet superconductive NbTi. These results show the possibility of creating a superconducting proximity gap in the quantum well and thus developing superconducting spintronics with semiconducting devices (Matsuo et al. 2018).

Theoretical investigations of triplet pairings in the anomalous Andreev reflection have shown that the reflection amplitude should be controllable in the experiment when tunable strain and Fermi level are incorporated in the nonsuperconductive material (Beiranvand et al. 2016). This finding is of special importance for the coupling across regions with Rashba spin–orbit interaction, for example, in experiments with graphene layers (Avsar et al. 2014; Dushenko et al. 2016; Avishai and Band 2022).

Even more complicated effects occur, for example, in Weyl semimetals with their broken time-reversal symmetry, in which at least two Weyl fermions of opposite chirality occur. Combining a Weyl semimetal in a normal conducting state now with a superconductor shows that the Andreev reflection at the interface between both materials must include a switch of the chirality; other reflection paths are blocked. This effect suppresses the proximity effect of the superconductor as long as the magnetization is parallel to the interface, which reduces the critical temperature of the superconductor, combined with the observation of weak superconductivity in the normal material due to the nonlocality of the metal electrons. This again results in an interface layer of certain thickness in the range of up to some hundred micrometers and can be regarded as the mesoscopic description of the microscopic Andreev reflection. The blockade can be overcome by Zeeman fields at the interface, which provide the desired switch of the chirality (Bovenzi et al. 2017).

After providing these insights into possible physical effects at interfaces between different layers in junctions, the next chapter will concentrate on transport effects within layers and over interfaces.

6 Transport in spintronic structures

Similar to the electrical current, which is based on moving electrons, a spin current necessitates moving (electron) spins. This means, however, that electrons have to be moved as in the case of the electrical current. While the charge of an electron will always be identical to the negative elementary charge, which is approximately -1.602×10^{-19} °C, the spin of the electron can vary. This, on the one hand, offers completely new degrees of freedom that can be utilized in spintronic applications. On the other hand, it means that effects like spin flips and dephasing of spin ensembles have to be taken into account when designing spintronic devices, in addition to the possible movement of the electrons carrying the spin. Especially, interfaces can modify the electron spins and are thus of special importance in such devices.

This chapter gives an overview of transport phenomena with a focus on interface effects, partly resulting in undesired variations of spins, but partly also necessary for spin injection and accumulation.

6.1 Diffusion and Drift of Spins

Let us assume that a certain spin polarization is injected into a layer of a spintronic device. This means that, at this point, the spin polarization is defined as the difference between up-spins and down-spins – that is, the deviation from equilibrium in which both sorts of spins would have the same probability – scaled with the overall amount of spins. But how long will this spin polarization survive, and how far can a spin current travel through the medium?

For metals, a relatively simple equation can be used to estimate the length scale L on which the spin polarization vanishes:

$$\nabla^2 \left(\mu_\uparrow - \mu_\downarrow \right) - \frac{\left(\mu_\uparrow - \mu_\downarrow \right)}{L^2} = 0, \tag{6.1}$$

with the electrochemical potentials $\mu_\uparrow$ and $\mu_\downarrow$ of the up-spin and down-spin electrons, respectively, and the Nabla operator ∇ denoting the derivative with respect to the three spatial directions (van Son et al. 1987). Spintronic devices, however, are often based on semiconducting devices where the electrical field – which can be assumed to be more or less screened in metals – may play an important role in the carrier motion. While this diffusion equation is nevertheless often used for semiconductors (Schmidt et al. 2000; Rashba 2000; Fert and Jaffrès 2002), experimental results have shown significant influence of electric fields on the spin diffusion in semiconductors (Kikkawa and Awschalom 1999; Malajovich et al. 2001). Thus, more general equations were developed, taking also into account the impact of an external electrical field (Yu and Flatté 2002, 2002a).

https://doi.org/10.1515/9783111383736-006

Several special systems were investigated in theory and experiment, aiming at describing the interaction of the spins with their environment. For III–V semiconductors, for example, such as the often-used GaAs/AlGaAs heterostructures, it is known that spin evolution is dominated by spin–orbit interaction (Dyakonov and Perel 1971; Dyakonov and Kachorovskii 1986). The spins are known to precess around a defined axis with a frequency of approximately 10^{10}–10^{11}/s (Miller et al. 2003; Mani et al. 2004). It should be mentioned that this precession can also be induced without an external magnetic field (Kato et al. 2004, 2004a). Scattering rates in such semiconductors, however, are dependent on the number of defects and are in the order of some 10^{12}/s (Shur 1987), showing that scattering happens several times per precession cycle, making such systems seem unsuitable for spintronics. On the other hand, a simple electric gate can control the coherent spin dynamics (Datta and Das 1990; Wang et al. 2002). Since in some cases the movement of the electrons and the spin evolution can be decoupled, using a special symmetry of the spin–orbit interaction, this diffusive transport regime can be used to create spintronic devices (Schliemann et al. 2003; Saikin et al. 2004).

Nevertheless, it should be mentioned that these properties also depend on the transport directions with respect to the crystallographic orientation of the involved semiconductors, making it necessary to define transport directions, respectively (Saikin 2004). Especially for GaAs, which has no inversion symmetry, spin splitting occurs for electrons with wave vectors along the [110] axes, but not along the [111] or [100] axes, resulting in the so-called D'yakonov–Perel' mechanism of electron spin relaxation (Crooker and Smith 2005). For two-dimensional (2D) heterostructures, an additional inversion asymmetry along the growth direction occurs, resulting in an additional Rashba term which works like an additional electric field and can thus also be controlled by an external magnetic field. Finally, the electron spins also couple to the strain induced by stress along a certain axis (Seiler et al. 1977; Cardona et al. 1984, 1988), working like a strain-induced effective magnetic field (Crooker and Smith 2005).

Measurements of spin transport can, for example, be performed all-optically, measuring the Kerr rotation with a weak probe laser pulse at a position different from the spot where the strong pump laser pulse with circular light polarization has initiated spin polarization of electrons and holes. While the holes relax relatively fast, the conduction electrons keep their net spin polarization for longer times and can drift away from the point of generation (Kikkawa and Awschalom 1999). Alternatively, they can be dragged in a desired direction by applying an electric field (Kikkawa and Awschalom 1999).

On the other hand, all-electrical methods can be used to inject, transport, and detect spin-polarized electrons, as shown, for example, in silicon-based circuits (Huang et al. 2007). Silicon (Si) is of great interest for spintronic applications, not only because conventional charge-based electronics are based on it, but also due to its low atomic weight, lattice inversion symmetry, and low nuclear spin (Žutić et al. 2004, 2006; Tyryshkin et al. 2003), resulting in a smaller spin–orbit and hyperfine interaction than the often examined GaAs (Kikkawa and Awschalom 1998, 1999; Lou et al. 2006) and thus expected longer spin lifetimes and spin coherence lengths (Žutić and Fabian 2007). Due to experimental difficulties, Si has nevertheless been examined only for about one decade. Lifetimes of approximately 500 ns at 60 K and nearly 100 ns at 150 K were indeed found about two orders of magnitude larger than those measured in metals or GaAs (Huang et al. 2007; Kikkawa and Awschalom 1998).

Another interesting material for spintronic devices is germanium (Ge) due to its high charge carrier mobility and compatibility with recent Si-based technology. Theoretical investigations also suggested long spin lifetimes and large transport lengths due to the lattice inversion symmetry, resulting in a weak spin–orbit interaction (Dyakonov 2017). In Ge, spin relaxation was suggested to occur via the Elliott–Yafet mechanism, a mechanism that can be found in elemental metals and semiconductors, which results in a certain spin relaxation probability in each spin-independent scattering event, meaning that spin relaxation rate and momentum relaxation rate are coupled (Elliott 1954; Yafet 1963). The latter depends on scattering at ionized impurities (dominating at low temperatures) and at phonons (mostly at higher temperatures, resulting in a stronger temperature dependence). This mechanism was verified experimentally for bulk n-type Ge, showing relatively short spin lifetimes, compared to measurements in Si, of several hundred picoseconds (Zhou et al. 2011).

Finally, another mechanism responsible for spin relaxation or dephasing of spin ensembles should be mentioned. While the Elliott–Yafet mechanism is caused by spin-flip scattering of electrons at impurities due to the spin–orbit coupling, and the D'yakonov–Perel' mechanism is caused by the momentum-dependent spin splitting in crystals without inversion symmetry, the Bir–Aronov–Pikus mechanism is caused by spin-flip electron–hole exchange interactions (Bir et al. 1975). The Elliott–Yafet mechanism dominates in bulk narrow bandgap semiconductors with high impurities, while the D'yakonov–Perel' mechanism is important in n-type semiconductors, and the Bir–Aronov–Pikus mechanism is essential in p-doped semiconductors (Song and Kim 2002; Aronov et al. 1983; Zerrouati et al. 1987). Finally, the hyperfine interaction can also induce spin relaxation (Pershin and Privman 2003). For thin-film samples, however, these rules may change, especially with the Bir–Aronov–Pikus mechanism often showing smaller effects than the D'yakonov–Perel' mechanism (Zhou and Wu 2008).

Besides these spin relaxation and dephasing mechanisms inside single materials, interfaces strongly influence the spin states of moving electrons; and in many cases the spin, on the contrary, influences the behavior of an electron at an interface, especially the electric resistance of this interface. While properties of some interfaces were already given in the last section, the following sections will describe several transport-related effects in detail.

6.2 Tunnel effect

If a potential barrier blocks a particle, in classical physics the particle cannot get over it. In quantum mechanics, however, there is a certain chance that the particle tunnels through the barrier, if it is not too thick or too high. Several fundamental ideas, but also technological devices, are based on the tunneling effect, such as the scanning tunneling microscope, which measures the tunneling current between a conducting tip and the surface under investigation to examine the surface morphology with atomic resolution.

In spintronics, tunnel junctions are of interest. Here, two conductors are separated by a very thin, but completely closed, insulating layer. In this way, tunnel diodes or magnetic tunnel junctions (cf. Section 7.3) can be created. Due to these important applications in spintronics, tunneling effects are often investigated in this research area.

Spin injection into graphene, a material under strong investigation in recent years, for example, was examined for different tunnel barriers. One of the tunnel barriers that has recently been established for spin injection into graphene is hexagonal boron nitride (Juma et al. 2021). Generally, spin accumulation occurs under the ferromagnetic contact with the graphene. To obtain high efficiency, it is necessary to overcome the problem of the strong conductivity mismatch between both materials (Filip et al. 2000; Han et al. 2009). This can theoretically be obtained by introducing a tunnel barrier between both materials (Rashba 2000; Fert and Jaffrès 2001). In reality, however, several problems were found with typical oxide-based tunnel barriers, such as MgO (Avsar et al. 2011), Al_2O_3, or TiO_2 (Tombros et al. 2007). On the other hand, an Al_2O_3 tunnel barrier with Co electrodes was found to result in large diffusion lengths above 10 μm at room temperature (Demirci et al. 2018).

Hexagonal boron nitride can be applied in single or few layers without pinholes or other problems, which typically caused issues in tunnel barriers prepared from other materials (Lee et al. 2011), enabling high spin polarization (Kamalakar et al. 2017), while the spin lifetimes still suffer from technical problems (Gurram et al. 2018).

Generally, this material enables high spin polarization and tunnel magnetoresistance if introduced in magnetic tunnel junctions. By directly growing hexagonal boron nitride on Co and Fe stripes using chemical vapor deposition, up to 50% tunnel magnetoresistance was found in magnetic tunnel junctions consisting of Co/hexagonal boron nitride/Fe (Piquemal-Banci et al. 2018). Even higher TMR values, up to 1200%, were found by integrating a graphene layer in the hexagonal boron nitride tunnel layer between two Ni(111) layers (Harfah et al. 2022).

With different tunnel barriers and direct contact, the spin injection in graphene was found to have a nonlinear dependence on the applied DC bias current, including a reversal of the sign of the spin signal. This sign change was correlated with a specific bias voltage range. These experiments showed that the bias could be used to tailor the desired spin polarization (Zhu et al. 2018).

Although problematic in combination with graphene, Fe/MgO is a typical material combination for magnetic tunnel junctions. While the system seems to be easily understandable at first glance, theoretical investigations have revealed complex behavior. The ground state structure at the interface, for example, was found to form a spin-spiral due to the frustrated interaction of ferromagnetic and antiferromagnetic moments. This also resulted in a reduced and position-dependent Curie temperature as well as temperature-induced spin reorientations at low temperatures (Cuadrado et al. 2018).

Using Fe/MgO as tunnel contacts on an n-type Si channel, giant spin accumulation could be created in the Si layer, combined with large spin splitting even at room temperature. These effects are based on the large tunnel spin polarization of the contacts and can be increased by using a spin injector with dimensions similar to the spin diffusion length in silicon (Spiesser et al. 2017).

An interesting effect happens in a Josephson junction, which is coupled over a spin precessing with the Larmor frequency. In this case, the boundary conditions are time-dependent, that is, tunneling amplitudes for both spin orientations are different. In addition, spin-flip scattering occurs due to the spin precession. This results in time-dependent spin currents which, on the other hand, create a torque acting on the coupling spin, shifting its precession frequency (Holmqvist et al. 2018).

Generally, coupling tunnel barriers with spinel structure, a special crystal structure, has been shown to result in high spin filtering efficiencies, but also to be problematic due to lattice mismatches between the barrier and electrodes. Recently, combining electrodes prepared from superconducting $LiTi_2O_4$ thin films with spin filter devices from spinel oxide $CoFe_2O_4$ has been shown to overcome this problem since both lattices have nearly identical lattice constants. Measurements at low temperatures, that is, in the superconducting state, have been examined to investigate the Josephson junction behavior as well as the tunneling process (Mesoraca et al. 2018).

Other special systems in which tunneling plays a role are three-dimensional (3D) so-called topological insulators (Dey et al. 2020, Fluckey et al. 2022). In these special systems, the spin is caught perpendicularly to the momentum. This means that a spin polarization is created by an unpolarized charge current; or conversely, a net spin injected into the topological insulator would result in a charge accumulation. For the topological insulator Bi_2Te_3, these effects were electrically detected via a tunnel contact (Li et al. 2018b).

In the similar system Bi_2Te_2Se (also a topological insulator) as part of a spin valve (cf. Section 7.2), the aforementioned material hexagonal boron nitride was investigated as a tunnel barrier, separating the topological insulator from the ferromagnetic Co contacts. The contact resistance was found to vary with the spin valve characteristics. The results were interpreted as tuning the spin exchange between the ferromagnetic metal and the topological insulator by the number of intermediate layers of hexagonal boron nitride (Vaklinova et al. 2018).

Another special tunnel junction is created by a ferromagnetic/antiferromagnetic/ normal metal junction. A voltage or a temperature difference can be used to obtain electron transport through this tunnel junction. Due to the combination of ferromagnet and antiferromagnet, the electron transport is spin-dependent. Additionally, spin torques were found to act on the antiferromagnet similar to the usual Slonczewski torque, which induced damping. On the other hand, electron transport through the antiferromagnet worked in a different way than in the ferromagnet. Finally, the relative orientation of the magnetic moments in the antiferromagnet and the ferromagnet controlled charge current and spin current flow (Yamamoto et al. 2018).

The so-called tunnel diodes have rectifying properties similar to normal diodes. In a system composed of a Co_2MnSi layer on a thin ZnO layer on a p-Si substrate, the current over this junction was found to decrease with a magnetic in-plane field. This effect became smaller at larger temperatures (Maji and Nath 2018).

In combination of inorganic and organic materials, tunnel barriers can also be used as spin filters. An MgO tunnel barrier, for example, was used to connect a Ni top electrode with a bacteriorhodopsin layer on a gold electrode functionalized with cysteamine to create a spin valve with antisymmetric magnetoresistance for the magnetic field applied along the current flow direction and a positive symmetric magnetoresistance for a magnetic field perpendicular to the current orientation (Varade et al. 2018).

Lastly, tunnel barriers can be used to create artificial antiferromagnets. If several ferromagnetic layers are stacked, separated by tunnel barriers as spacer layers, a synthetic antiferromagnet can be created, which can also be used for diverse spintronic applications (Duine et al. 2018).

Nanoelectronic transport properties are dimensionally dependent. For classical macroscopic devices, the well-known property of electric current flow, also known as Ohm's law, holds very well, and the current is proportional to the applied voltage and a material factor represented by a given specific conductance. However, going into the meso- and nanoscales where magnetoelectronic devices operate, the mechanism of conduction is different. First of all, the quantum properties of matter do have an influence in these cases.

To be more precise, what will be presented here are considerations related to a quite elementary type of conductive electronic device element made from a central part (an island) connected with two terminals (leads), thus consisting of two tunnel junctions where the spin properties of carriers are significant for understanding the underlying effects (Fig. 6.1). For such types of physical systems with reduced spatial dimensions, the mechanism of conduction can be ballistic, and this is the significant property of such devices. Also, what is very fundamental, such devices cannot be treated only as totally separated from the surrounding environment since the transport performance can be influenced by the electromagnetic fields coming from other external devices and materials. On the other hand, from the quantum perspective, the underlying effects can be associated also with the so-called Coulomb blockade (cf. Section 6.3).

Tunnel barier

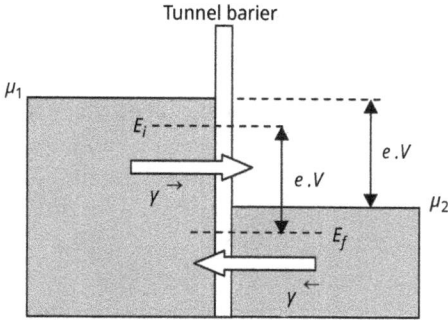

Fig. 6.1: The tunnel barrier between two materials of different electrochemical potentials μ_1 and μ_2 separated by the use of an externally applied voltage V.

Electric current is defined as the amount of electric charge transferred per unit time, that is $I = Q/t$ or, in a naive imagination of a single particle, $I = e/t$. The transport of single electrons at the mesoscale does not distinguish any moment of time or the period of it, which results from the quantum nature of underlying effects. Assuming only that the time of flight in the so-called ballistic regime is larger than the characteristic time of scattering events, then the one-by-one transport of electrons can be statistically expressed as $I = e \cdot y$, with y being the time rate of the transport phenomena – in the case of tunneling, it is the tunneling rate from one material to another.

The following three types or regimes of tunneling via the above generally sketched scheme can be realized (to be more precise, we talk about tunneling through the interface between two materials, a terminal and an island, and we do not care about whether these materials are magnetic or nonmagnetic – but the materials are conductors):

(a) The sequential regime with weak coupling between a lead and an island, when the tunnel junction resistance is significantly larger than the so-called resistance quantum $R_Q = h/(2e^2)$ – this term will be explained later.

(b) The cotunneling, or the second-order tunneling regime, occurs when an electron can travel from the left to the right via available intermediate states at the interface.

(c) The strong coupling regime, in which, at low enough temperatures, the Kondo effect takes place.

Nevertheless, the main goal of this chapter is to provide adequate formulas for the tunneling rates.

In the subsequent considerations, we will concentrate mainly on the sequential type of transport. Importantly, in our analysis, the spin will be included. Thus, we start with the derivation of the formula for the energy state density function $g(E) = dn/dE$, comparing 3D and 1D (one-dimensional) cases, while assuming that our electronic device reveals 1D transport. The reason for this approach is to express the underlying effects using the energy scale. Also, we will carefully discuss first the one-interface system and later the three-terminal device consisting of a two-interface device composed of a centrally located island and two terminals.

6.2.1 The single tunnel junction

In the derivation shown below in tabular form, we start by writing down the minimum unit of the available phase space for a single particle, governed by the Heisenberg principle of uncertainty, followed by the introduction of the momentum state density function $g(p)$ and finishing with the energy-related $g(E)$ equivalent quantity. At the end, the current density and the respective transport parameters are derived for the 1D case.

The 3D Cartesian space and the 6D phase space	**The 1D Cartesian space and the 2D phase space**
The elementary volume of the 6D phase space:	The elementary volume of the 2D phase-space:

$$dV \cdot dV_p = \frac{h^3}{2} \qquad (6.2a)$$

$$dx \cdot dp = \frac{h}{2} \qquad (6.2b)$$

$$dV_p = \frac{h^3}{2} \cdot \frac{1}{dV} \qquad (6.3a)$$

$$dp = \frac{h}{2} \cdot \frac{1}{dx} \qquad (6.3b)$$

$$\frac{4\pi p^2 dp}{dV_p} = \frac{4\pi p^2 dp}{\frac{h^3}{2}} \cdot dV \qquad (6.4a)$$

$$\frac{dp}{dV_p} = \frac{dp}{\frac{h}{2}} \cdot dx \qquad (6.4b)$$

Since the elementary volume for the momentum 3D space is $4\pi p^2 dp$, thus

Since the elementary volume for the momentum 1D space is d_p, thus

$$g(p)dp = \frac{8\pi p^2}{h^3} dp \qquad (6.5a)$$

$$g(p)dp = \frac{2}{h} dp \qquad (6.5b)$$

After regular common derivations, we get

$$g(p)dp \rightarrow g(E)dE \qquad (6.6)$$

$$E = \frac{p^2}{2m} \qquad (6.7)$$

$$dE = \frac{p}{m} dp \qquad (6.8)$$

$$dp = \frac{m}{p} dE \qquad (6.9)$$

$$\frac{8\pi p^2}{h^3} dp \rightarrow \frac{8\pi}{h^3} \cdot 2mE \cdot \frac{m}{\sqrt{2mE}} dE = g(E)dE \quad (6.10a)$$

$$\frac{2}{h} dp \rightarrow \frac{2}{h} \cdot \frac{m}{p} dE = \frac{2}{hv} dE = g(E)dE \quad (6.10b)$$

The energy density function:

$$g(E) = \frac{dn}{dE} = \frac{4\pi}{h^3} \cdot (2m)^{3/2} \cdot \sqrt{E} \qquad (6.11a)$$

$$g(E) = \frac{dn}{dE} = \frac{2}{hv} \qquad (6.11b)$$

(continued)

The 3D Cartesian space and the 6D phase space	The 1D Cartesian space and the 2D phase space

The current density is

$$j = (-e)nv \qquad (6.12)$$

Uniformity condition for the physical state distribution with energy gives us

$$\frac{dn}{dE} = \frac{n}{\mu_2 - \mu_1} \qquad (6.13)$$

The junction polarized by the external voltage V equals:

$$(-e)V = \mu_2 - \mu_1 \qquad (6.14)$$

$$
\begin{aligned}
j &= (-e)(-e)v \cdot \tfrac{4\pi}{h^3}(2m)^{3/2}\sqrt{E} \cdot \tfrac{\mu_2 - \mu_1}{(-e)} \\
&= e^2 v \cdot \tfrac{4\pi}{h^3}(2m)^{3/2}\sqrt{E} \cdot V
\end{aligned}
\qquad (6.15a)
$$

$$j = (-e)(-e)v \cdot \frac{2}{hv} \cdot \frac{\mu_2 - \mu_1}{(-e)} = \frac{2e^2}{h}V \qquad (6.15b)$$

$$\frac{j}{V} = G_Q = \frac{2e^2}{h} \qquad (6.16)$$

$$R_Q = \frac{h}{2e^2} \qquad (6.17)$$

To conclude shortly, the obtained conductance quantum G_Q, for the 1D case, is expressed in the unit Ω^{-1}, since $C^2/(Js) = AC/J = AC/(CV) = A/V = \omega^{-1}$, while the unit of the classical bulk conductance is $(\Omega \cdot m)^{-1}$. This obvious information results from the fact that in 1D space there are no transverse dimensions. The classical current density is represented by the unit $[j] = A/m^2$ and has to be derived from the above 1D current–voltage dependence. Once again, since our 1D considerations are related to a single electron, we can conclude that the conductance phenomenon can be referenced to the "elementary" $h/2e^2$ level.

Thus, the mesoscopically understood Ohm's law, written in the quantum style as

$$I = \frac{V}{R_0} = \frac{V}{h} \cdot 2e^2, \qquad (6.18)$$

can be equivalently expressed in the statistical manner as

$$I = e \cdot \gamma, \qquad (6.19)$$

thus as the product of the elementary charge and the rate expressed in the units of s^{-1}. These two simple equations yield the very intuitive expression for the rate:

$$\gamma = \frac{2Ve}{h} = \frac{Ve}{\pi\hbar} = \frac{energy}{\pi\hbar}. \qquad (6.20)$$

The important point is that a rate, in general, is a quantum-type parameter dependent, however, on material parameters like the concentration of spin-current carriers

n treated as a parameter continuous in space and time. Thus, further formulas should also contain the probability, which was preliminarily mentioned in Section 1.7. Let us remind and compare the adequate classical versus quantum formulas for electrons propagating in 1D. Classically, it was

$$n_\sigma(x, t) = P_\rightarrow \cdot n_\sigma(x - \Delta x, t - \Delta t) + P_\leftarrow \cdot n_\sigma(x + \Delta x, t - \Delta t), \tag{6.21}$$

where at a given point and moment of time (x, t), the output balance of the spin concentration $n_\sigma(x, t) = n_{up}(x, t) - n_{down}(x, t)$ was tailored by the spins coming from the left with the corresponding probability $P_\rightarrow$, and by the spins coming from the right with the corresponding probability $P_\leftarrow$. In the contemporary quantum picture, the relevant formula is

$$P(n) \cdot \left(\gamma_\sigma^\rightarrow(n) + \gamma_\sigma^\leftarrow(n)\right) = P(n-1) \cdot \gamma_\sigma^\rightarrow(n-1) + P(n+1) \cdot \gamma_\sigma^\leftarrow(n+1) \tag{6.22}$$

where concentrations are not included in an evident form, but instead are substituted by rates.

There is an important modification in the quantum approach seen on the left side of the above equation, namely, the physical state depends not only on a given spin state σ, controlled by the factor $\left(\gamma_\sigma^\rightarrow(n) + \gamma_\sigma^\leftarrow(n)\right)$, but first of all on $P(n)$, the probability of finding a given number (the concentration) of the excess electrons. The quantum approach distinguishes the excess electrons from those that are there in general. The classical approach, based on the statistical rate equations described in Section 1.7, counts all the contributions together, providing the classical concentration n.

The significant question is, what is the physical meaning and the origin of the probabilities $P(n)$, $P(n-1)$, and $P(n+1)$? All this results from the availability and non-availability of physical states for the spin carriers, which is a very direct material-dependent property. And again, in other words, it can be said that $P(n)$ is the reason, while the γ rate is the consequence. $P(n)$, however, is also a consequence of the external voltage usage – in the practical device realization. Thus, what is really important, the probability of the occurrence of excess electrons can be tailored and modified by the external load (voltage) in the lead–island–lead three-region basic device.

Apart from that, the statistical properties of materials are tailored by the electro-chemical potentials, the Fermi levels, involved in the well-known Fermi probability-density distribution function. Thus, the very operative formula for the transmission rate between two materials should look as follows:

$$\gamma_Q^\rightarrow \sim \frac{\text{energy}}{\pi\hbar} \cdot f(E_i) \cdot \left(1 - f(E_f)\right) \cdot P(E). \tag{6.23}$$

The above formula is valid for the single energy-level transport, where the subscript Q was temporarily introduced to emphasize the quantum nature of the underlined formalism, where the factor $f(E_i)$ informs about how many electrons are on the left side, while the factor $(1-f(E_f))$ counts how many places are not occupied by electrons and are ready to adopt electrons tunneling from the left (Fig. 6.1).

The above-described proportionality can be transformed into the rate equation using the following simple postulate: the full quantum description uses probability language and deals with the excess electrons. On the other hand, what is accessible during laboratory measurements of tunneling are the classical, macroscopic electric parameters, like, for example, currents. In other words, we can assume that the ratio of the normal conductance to the quantum conductance should be proportional to the corresponding ratios of current densities and rates; thus,

$$\frac{j_T}{j_Q} = \frac{\gamma_T}{\gamma_Q} = \frac{G_T}{G_Q} \tag{6.24}$$

and the tunneling rate then equals

$$\gamma_T = \frac{G_T}{G_Q} \cdot \gamma_Q, \tag{6.25}$$

which finally leads to the equation describing the adequate rate:

$$\overrightarrow{\gamma_T} = \frac{G_T}{G_Q} \cdot \frac{energy}{\pi\hbar} \cdot f(E_i) \cdot (1-f(E_f)) \cdot P(E). \tag{6.26}$$

Going even beyond this simple physical scheme, in naturally met situations we deal with the whole spectrum of events being in agreement with energy conservation $E_i - E_f = -e \cdot V$, where V is the aforementioned applied voltage. In this case, in order to include all the relevant events, we should perform integration. The integration, due to mathematical reasons, is formally carried out from $-\infty$ to $+\infty$ in the energy domain. Thus, the final solutions for the tunneling rate (the T subscript was omitted for simplicity and clarity, also in order to release a place for the proper subscript related to the spin degree of freedom) from the left to the right side and backward, respectively, are equal to

$$\overrightarrow{\gamma} = \frac{1}{\pi\hbar}\frac{G_T}{G_Q} \int\limits_{-\infty}^{+\infty} dE_i \int\limits_{-\infty}^{+\infty} dE_f f(E_i)\left[1-f(E_f)\right] \cdot P(E_i - E_f + eV) \tag{6.27}$$

and

$$\overleftarrow{\gamma} = \frac{1}{\pi\hbar}\frac{G_T}{G_Q} \int\limits_{-\infty}^{+\infty} dE_i \int\limits_{-\infty}^{+\infty} dE_f [1-f(E_i)]f(E_f) \cdot P(E_f - E_i - eV), \tag{6.28}$$

where the energy conservation rule was directly included within the probability function argument, the function being a consequence of something beyond purely classical description, as it was mentioned. The electric current can then be derived from

$$I = e(\gamma^{\rightarrow} - \gamma^{\leftarrow}), \tag{6.29}$$

or, in a more direct form, using the following obvious expression:

$$I = \frac{e}{\pi \hbar} \frac{G_T}{G_Q} \int_{-\infty}^{+\infty} dE_i \int_{-\infty}^{+\infty} dE_f \left(f(E_i) \left[1 - f(E_f)\right] \cdot P(E_i - E_f + eV) - \left[1 - f(E_i)\right] f(E_f) \cdot P(E_f - E_i - eV) \right). \tag{6.30}$$

This leads us almost to the final practical result; however, under the assumption that the concrete shape of the $P(E)$ function will be known. This requires some additional explanation.

First of all, the $P(E)$ function has the following two properties:

$$\int_{-\infty}^{+\infty} P(E)dE = 1, \tag{6.31}$$

what seems to be quite elementary for a probability, and the property enabling the obtainment of the average measure of energy, thus

$$\int_{-\infty}^{+\infty} P(E)EdE = E_Q = \frac{e^2}{2C}. \tag{6.32}$$

This averaged energy, in the quantum approach, equals the elementary single-electron charging energy and is related to the tunnel junction capacity C. As mentioned earlier, the character of $P(E)$ reveals the material properties of the tunnel junction and, importantly, the coupling with the external electrical environment (power supply and terminals). There are two limiting cases to be considered from that perspective (Devoret et al. 1990, Ingold et al. 1991, Grabert et al. 1991):

a) When an electron tunnels through the barrier, it does not exchange energy with the hosting material. This case is equivalent to the sequential tunneling mentioned earlier. In this case, $P(E)$ reduces to Dirac's delta function: $P(E) = \delta(E)$. In practical applications, this regime can be narrowed to situations when control of a junction is done only by the voltage bias and where the following expressions for the "left-to-right" and "right-to-left" rates are valid (Dirac's delta cancels integration):

$$\gamma^{\rightarrow} = \frac{1}{\pi \hbar} \frac{G_T}{G_Q} \frac{eV}{1 - \exp(-(eV/k_B T))}, \tag{6.33}$$

and

$$\gamma^{\leftarrow} = \exp\left(-\frac{eV}{k_B T}\right) \cdot \gamma^{\rightarrow} = \frac{1}{\pi \hbar} \frac{G_T}{G_Q} \frac{eV}{1 - \exp(-(eV/k_B T))} \cdot \exp\left(-\frac{eV}{k_B T}\right), \tag{6.34}$$

which leads us finally to the simple formula for the current

$$I(V) = e \cdot \frac{eV}{\pi\hbar} \frac{G_T}{G_Q} \left(\frac{1}{1-\exp(-(eV/k_BT))} - \frac{\exp(-(eV/k_BT))}{1-\exp(-(eV/k_BT))} \right) = \frac{e^2 V}{\pi\hbar} \frac{G_T}{G_Q}. \tag{6.35}$$

b) When an electron tunnels and exchanges energy with the hosting materials, $P(E)$ depends exponentially on the energy as well as on the single-electron charging energy (there is, in principle, a normal distribution around the mean value $E_Q = e^2/2C$), namely

$$P(E) = \frac{1}{\sqrt{4\pi k_B T \cdot E_Q}} \exp\left(-\frac{(E-E_Q)^2}{4 k_B T \cdot E_Q} \right), \tag{6.36}$$

or by modification of the variable $E \to eV - E$ (the argument of the function $P(E)$ informs us how much energy E a single electron loses during tunneling), we get

$$P(eV - E) = \frac{1}{\sqrt{4\pi k_B T \cdot E_Q}} \exp\left(-\frac{(eV - E - E_Q)^2}{4 k_B T \cdot E_Q} \right) \tag{6.37}$$

and the following expressions for the equivalent rates are now valid:

$$\overrightarrow{\gamma} = \frac{1}{\pi\hbar} \frac{G_T}{G_Q} \int_{-\infty}^{+\infty} \frac{E}{1-\exp(-(E/k_BT))} \cdot \frac{\exp\left[-(eV - E - E_Q)^2/(4k_BT \cdot E_Q) \right]}{1-\exp(-(E/k_BT))} dE. \tag{6.38}$$

for the left-to-right rate, and

$$\overleftarrow{\gamma} = \exp\left(-\frac{eV}{k_BT} \right) \cdot \overrightarrow{\gamma} = \frac{1}{\pi\hbar} \frac{G_T}{G_Q} \int_{-\infty}^{+\infty} \frac{E}{1-\exp(-(E/k_BT))} \cdot \exp\left(-\frac{eV}{k_BT} \right) \cdot$$
$$\frac{\exp\left[-(eV - E - E_Q)^2/(4k_BT \cdot E_Q) \right]}{1-\exp(-(E/k_BT))} dE, \tag{6.39}$$

for the right-to-left rate, respectively. This leads us finally to the equivalent formula for the electric current:

$$I(V) = e(\overrightarrow{\gamma} - \overleftarrow{\gamma}) = \frac{1}{\pi\hbar} \frac{G_T}{G_Q} \left[1 - \exp\left(-\frac{eV}{k_BT} \right) \right] \int_{-\infty}^{+\infty} \frac{E}{1-\exp(-(E/k_BT))} \cdot$$
$$\frac{\exp\left[-(eV - E - E_Q)^2/(4k_BT \cdot E_Q) \right]}{1-\exp(-(E/k_BT))} dE. \tag{6.40}$$

It is also worth mentioning that the formulas for the rates, as derived earlier, can be modified straightforwardly by the inclusion of equivalent subscripts relevant to the spin degree of freedom. The general mathematical shape remains the same.

6.2.2 The lead–island–lead basic three-component system

What was discussed earlier was related to the single tunnel barrier being a component of our elementary three-region magnetoelectronic structure. Now, we will extend this picture by including another junction in order to analyze the elementary case of a spintronic device consisting of a central island and two leads (terminals). The theory provided was introduced by Hermann Grabert and coworkers at the beginning of the 1990s and published in several scientific articles (Devoret et al. 1990; Ingold et al. 1991; Grabert et al. 1991). The approach, even if it seems to be quite complex in several points, as each advanced theory in the discipline of solid-state physics, has, however, very practical consequences, since it couples the quantum picture level with measurable quantities, realistic material parameters and devices and, what should be emphasized, it takes into account the influence of an external load on the tunneling effects.

Before we take a step into the main topic of this section, let us make some general quantum-mechanical considerations in order to remind ourselves of some fundamental aspects of the issue.

In quantum mechanics, the commonly met exponent factor, being a component of the wave function or treated as an operator, is equipped with the generally understood wave phase φ, namely

$$e^{i\varphi} = e^{i(E/\hbar)t}, \tag{6.41}$$

where the phase is the product of energy E and time t over the Planck constant $\hbar$. In the case of magnetoelectronic devices, we have, however,

$$\varphi = \frac{\text{energy}}{\hbar} t = \frac{e \cdot U}{\hbar} t = \frac{e}{\hbar} \frac{Q}{C} t \tag{6.42}$$

or, in equivalent form,

$$\frac{\hbar}{e} \varphi = \frac{Q}{C} t, \tag{6.43}$$

which finally enables calculations of the time derivative of the phase, thus

$$\frac{\hbar}{e} \dot{\varphi} = \frac{Q}{C}, \tag{6.44}$$

where in the traditional manner the $\dot{\varphi} = \partial \varphi / \partial t$ derivative was marked.

All the above means that the phase φ and the charge Q – expressed with the time derivative of $\dot{\varphi}$ – create a pair of conjugate dynamical variables represented by the operators that do not commute, in the same way like quantum-mechanical operators of the position x and the momentum $p = m\dot{x}$ do not commute since $[x, m\dot{x}] = i\hbar$. Similarly, we have

$$\left[\frac{\hbar}{e}\varphi, Q\right] = i\hbar \tag{6.45}$$

or equivalently

$$[\varphi, Q] = ie. \tag{6.46}$$

The practical meaning of the phase, from the electronic science perspective, becomes more obvious if one integrates the time derivative of the phase, as a consequence of the dynamical change of the voltage imposed onto the tunnel junction, namely

$$\frac{d\varphi}{dt} = \frac{e}{\hbar}\frac{Q}{C} = \frac{e}{\hbar}U(t) \Rightarrow \varphi(t) = \frac{e}{\hbar}\int_{-\infty}^{t} U(t)dt. \tag{6.47}$$

The physical meaning of the phase becomes even clearer since the quantum approach for the tunneling phenomenon counts excess electrons, and this fact can be expressed directly by the use of phase operators acting from the left and from the right onto the junction charge, namely

$$e^{i\varphi}Qe^{-i\varphi} = Q + (-e) = Q - e. \tag{6.48}$$

Now, going forward, the classical tunneling Hamiltonian

$$H_T = \sum_{k_1 k_2 \sigma} T_{k_2 k_1} c^\dagger_{k_2 \sigma} c_{k_1 \sigma} + h.c., \tag{6.49}$$

with *h.c.* meaning higher coefficients, it describes a tunneling event which annihilates an electron with the wave vector $\vec{k}_1$ in the lead material number 1 and creates an electron with the wave vector $\vec{k}_2$ in the island material number 2. This expression distinguishes between spins of an electron. In the case of voltage application across the junction, we have

$$H_T = \sum_{k_1 k_2 \sigma} T_{k_2 k_1} c^\dagger_{k_2 \sigma} c_{k_1 \sigma} \exp(-i\varphi) + h.c. \tag{6.50}$$

In this way, the charge-shift operator $\exp(-i\varphi)$ can be interpreted as the modification of the pure tunneling amplitude $T_{k_2 k_1}$ by the externally applied voltage source.

For the double-junction system considered here, we are dealing with the connection of the two, in general different, capacitances in series. The relevant quantum-like expressions for such a device are quite straightforward. The island, the central part, collects the charge $Q_1 - Q_2$ adopted from both junctions if no tunneling takes place. Also, the total capacitance C is

$$\frac{1}{C} = \frac{1}{C_1} + \frac{1}{C_2}, \tag{6.51}$$

and the total electric charge Q kept in this set of capacitors equals

$$Q = U\frac{C_1 C_2}{C_1 + C_2} = (U_1 + U_2)\frac{(Q_1/U_1)(Q_2/U_2)}{C_1 + C_2} = \frac{(1 + U_2/U_1)Q_1(Q_2/U_2)}{C_1 + C_2} =$$
$$= \frac{Q_1(Q_2/U_2) + Q_2(Q_1/U_1)}{C_1 + C_2} = \frac{C_1 Q_2 + C_2 Q_1}{C_1 + C_2} = \frac{C_2}{C_1 + C_2}Q_1 + \frac{C_1}{C_1 + C_2}Q_2. \tag{6.52}$$

However, since the quantum-mechanical phase is associated with the given capacity, $\dot{\varphi} = [e/(\hbar C)]Q$ the variable Ψ canonically conjugated to the charge $Q_1 - Q_2$ kept at the island equals

$$\Psi = \frac{C_1 \varphi_1 - C_2 \varphi_2}{C_1 + C_2}, \tag{6.53}$$

and

$$\dot{\Psi} = \frac{C_1 \dot{\varphi}_1 - C_2 \dot{\varphi}_2}{C_1 + C_2} \tag{6.54}$$

with

$$\dot{\Psi} = \frac{e}{\hbar(C_1 + C_2)}(Q_1 - Q_2) \tag{6.55}$$

and

$$[\Psi, (Q_1 - Q_2)] = [\Psi, q] = ie, \tag{6.56}$$

where the charge on the island is an integer multiple of the elementary charge, that is $q = ne$.

All the above, derived here, enables us to directly rewrite the formulas valid for a single tunnel junction elaborated in the previous section.

The total Hamiltonian of the recent system consists of the term H_0, describing the state in the absence of tunneling, and the two terms H_1 and H_2, reflecting the tunneling for the left and right junctions, respectively.

For example,

$$H_1 = \sum_{k_1 k_2 \sigma} T^1_{k_2 k_1} c^\dagger_{k_2 \sigma} c_{k_1 \sigma} \exp(-i\varphi_1) + h.c., \tag{6.57}$$

where we should provide the relation between the charge-shift operator phase for the first junction, the total phase $\varphi = \varphi_1 + \varphi_2$ as and the island phase Ψ, that is $\varphi_1 = \varphi_1(\varphi, \Psi)$. The derivation is quite straightforward, namely

$$\Psi = \frac{C_1 \varphi_1}{C_1 + C_2} - \frac{C_2 \varphi_2}{C_1 + C_2}, \tag{6.58a}$$

$$\Psi + \frac{C_2 \varphi_2}{C_1 + C_2} = \frac{C_1 \varphi_1}{C_1 + C_2}, \tag{6.58b}$$

$$\Psi + \frac{C_2}{C_1 + C_2}(\varphi - \varphi_1) = \frac{C_1 \varphi_1}{C_1 + C_2}, \tag{6.58c}$$

$$\Psi + \frac{C_2}{C_1 + C_2}\varphi = \left(\frac{C_2}{C_1 + C_2} + \frac{C_1}{C_1 + C_2}\right)\varphi_1, \tag{6.58d}$$

$$\Psi + \frac{C_2}{C_1 + C_2}\varphi = \varphi_1 \quad \Rightarrow \quad \Psi + \frac{C}{C_1}\varphi = \varphi_1. \tag{6.58e}$$

Similarly,

$$\Psi + \frac{C}{C_2}\varphi = \varphi_2, \tag{6.59}$$

and finally,

$$H_1 = \sum_{k_1 k_2 \sigma} T^1_{k_2 k_1} c^\dagger_{k_2 \sigma} c_{k_1 \sigma} \exp\left(-i\frac{C}{C_1}\varphi - i\Psi\right) + \text{h.c.}, \tag{6.60a}$$

$$H_2 = \sum_{k_1 k_2 \sigma} T^2_{k_2 k_1} c^\dagger_{k_2 \sigma} c_{k_1 \sigma} \exp\left(-i\frac{C}{C_2}\varphi - i\Psi\right) + \text{h.c.}. \tag{6.60b}$$

Thus, to summarize, the $\exp(-i\Psi)$ operator counts electrons coming onto and moving out of the island due to tunneling. The operator $\exp(-i\varphi)$, weighted by the respective capacitance, describes the influence of the external load on the tunneling events.

If we compare the tunneling rate formula for the single junction presented in the previous section, where the golden rule (energy conservation principle) of the tunneling event was included as the argument of the probability $\gamma^\rightarrow \sim P(E_i - E_f + eV)$, then the probability is the two-argument function $P(C/C_1, E_i - E_f + E_1(V, q))$ for the first junction and the tunneling from the lead into the island, and $P(C/C_1, E_f - E_i - E_1(V, q))$ for the back-tunneling events from the island to the first lead, and where

$$E_1(V, q) = \frac{C}{C_1}eV + \frac{q^2}{2(C_1 + C_2)} - \frac{(q-e)^2}{2(C_1 + C_2)}. \tag{6.61}$$

The energy consists of, from left to right, the energy of the external load imposed onto a junction, the island steady-state charging energy, and the negative energy contribution relaxing the device energy stemming from the additional electron tunneling through the barrier. Also, for the special case when $E_i = E_f/2$, which can be interpreted as the statistically averaged level, $E_i - E_f + E_1(V, q) = E_1(V, q) - E_i$, the following equations connect the flow and antiflow rates:

$$\gamma^\rightarrow_1(V, q) = \gamma^\leftarrow_1(-V, -q), \tag{6.62}$$

$$\gamma_1^{\leftarrow}(V, q - e) = \gamma_1^{\leftarrow}(V, q) \exp(-E_1(V, q)/kT). \tag{6.63}$$

Symmetrically, for the second junction, we get

$$\gamma_2^{\rightarrow}(V, q) = \gamma_2^{\leftarrow}(-V, -q), \tag{6.64}$$

$$\gamma_2^{\rightarrow}(V, q - e) = \gamma_2^{\leftarrow}(V, q) \exp(-E_2(-V, q)/kT), \tag{6.65}$$

with the system energy at the second junction equivalently written as

$$E_2(V, q) = \frac{C}{C_2} eV + \frac{q^2}{2(C_1 + C_2)} - \frac{(q - e)^2}{2(C_1 + C_2)}. \tag{6.66}$$

At the end of this section, similarly to the single junction case, the two limiting cases of low-impedance and high-impedance electromagnetic environments will be presented.

The general rate formula for the first tunneling junction is

$$\gamma_1^{\rightarrow} = \frac{1}{\pi\hbar} \frac{G_1}{G_Q} \int_{-\infty}^{+\infty} dE_i \int_{-\infty}^{+\infty} dE_f f(E_i) \left[1 - f(E_f)\right] \cdot P\left(\frac{C}{C_1}, E_1(V, q) - E_i\right). \tag{6.67}$$

For a low-impedance environment, or in other words, when the phase fluctuations $\delta\varphi$ are very limited, the probability is the very narrow Dirac-delta function, and the transport is well defined for the given energy, $p = \delta(E)$. In this case,

$$\gamma_1^{\rightarrow} = \frac{1}{\pi\hbar} \frac{G_1}{G_Q} \frac{E_1(V, q)}{1 - \exp(-(E_1(V, q)/k_B T)))} \tag{6.68}$$

For the opposite case, when the phase fluctuations follow a Gaussian distribution, the probability function is

$$P\left(\frac{C}{C_1}, E\right) = \frac{1}{\sqrt{4\pi k_B T \cdot (C/C_1)^2 eV}} \exp\left(-\frac{\left(E - (C/C_1)^2 eV\right)^2}{4k_B T \cdot (C/C_1)^2 eV}\right), \tag{6.69}$$

$$\gamma_1^{\rightarrow} = \frac{1}{\pi\hbar} \frac{G_1}{G_Q} \int_{-\infty}^{+\infty} \frac{E}{1 - \exp(-(E/k_B T))} \cdot \frac{1}{\sqrt{4\pi k_B T \cdot (C/C_1)^2 eV}} \exp\left(-\frac{\left(E - (C/C_1)^2 eV\right)^2}{4k_B T \cdot (C/C_1)^2 eV}\right) dE, \tag{6.70}$$

where the repeated term $(C/C_1)^2 eV$ comes from the charging energy of the external load.

In order to summarize the above considerations, the formula for the measurable current should be provided. This is, however, not the case for the simple $I(V) = e(\gamma^{\rightarrow} - \gamma^{\leftarrow})$ formula. For the double junction, and especially for a temperature above absolute zero, the stochastic nature of current flow results from uncorrelated tunneling into an island, from left to right, including not only single electron events but also packets of electrons. In this case, the formula for the current is

$$I(V) = e \sum_{-\infty}^{+\infty} p_n \left(\gamma_1^{\rightarrow}(n) - \gamma_1^{\leftarrow}(n) \right) = e \sum_{-\infty}^{+\infty} p_n \left(\gamma_2^{\rightarrow}(n) - \gamma_2^{\leftarrow}(n) \right), \tag{6.71}$$

where the charge at the island is proportional to the elementary charge $q = ne$ and p_n is the probability that this charge is kept on the island. It is obvious then that we turn back to the physical problem of solving rate equations presented before in the following form:

$$P(n) \cdot \left(\gamma_\sigma^{\rightarrow}(n) + \gamma_\sigma^{\leftarrow}(n) \right) = P(n-1) \cdot \gamma_\sigma^{\rightarrow}(n-1) + P(n+1) \cdot \gamma_\sigma^{\leftarrow}(n+1). \tag{6.72}$$

To avoid misunderstandings, we will no longer use the capital letter $P(n)$ to mark the stochastic probability – the symbol used in this chapter to describe probabilities for the transport rates – and we will use lowercase $p(n)$ instead. Thus, the last but not least problem to be solved is the other equation for $p(n)$ using the so-called master equation approach:

$$\frac{\partial p(n)}{\partial t} = \sum_{n=-\infty}^{n=+\infty} [\gamma(n, n+1)p(n+1) + \gamma(n, n-1)p(n-1) - \gamma(n+1, n)p(n) - \gamma(n-1, n)p(n)] \tag{6.73}$$

along with the clear condition

$$\sum_{n=-\infty}^{n=+\infty} p_n = 1. \tag{6.74}$$

The master equation describes the time evolution of the $p(n)$ balanced by equivalent events. In the formula, the meaning of the term γ is γ(later, before). For example, $\gamma(n+1, n)$ means the island charge evolves from the $q = ne$ state into the $q = (n+1)e$ one. Obviously, the γ term can be completed with the spin index, such as $\gamma_\sigma(n+1, n)$. Hence, the direction of flow is expressed in the following elementary equations for rates introduced before:

$$\gamma(n+1, n) = \gamma_1^{\leftarrow} + \gamma_2^{\rightarrow}, \tag{6.75a}$$

$$\gamma(n-1, n) = \gamma_1^{\rightarrow} + \gamma_2^{\leftarrow}. \tag{6.75b}$$

The stationary case of the master equation implies the following identities:

$$\gamma(n, n+1)p(n+1) = \gamma(n+1, n)p(n) \tag{6.76a}$$

and

$$\gamma(n, n-1)p(n-1) = \gamma(n-1, n)p(n). \tag{6.76b}$$

Equation (6.76a) can be written sequentially starting from $n = 0$ (there is no charge at the island) state,

$$n = 0, \gamma(0, \ 1)p(1) = \gamma(1, \ 0)p(0), \tag{6.77a}$$

$$n = 1, \gamma(1, \ 2)p(2) = \gamma(2, \ 1)p(1), \tag{6.77b}$$

$$n = 2, \gamma(2, \ 3)p(3) = \gamma(3, \ 2)p(2), \tag{6.77c}$$

. . .

$$n = n, \gamma(n, n + 1)p(n + 1) = \gamma(n + 1, n)p(n), \tag{6.77d}$$

and consequently

$$p(2) = \frac{\gamma(2, \ 1)}{\gamma(1, \ 2)}p(1) = \frac{\gamma(2, \ 1)}{\gamma(1, \ 2)}\frac{\gamma(1, \ 0)}{\gamma(0, \ 1)}p(0), \tag{6.78a}$$

$$p(3) = \frac{\gamma(3, \ 2)}{\gamma(2, \ 3)}p(2) = \frac{\gamma(3, \ 2)}{\gamma(2, \ 3)}\frac{\gamma(2, \ 1)}{\gamma(1, \ 2)}\frac{\gamma(1, \ 0)}{\gamma(0, \ 1)}p(0), \tag{6.78b}$$

which finally leads to

$$p(n) = \frac{\gamma(n, \ n - 1)}{\gamma(n - 1, \ n)}p(n - 1) = \frac{\gamma(n, \ n - 1)}{\gamma(n - 1, \ n)}\frac{\gamma(n - 1, \ n - 2)}{\gamma(n - 2, \ n - 1)} \cdots \frac{\gamma(1, \ 0)}{\gamma(0, \ 1)}p(0)$$

$$= p(0) \prod_{m=0}^{m=n-1} \frac{\gamma(m + 1, \ m)}{\gamma(m, \ m + 1)}. \tag{6.79}$$

Similarly, calculating the negatively indexed $p(-n)$ values of the first identity, one obtains

$$n = 0 \ \gamma(0, \ 1)p(1) = \gamma(1, \ 0)p(0), \tag{6.80a}$$

$$n = -1 \ \gamma(-1, \ 0)p(0) = \gamma(0, \ -1)p(-1), \tag{6.80b}$$

$$n = -2 \ \gamma(-2, \ -1)p(-1) = \gamma(-1, \ -2)p(-2) \tag{6.80c}$$

and

$$p(-n) = \frac{\gamma(-n, \ -n + 1)}{\gamma(-n + 1, \ -n)}\frac{\gamma(-n + 1, \ -n + 2)}{\gamma(-n + 2, \ -n + 1)} \cdots \frac{\gamma(-1, \ 0)}{\gamma(0, \ -1)}p(0) = p(0) \prod_{m=-n+1}^{m=0} \frac{\gamma(m - 1, m)}{\gamma(m, m - 1)}. \tag{6.81}$$

This set of equations enables calculations of the current–voltage characteristics.

The typical values for mesoscopic systems are as follows: The tunnel junction resistance falls in the range of 10^6–10^8 Ω, and the capacitance can be of the order of 10^{-18} F. Applying a voltage of some millivolts, the currents flowing through the two-junction systems are of the order of 0.01–0.05 nA, assuming a circular cross-section of the tunnel junction with a diameter of 10 nm.

The characteristics of such junction systems are associated with the discretized charging energy effect. If the energy provided from an external source exceeds the two-junction charging energy

$$E_C = \frac{e^2}{2(C_1 + C_2)}, \tag{6.82}$$

a tunnel current occurs. In other words, if the applied voltage exceeds the value

$$V_C = \frac{E_C}{e} = \frac{e}{2(C_1 + C_2)}, \tag{6.83}$$

then the current can flow until the next so-called Coulomb barrier is overcome (cf. Section 6.3). The current–voltage dependence then evolves through a characteristic staircase shape. The shape is clearly visible in situations where there exists a high contrast between the capacitances and resistances of the two junctions, thus, for example $C_1 \gg C_2$ and $R_1 \gg R_2$, when the situation is more monolithic. Assuming $C_1 = 100$ aF and $C_2 = 1$ aF, we get

$$V_C = = \frac{1.6 \cdot 10^{-19} C}{202 \cdot 10^{-18} F} = 7.9 \cdot 10^{-4} V = 0.79 \, \text{mV}, \tag{6.84}$$

which can be measured rather easily. The equivalent current jumps are on the order of 0.01 nA.

At the end of this chapter, let us mention another important case of magnetoelectronic devices, which is the mesoscopic transistor. By providing an extra voltage through the additional capacitance C_3 to the island, we have a modified charging energy which equals

$$E_C = \frac{e^2}{2(C_1 + C_2 + C_3)}. \tag{6.85}$$

Another interesting issue, which is out of the scope of this book, is the detailed analysis of energy-level spectra of the island and the external terminals – there are two limiting cases considered, namely, continuous-level systems and systems with discretized energy levels.

This overview has shown the broad bandwidth of possible phenomena in spintronics, which are related to tunneling. Another effect, in which typically a tunnel contact is used for measurements, is described in the next section.

6.3 Coulomb blockade

The Coulomb blockade is an effect that may occur in very small objects, for example, in quantum dots or conductive islands in nonconductive materials. It describes a phenomenon that can be observed if the contact resistances are very high – practically, this is realized by tunnel barriers, as described in the last section. In this case, the small object connected by tunnel barriers has a certain capacitance. If an electron is added to this small conductor, the voltage of the latter is increased by $\Delta U = e/C$ with the elementary charge e and the capacitance C. In very small objects, the capacitance can be very small too, making ΔU so high that tunneling of an electron onto the object is impeded. This effect is typically visible for small temperatures – since the thermal energy must be small enough to avoid enabling the electron to pass the small object by thermal excitation – and small voltages driving the electron flow; however, for very small objects on the nanoscale, it may become visible even at room temperature. This effect is necessary for the single-electron transistor (Section 7.6) (Averin and Likharev 1985).

Not only due to its application in the single-electron transistor has the Coulomb blockade been investigated by many researchers. An interesting idea applying the Coulomb blockade regime was reported by Ilinskaya et al. (2018). It is based on the idea to move electrons by mechanical procedures, for example, by a grain moving between two connection lines. This possibility was predicted much earlier (Gorelik et al. 1998) and examined by different groups since (Shekhter et al. 2013, Kulinich et al. 2014). Ilinskaya and coworkers investigated the possibility to "shuttle" an electron on a movable metallic grain between two magnetic leads at different temperatures. The latter are fully antiparallel spin-polarized. Shuttling was indeed found for a certain external magnetic field range, depending on the upper and lower temperatures of the magnetic connections.

In carbon nanotubes coupled to conductive leads over tunnel barriers, giant magnetoresistance (GMR, cf. Section 7.2) values up to 1,000% were found (Urdampilleta et al. 2011; Urdampilleta et al. 2013), which means that the resistance through this spin valve depended strongly on the relative orientation (ferromagnetic or antiferromagnetic) of two ferromagnetic layers, coupled to an intermediate nonmagnetic spacer layer. Here, the carbon nanotube corresponds to the nonmagnetic spacer, while single-molecule magnets were attached to both ends. In a theoretical model by Krainov et al. (2017), this large GMR was attributed to the Coulomb blockade in combination with the so-called Fano effect (for a more detailed description, cf. Section 7.14). While the single-molecule magnets originally couple antiferromagnetically, a large external magnetic field can align their orientation, resulting in one spin orientation not being scattered in any of the molecules and thus having a significantly reduced resistance.

An effect similar to the Coulomb blockade was found in quantum dots, allowing for controlling the spin-dependent tunneling through defect states at the interface by external magnetic fields. While this magnetoresistance effect was previously attributed to ensemble dephasing of a spin accumulation, Swartz et al. (2016) showed that spin accumulation in the nonmagnetic spacer layer or along the interfaces between nonmagnetic and ferromagnetic materials was unlikely; instead, the new explanation by a Coulomb blockade was consistent with the experimental results in these experiments. In a quantum well separated from two ferromagnetic connection lines by tunnel barriers, Sverdlov et al. (2016) found resonant tunneling besides the Coulomb blockade, resulting in a similar "spin blockade."

In a magnetic quantum dot as part of a single-electron transistor, magnetic polarons can be created, fermionic quasiparticles that describe the interaction between electrons and atoms in condensed matter. More exactly, the electron can spin-polarize the quantum dot atoms, while at the same time, the electrons are more tightly bound to the quantum dot. Interestingly, the temperature range in which stable polarons occur is controlled by the gate voltage of the single-electron transistor and depends on the external magnetic field. The stability of this magnetic polaron was attributed to the Coulomb repulsion between electrons in the Coulomb blockade regime, which separates spin-up and spin-down levels (Lebedeva et al. 2012).

Things are more complicated in double quantum dots. Here, Coulomb blockade and spin blockade could be utilized in spintronic applications since they enable sophisticated single-spin operations. The Coulomb blockade in double quantum dots, however, is typically observed in very low temperature regions below 1 K, that is, in temperatures that cannot be reached by the relatively simple cooling with liquid helium. In such systems, Rossella et al. (2014) managed to control a double quantum dot device of arbitrary dimensions via the so-called Stark effect (splitting degenerate energy levels in an electric field) which also led to a spin blockade, in this case up to 10 K. While this temperature is still far below room temperature, it can be reached at least easier than the aforementioned sub-kelvin range.

Even a triple quantum dot molecule was suggested by Chi and Yuan (2009) as a spin splitter. Using the spin-dependent chemical potentials on two drains as driving forces for spin-polarized transport, spin-up and spin-down electrons could be separated into two different drains.

In such double quantum dots with ferromagnetic electrodes, the coupling between the dots typically destroys the Coulomb blockade and leads to co-tunneling as the major electron transport mechanism. Nevertheless, tuning the external magnetic field can significantly influence the currents in the quantum dot, especially in the case of different magnetic fields (Yuan 2012). By applying a small external magnetic field and optimized gate voltages, Bordoloi et al. (2020) found a spin polarization of 80% in double quantum dot spin valves.

Coupling a quantum dot to a normal metal and a ferromagnetic metal lead was found to result in 100% spin polarization in a strong enough magnetic field. This means that on the quantum dot, only one of the two possible spin orientations was found. This system can thus be used as a spin injector, injecting spins of the desired orientation from a normal metal into a semiconductor for information processing in spintronic applications (Li et al. 2012; Chi et al. 2010). Combining an InAs quantum dot with a tunneling-coupled GaNAs spin filter, Huang et al. (2021) showed more than 90% spin polarization at room temperature.

The spin population on a quantum dot, connected by two ferromagnetic leads in the Coulomb blockade regime, can also be influenced by either adjusting the energy level of the quantum dot by a temperature gradient or by inverting the temperature gradient between the leads while the dot energy level is kept constant. In this way, spin accumulation in the quantum dot can be increased by the leads' spin polarization as well as by the asymmetry of the coupling between dots and leads (Bai et al. 2012).

The Coulomb blockade was even found in molecular junctions. Using a single benzene junction between platinum electrodes, a Coulomb blockade was found at the interfaces between the molecule and electrodes, connected with high spin polarization. This makes such molecular junctions potentially interesting for spintronic applications (Zhu et al. 2014a).

A spin transistor based on a magnetic molecule, for example, is connected with two electrodes along which a bias voltage is applied. The gate electrode controls the molecular levels to enable resonant tunneling and oxidation or reduction of the molecule. This gate field can be used to significantly modify the magnetic properties of the molecule in the different redox states (Burzuri and von der Zant 2014). On the other hand, even the leads were found to influence the anisotropy of the charged molecule states (Nossa et al. 2013).

Oppositely, connecting iron nanoparticles via organic barriers also resulted in spin-dependent tunneling and typical Coulomb blockade behavior. Apparently, the organic ligands could be used as spin-conservative tunnel barriers, allowing for using such systems for spintronic applications at room temperature (Dugay et al. 2011).

After this short overview of possible applications of the Coulomb blockade in spintronics, the next section will introduce another effect that is also typically observed in quantum dots.

6.4 Kondo effect

The Kondo effect describes the finding that the resistance of metals typically has a minimum at temperatures larger than 0 K. The effect was first defined by Jun Kondo who modeled the scattering rate of conduction electrons at low temperatures using third-order perturbation theory and found that scattering at magnetic impurities diverged near 0 K (Kondo 1964). According to Kondo, the resistivity shows the following temperature dependence:

$$\rho(T) = \rho_0 + aT^2 + bT^5 + c\ln(\mu/T). \tag{6.86}$$

Here, the first term describes the residual resistance at $T = 0$ K – if the diverging logarithmic term is ignored – the square term is derived for a Fermi liquid, and the third one stems from lattice vibrations. The logarithmic term is the one introduced by Kondo. It can result in electrons behaving as if they had masses up to three orders of magnitude higher than usual electrons, as it was found in some rare earth and actinide elements. This means that electron speeds can be correspondingly smaller due to the Kondo effect.

Interestingly, the Kondo effect can also occur in quantum dots where an unpaired electron serves as a magnetic impurity, scattering conduction electrons from a coupled metallic material, or in a single molecule (Kouwenhoven and Marcus 1998; Cronenwett 1998; Martinek et al. 2007; Sanchez et al. 2005). This indicates already that the Kondo effect may be of interest for spintronic applications with their typically very small dimensions. Especially for systems with two or three coupled quantum dots, interesting effects can occur, enabling the control of molecular electronics and spintronics (Xiong et al. 2017; Xiong et al. 2017a; Andrade et al. 2017; Yi et al. 2020; Mantsevich and Smirnov 2023). The Kondo effect was even found in ferromagnetic dots in single-electron devices, making such dots possible candidates for spintronic elements which use the quantum properties of an electron instead of the average spin polarization of an ensemble of electrons. In this way, solid-state spin-based qubits could be created (Dempsey et al. 2011).

Research on the Kondo effect in spintronics-related components can indeed be found in recent scientific literature. Yan et al. (2018), for example, studied the Kondo effect in graphene with magnetic impurities, produced by magnetic molecules decorating the graphene, using transport measurements at low temperatures. The typical logarithmic behavior described in the Kondo effect was found in some samples up to temperatures of 20 K, indicating a strong coupling of the electrons to the magnetic impurities. The Kondo effect could be tailored in this system by a gate voltage that might be useful for spintronic applications.

Semiconducting graphene nanoribbons, on the other hand, have large band gaps and thus no free electrons, which should result in decoupling magnetic molecules on top from a metal below them, that is, no Kondo effect. Unexpectedly, the Kondo effect was not reduced by placing a semiconducting graphene nanoribbon between the magnetic molecules and the metal, that is, the graphene nanoribbons mediate spin coupling (Li et al. 2017b).

Magnetic doping of the 3D topological insulator Bi_2Se_3 in the shape of a nanoribbon with less than 2% iron was already sufficient to show a Kondo effect up to temperatures of 30 K, making such doped topological insulators also interesting for spintronic applications (Cha et al. 2010).

In carbon quantum dots that were realized as single-atom junctions in patterned graphene nanoflakes, the interplay between orbital and spin degrees of freedom, as well as the orbital-selective Kondo effect, was investigated, finding an unusual electronic structure unlike larger graphene layers (Craco et al. 2016).

A possibility to tailor molecular behavior, including the Kondo effect, mechanically was found for metal benzene sandwich molecules, spanned between two Cu nanocontacts. The electronic transport properties depended on the molecule's geometry, which could be modified by the nanocontact tips, switching half-metallic behavior on and off, making this system interesting for spintronic applications. For some of these molecule geometries, an orbital Kondo effect was observed (Karolak and Jacob 2016).

Tailoring the Kondo effect by hydrogen absorption was found possible for metal phthalocyanines and their derivatives, placed on a gold substrate. Absorbing a hydrogen atom on the central Mn ion of the molecule resulted in a spin decrease, weakening of the molecule–substrate coupling, and a change of the preferred adsorption sites of the molecules on the gold surface. The modified coupling resulted in quenching the Kondo effect, which could be restored by a voltage pulse or sample heating. In this way, the Kondo effect and other electronic properties can be tailored for single molecules (Xiao et al. 2015).

On the other hand, a stable Kondo effect was found in organic radicals having an unpaired electron in the ground state and thus an intrinsic magnetic moment. Frisenda et al. (2015) used the Kondo effect to verify that the paramagnetic ground state of polychlorotriphenylmethyl radical molecules was preserved in solid-state devices with two or three terminals, independent of the electrode position and the electrostatic environment, making such radical molecules useful for spintronic devices.

Another interesting molecular magnet system was investigated by Ormaza et al. (2016). They placed co-ferrocene molecules on metallic substrates and used the Kondo effect to prove that co-ferrocene molecules, opposite to pure ferrocene, have a magnetic moment. By chemically tailoring nanowires based on metallocene, new approaches for molecular spintronics can be enabled.

The Kondo effect was also investigated in multiwalled carbon nanotubes filled with magnetic nanoparticles. Such systems are of interest due to their special magnetic and electric properties. Ncube et al. (2018) found a competition between the Kondo effect and cotunneling in the superparamagnetic limit, combined with magnetic coherence and the spin-flip effect, enabling tailoring magnetic interactions in carbon nanotube-based systems. Magnetoresistance measurements even exhibited spin-valve switching behavior, which may be used for spin-based quantum computing using 1D channels (Ncube et al. 2018a). In single-walled carbon nanotubes ligated by a magnetic impurity, Lobo et al. (2020) found a severe impact of the hydrostatic pressure on the conductance, which the authors attributed to the displacement of the Kondo peak from the chemical potential. In a carbon nanotube/lanthanide hybrid molecular complex, Mosse et al. (2021) found deviations of the temperature dependence of the magnetic susceptibility from a linear behavior at low temperatures, which they attributed to the onset of the Kondo effect below the Kondo temperature.

In addition, a similar Kondo-like effect was found in diluted magnetic semiconductors. In these materials, correlations between electric carriers and impurities result in a spin transfer between both and a modification of the spin dynamics, similar to the Kondo effect in metals (Cygorek et al. 2016).

Finally, it should be mentioned that one of the recent approaches to the Kondo effect is based on the Numerical Renormalization Group (NRG) method. The method was developed with great success in recent years and has its origin in a work by K. G. Wilson (1975). It treats the physical system from the quantum perspective, or more precisely, it uses the methods of quantum field theory, including quantum electrodynamics.

It is worth mentioning that this is quite a general approach found in a variety of physical sub-disciplines, where we can recognize dominating local degrees of freedom and continuous baths of excitations. The not-so-direct similarity concerns, for example, the parton model of the elementary particle, in which the particle consists of a few quarks and the quark-antiquark sea, in order to describe the deep inelastic scattering collisions.

Within the NGR approach, the system under consideration is represented by selected fundamental physical states and quantum impurities. In other words, the Kondo problem is an example of a quantum impurity model in which a local spin interacts with a bath of fermions. Renormalization, in the meaning of transformation of the physical problem, i.e., the continuous energy scale of the bath, into a quantized energy scale, relies on introducing a set of logarithmically spaced discrete frequencies (energies) $\omega_n = \Lambda^{-n}$ ($n = 0, 1, 2, \ldots$) with the discretization parameter $\Lambda > 1$. The width of the intervals equals $\Delta \omega_n = \Lambda^{-n}(1 - \Lambda^{-1})$. In the well-arranged procedure, the continuum limit should be obtained for the limit $\Lambda \to 1$ (Peters et al. 2006).

Generally, the physics of a quantum impurity system can be represented by the following Hamiltonian:

$$H = H_{\text{bath}} + H_{\text{imp}} + H_{\text{int}}, \tag{6.87}$$

where H_{bath} is the continuum energy part of electrons, H_{imp} is the impurity energy, and H_{int} represents the interaction between bath and impurities. For a single impurity, the three Hamiltonians can be written as follows (Anderson 1961)

$$H_{\text{bath}} = \sum_{k,\sigma} \varepsilon_k c_{k\sigma}^\dagger c_{k\sigma}, \tag{6.88a}$$

$$H_{\text{imp}} = \sum_{\sigma} \varepsilon_f f_\sigma^\dagger f_\sigma + U f_\uparrow^\dagger f_\uparrow f_\downarrow^\dagger f_\downarrow, \tag{6.88b}$$

$$H_{\text{int}} = \sum_{k,\sigma} V_k \left(f_\sigma^\dagger c_{k\sigma} + c_{k\sigma}^\dagger f_\sigma \right), \tag{6.88c}$$

where $c_{k\sigma}$, $c_{k\sigma}^\dagger$ are the fermionic operators for bath space with equivalent energy, ε_k, f_σ, $f_\sigma^\dagger$ are the fermionic operators for impurities with energy, ε_f, U is the Coulomb interaction energy between two impurity levels, and V_k is the measure of the coupling energy between the impurity and a given electron level.

In the strict Kondo model, an impurity can be represented by the spin $\vec{S}_{\text{imp}}$, allocated at a single energy level, and antiferromagnetically exchange-coupled to a conduction electron-spin from the bath at the impurity location $\vec{s}(0)$. The Hamiltonian for such a case reads

$$H = \sum_{k,\sigma} \varepsilon_k c_{k\sigma}^\dagger c_{k\sigma} + J\vec{S}_{\text{imp}} \cdot \vec{s}(0), \qquad (6.89)$$

where the coupling is represented by the exchange constant J.

The NRG and the other second quantization methods were used in the analysis of several important spintronic topics, such as single-level molecular quantum dots (Misiorny et al. 2012), thermoelectric effects coupled to ferromagnetic leads (Weymann and Barnaś 2013), magnon transport through a quantum dot (Karwacki et al. 2015), or the shot noise in tunneling structures (Szczepański et al. 2016), to mention only a few.

The open numerical code (GNU license) of the NRG approach is developed and achievable from the research group of the Department of Theoretical Physics, Budapest University of Technology and Economics (Tóth et al. 2008), (https://www.phy.bme.hu/~dmnrg/). The codes are also available from NRG Ljubljana open source numerical renormalization group (Žitko and Pruschke 2009) (http://nrgljubljana.ijs.si/) as well as from University College Dublin, School of Physics, Theoretical Nanoelectronics Group (https://www.ucd.ie/nanoelectronics/nrg.html).

While the Kondo effect has to be taken into account for interactions of electrons with magnetic impurities at low temperatures, electrons and other carriers are also influenced by interfaces and other effects inside materials. The following sections discuss some of these effects.

6.5 Reflection and scattering of carriers

Electron transport in materials, and especially over interfaces, can be described by several theories, some of which will be described briefly here. In the most detailed case, quantum transport is taken into account; that is, transport is described by the Schrödinger equation. Specifically for transport in semiconductors, Wigner (1932) introduced a special form of the Schrödinger equation, thus called the Wigner function, in which the electron momentum is related to its position and wave vector. If many particles are involved, the Pauli principle and the electron–electron interactions have to be included. In the Wigner equation, quantum transport is defined, taking into account a potential energy which includes not only the macroscopic long-range electrostatic potential but also the short-range potential due to scattering at phonons and impurities. It can describe the situation of a semiconductor in contact with metallic contacts.

Going one step back, a semiclassical transport description can be used. By applying several approximations to the Wigner equation, the Boltzmann equation is reached. These approximations are only allowed for systems with small electron scattering rates, in which the electrostatic potential changes only slowly in time and space, and in which the electron distribution changes only on large scales. Thus, applying the Boltzmann equation inside the bulk of a material is often sufficient, while for transport across interfaces usually this is not the case. It should be mentioned that nevertheless it is often possible to regard the interface as a layer of negligible thickness so that most of a device can be modeled using the Boltzmann equation, while quantum effects around the interface are included in the special interface model (Schroeder 2013).

Since it is often not necessary to treat all single carriers separately, further approximations are in many cases possible, decreasing the accuracy level as well as the necessary computing power. For this, different transport models can be used for different situations.

A special case is a tunnel barrier, which has to be treated very carefully. Here, the density gradient is not necessarily continuous, resulting in apparent charge storage in the tunnel barrier, which does not make sense. This problem is based on the fact that tunneling is a nonlocal effect, and thus reducing the tunnel barrier to zero extension and imagining the neighboring layers to fulfill the preconditions for applying the Boltzmann approximations are not allowed (Ferry et al. 2003). This is one of the situations for which diverse extensions and other approximations of transport in materials and over interfaces have been developed during the last years (e.g., Gamba 2014).

A more detailed description of the spin-dependent transport, especially from a ferromagnetic material into a semiconductor, a situation that is often used for spin injection, is given in the next section.

6.6 Spin injection and spin accumulation

In most spintronic devices, it is necessary to inject charge carriers with a defined spin polarization into nonmagnetic materials, often semiconductors. First, spin injection experiments showed the possibility of creating a spin polarization in a ferromagnetic material and transferring it by a current into a normal metal (Johnson and Silsbee 1987). To inject spins electrically into a semiconductor, the Ohmic contact between the ferromagnet and the semiconductor was found to be important. Good contact properties – in the sense of the Fermi level of the semiconductor near the metal–semiconductor interface being located in the conduction band of the metal – were found, for example, in InAs and similar semiconductors with small band gaps. Nevertheless, first experimental results of spin injection into such semiconductors showed disappointing results (Gardelis et al. 1999; Monson and Roukes 1999). Schmidt et al. (2000) showed theoretically that better results could only be expected for nearly 100% spin polarization in the ferromagnet due to the severely different densities of states in both materials, which can only be expected for Heusler alloys and some other materials, but not for common ferromagnets like iron, cobalt, or nickel. The problem is based on the fact that the resistance of the semiconductor is much larger than one of the ferromagnetic metals, which is why the semiconductor is mostly responsible for the total interface resistance, and thus the magnetoresistance is nearly negligible (Albrecht and Smith 2002).

Thus, next approaches focused on creating a spin polarization directly in a semiconductor, which is possible due to the spin–orbit coupling (Johnson 2001). On the other hand, using ballistic spin transport over the interface showed, first in theory, promising results (Tang et al. 2000).

Experimentally, tunnel barriers were shown to enable spin injection from a ferromagnetic material into a semiconductor (Alvarado and Renaud 1992; Motsnyi et al. 2003; Ikeda et al. 2010). Generally, tunnel junctions with a high TMR can maintain spin orientations during the spin injection from a ferromagnet to a semiconductor through the insulating tunnel barrier. Spin injection over Schottky tunnel contacts was investigated by Lou et al. (2006).

Another possibility to overcome this problem, as shown by Hamaya et al. (2018) for n-Ge as the semiconductor, is decreasing the contact resistance in comparison to tunnel barriers by using a more "continuous" interface, for example, by creating a ferromagnet/semiconductor heterointerface with a phosphorus delta-doped germanium layer to have the atomic arrangements of both materials match at the interface. The atomic order or disorder at the interface was also found to be crucial for spin injection from the Heusler alloy $Co_2FeAl_{0.5}Si_{0.5}$ into n-Ge, leading to a decrease of the spin injection efficiency by one order of magnitude when the system was post-annealed. This effect was attributed to intermixing of phases at the interface by the temperature treatment (Kuerbanjiang et al. 2018).

A special material combination is given by MoS_2 and the half-metallic ferromagnet CrO_2. Kumar and Choudhary (2018) found in simulations not only a very high magnetoresistance of approximately 860% for a MoS_2 monolayer with CrO_2 electrodes but also a large spin injection efficiency of nearly 100% for high bias voltages, making this material combination highly interesting for spintronic applications.

Another possibility to overcome the conductivity mismatch between a ferromagnet and a semiconductor is using the so-called hot electrons, that is, electrons with much higher energies, which are transported from a normal metal with relatively high Fermi energy through a tunnel barrier into a ferromagnet and further into the semiconductor in contact with the ferromagnet. This process was first proposed by Jansen (2003) and verified experimentally by Appelbaum et al. (2007) and Huang et al. (2007).

Interestingly, it is also possible to use a graphene layer at the interface between a ferromagnet and the semiconductor silicon to increase the spin injection efficiency (van't Erve et al. 2012). Oppositely, Co/TiO_2 can serve as a spin injector for graphene flakes (Dankert et al. 2014). One of the largest problems of working on graphene is the missing "wettability" by several oxidic dielectric materials, resulting in undesired pinholes and discontinuities of Al_2O_3 or MgO tunnel barriers on graphene (Tombros et al. 2007; Wang et al. 2008a). These problems were overcome by magnetron sputtering of these materials as homogeneous tunnel barriers on graphene (Dlubak et al. 2012; Cubukcu et al. 2015). Another interesting approach is based on chemically functionalizing the top layer in a graphene bilayer or multilayer so that it becomes a single-layer tunnel barrier, while the lower layers serve as a high-mobility transport channel (Friedman et al. 2014). Possible chemical treatments include fluorinating and hydrogenating the graphene top layer, with the latter process resulting in lower spin lifetimes (Lee et al. 2015). A very high magnetoresistance of approx. $344 \cdot 10^3$% was found in functionalized H-graphene-F (i.e., containing one F and one H atom per unit cell) with CrO_2 electrodes (Pandey et al. 2020).

Leutenantsmeyer et al. (2018) found that the spin injection efficiency from the ferrimagnetic insulator yttrium iron garnet (YIG) into single-layer or bilayer graphene through a hexagonal boron nitride tunnel barrier could be tailored in a broad range by modifying the applied DC bias current. This makes hexagonal boron nitride single layers interesting for graphene spintronic applications.

Zhu et al. (2018) also investigated hexagonal boron nitride and, in addition, MgO as tunnel barriers between a ferromagnet and graphene. For both tunnel barriers, a strong dependence on the DC bias current at the injector electrode was found, including even a sign change of the nonlocal spin signal. They attributed this finding to the energy-dependent spin-polarized electronic structure of the spin-injecting ferromagnet.

Another interesting effect was investigated by Chan et al. (2018). When they applied Co nanoparticles on graphene, the originally superparamagnetic nanoparticles became ferromagnetic. This change was attributed to the dipolar interaction between neighboring nanoparticles. Ferromagnetism is supportive of high-spin injection efficiency into the graphene layer.

Besides changing the interface, it is also possible to change the spin injection method. Using ferromagnetic resonance (FMR), it is possible to create a pure spin current by spin pumping, without creating a charge current simultaneously. According to this method, Indolese et al. (2018) used a tunable magnetic field to create FMR in small permalloy pads. These pads were in contact with a single layer of graphene. Under resonance conditions, a spin current was injected into the graphene layer.

Generally, 2D materials such as graphene are of large interest for spintronic applications. Most of these materials, however, are nonmagnetic. Making them magnetic can be achieved by doping (Wang et al. 2014). For example, graphene becomes a diluted magnetic semiconductor when doped with transition metal atoms (Krasheninnikov et al. 2009). Doping graphene with chromium, iron, or cobalt, on the other hand, should theoretically result in a half-metallic ferromagnet (Pan et al. 2013). Diluted magnetic semiconductor behavior was also predicted theoretically for blue phosphorene doped with specific transition metal atoms (Sun et al. 2016; Yu et al. 2016), while doping blue phosphorene with other transition metal atoms should result in a half-metallic state, making these 2D materials feasible as spin injectors (Su and Li 2019).

Besides current-induced spin injection from a ferromagnet to a nonmagnetic material, light can also be used to trigger spin injection. The efficiency of this process is usually attributed to the spin–orbit coupling of the nonmagnetic layer, claimed to be responsible for the spin injection itself as well as for an increase in the demagnetization behavior of the ferromagnetic layer (von Korff Schmising et al. 2015; Hofherr et al. 2017). Recently, Dewhurst et al. (2018) showed by an ab initio calculation of diverse material pairs that spin injection starts with a short (<20 fs) spin-transfer phase in which majority spins flow from the ferromagnet into the nonmagnetic material and minority spins in the opposite direction, while the spin–orbit coupling resulted in spin-flip scattering and thus supported the demagnetization process in the whole system. Both processes were mainly controlled by the density of states at the interface, allowing for designing interfaces with high spin injection properties.

Finally, by bringing a ferromagnet in contact with a nonmagnetic conductor, it is possible to create a nonequilibrium spin configuration at the interface between both materials by an in-plane charge current. These spins can diffuse from the interface into the ferromagnet. If a radio-frequency current is applied to the conductor, then the spins will start a precession around the effective magnetic field direction, which is a superposition of a possible external magnetic field and internal fields in the ferromagnetic layer. In this way, a charge current can be transferred into a spin current. Measuring the spin current is possible, for example, by the FMR (Wang et al. 2018a).

One aspect correlated with the spin injection from a ferromagnet into a nonmagnetic metal or another nonmagnetic material is the spin accumulation in both materials. Electrons with preferred spin orientations accumulate within a certain range on both sides around the interface (van Son et al. 1987). Quantitatively, the spin accumulation can be described by the splitting of the spin-dependent electrochemical potentials for up and down spins. It is normally measured in a nonlocal geometry, as described by Jedema et al. (2001) for room temperature measurements of lateral permalloy/copper/permalloy structures or a cobalt/aluminum/cobalt device with Al_2O_3 tunnel barriers between the adjacent layers (Jedema et al. 2002). The general idea of such nonlocal measurements is to separate the positions of spin injection and spin detection to avoid probing the influence of the charge (Takahashi and Maekawa 2008).

An important aspect of spin accumulation in three-layer structures was investigated by Taniguchi (2018). He calculated the spin current flowing in the second ferromagnetic layer, triggered by a spin current in the first ferromagnetic layer by spin accumulation in the intermediate nonmagnetic material, if the latter has a large enough spin diffusion length. This means that another spin current flows, dependent on the different spin accumulations in both ferromagnets. This, in turn, creates an additional spin torque, which leads to a parallel or antiparallel synchronized motion of the magnetization in both ferromagnetic layers.

Another possibility to generate spin accumulation in a ferromagnet is by applying a thermal gradient, which results in the longitudinal spin Seebeck effect. The thermally achieved spin accumulation in YIG/platinum heterostructures was measured by Wang et al. (2018c) by detecting the charge current in platinum using the inverse spin Hall effect. In this way, they found hints of a spin-glass state at the interface and proved that the longitudinal spin Seebeck effect in this system could not be explained by a magnon spin current across the YIG bulk, again indicating the importance of the interface layer.

For spin accumulation in *p*-Si, a Co_2CrAl/SiO_2 tunnel heterojunction was evaluated in a broad temperature range between 10 and 300 K. Below 100 K, spin accumulation was found in *p*-Si, showing the possibility of using such heterostructures for spin injection (Kar et al. 2018).

Finally, it should be mentioned that spin accumulation is not restricted to thin films or nanodots. In chiral molecules, Kumar et al. (2018) found a spin-selective electron transmission based on chiral-induced spin selectivity. This means that a magnetic field can be applied to create spin-polarized currents from ferromagnetic electrodes, or vice versa, spin accumulation results in a magnetic field.

6.7 Spin-transfer torque

Electron spins can influence the magnetization orientation in a ferromagnetic sample. More exactly, it is necessary to have a nonconstant spin flow through a sample; that is, the spin flow has sources or sinks in the sample (Ralph and Stiles 2008). This can occur when a spin current flows through a magnetic film with a different magnetization orientation than the one which created the spin current, or when a spin current flows through a magnetic domain wall, a magnetic vortex, a magnetic skyrmion, or any other area of nonuniform magnetization. Since the spins of the electron current have to change their spin-angular momentum by interacting with the ferromagnetic spin, conversely, the ferromagnetic spins also experience a torque due to the interaction with the electron current.

While the first experimental proofs of this effect, carried out in macroscopic magnetic samples, necessitated very large currents (Freitas and Berger 1985; Hung and Berger 1988), recent nanostructure devices need only some mA to trigger a spin-transfer torque (Gan et al. 2000; Grollier et al. 2002; Tsoi et al. 2003; Ralph and Stiles 2008). For a system built from two ferromagnets, separated by a tunnel barrier, Slonczewski (1989) first calculated the spin-transfer torque due to a current perpendicular to the plane, before he and Berger at the same time showed theoretically that such a perpendicular current could be sufficient to switch the magnetization in one of the ferromagnetic layers of a GMR (cf. Section 7.2) system (Slonczewski 1996; Berger 1996). Measurements of the spin-transfer torque were performed in nanostructure samples (Katine et al. 2000) and showed magnetization switching as well as precessions of the magnetization (Rippard et al. 2004; Krivorotov et al. 2005). In the meantime, not only GMR but also TMR systems could be produced sufficiently small and exact to allow for using the spin-transfer torque to switch the magnetization in one of the ferromagnetic layers (Huai et al. 2004).

Besides these technical applications, the spin-transfer torque can also be used to trigger FMR measurements, enabling measurements of the normal modes of spin waves in nanostructures (Tulapurkar et al. 2005; Sankey et al. 2006). Using micromagnetic simulations and experiments, the FMR in a nanoring magnetic tunnel junction, triggered by spin-transfer torque, was found to contain two resonance modes with different frequencies as well as second and third harmonics of these modes (Chen et al. 2018c).

Recently, Houshangh demonstrated the possibility of driving higher-order spin waves in magnetic tunnel junctions with nanocontacts using spin-transfer torque. They found large-frequency steps that they attributed to second- and third-order spin waves with wavelengths smaller than the nanocontact. In this way, not only could higher frequencies be reached in magnonic devices, but also higher transmission rates and spin wave propagation lengths.

Solyom et al. (2018) used a nitrogen vacancy center in an artificial diamond to probe magnetic oscillations with large amplitude, driven by the spin-transfer torque in a ferromagnetic nanowire. In addition, they found strong spin-transfer torque damping of the stray field noise, similar to cooling down the sample from room temperature to half this temperature.

Due to its great importance in technical applications and for basic research of spintronic systems, several research groups concentrate on investigating this effect in theory and experiment.

Generally, the possibility to switch magnetization by the spin-transfer torque is of high importance for spintronic devices. Bainsla et al. (2018) pointed out the importance of finding materials with low Gilbert damping and small magnetization to produce fast and energy-efficient spintronic devices working with the spin-transfer torque. They found very low Gilbert damping and soft magnetic properties in the Heusler alloy Co-FeMnSi, making this material possibly useful for spintronic devices. It should be mentioned, however, that the switching time for such structures is not only inversely proportional to the applied field but also to the damping coefficient, meaning that a low damping factor negatively influences the switching time (Moon et al. 2018).

The necessity to use materials with low damping impedes using materials like FePt or Co/Pd, which were first investigated for magnetoresistive random access memory (MRAM) applications (cf. Section 7.11). Another magnetic material with a low damping factor is CoFeB (Mizukami et al. 2009), which can be expected to be used in magnetic tunnel junctions to the order of magnitude of 20 nm. However, below this magnitude, thermal stability necessitates materials with a higher anisotropy, again leading to higher damping (Bhatti et al. 2017). One of the materials suggested for high-density MRAM applications, showing a large perpendicular magnetic anisotropy, is epitaxial MnAl thin films in which high anisotropy and small surface roughness could be combined by post-annealing the sputtered films (Parvin et al. 2018). Similar problems arise for larger temperatures – if an MRAM is to be used up to approximately 150 °C, the decrease of the magnetic anisotropy at higher temperatures necessitates correspondingly higher anisotropies at room temperature and below. Iwata-Harms et al. (2018) suggest CoFeB, diluted with non-magnetic metallic impurities, as a free layer. By changing the magnetization in this material due to changing the material and concentration of the impurities, they could predict and tailor the thermal stability of MRAMs working with spin-transfer torque.

To overcome the problem of the necessarily higher anisotropies of materials used for high-density storage devices, a higher efficiency of the spin-transfer torque is necessary, that is, a higher spin polarization. Here, Heusler alloys have been investigated (Jourdan 2014). Alternatively, the geometry of the magnetic tunnel junction can be improved by using two reference layers in combination with one free layer (Diao et al. 2007; Khvalkovskiy et al. 2018), using point contacts instead of more extended contacts (Sbiaa et al. 2013), or tilting the free layer with respect to the reference layer (Sbiaa 2013).

The spin-transfer torque allows for manipulating magnetic tunnel junctions purely electrically, making them useful as possible parts of nonvolatile storage devices (Gajek et al. 2012; Khvalkovskiy et al. 2013; Sverdlov and Selberherr 2018).

An important point in such data storage devices is the probability of write errors. Pramanik et al. (2018) investigated write error rates in random access memories by micromagnetic simulations, performed for temperatures above 0 K with a special technique enhancing rare events. In this way, they found different spatially incoherent switching modes which reduced switching speeds and could distinguish between different switching mechanisms for smaller and larger currents.

Another technical application of the spin-transfer torque is its utilization in microwave-assisted magnetic recording devices, in which a spin-transfer torque is applied by the microwave oscillator (Nakagawa et al. 2023). Liu et al. (2018a) showed that the AC field strength, the geometry of the writing head, as well as the oscillation frequency, influenced the resulting spin torque. With this technique, switching times of the spin-torque layer below 300 ps can be reached, which are strongly influenced by the dimensions of this spin-torque layer (Ding et al. 2022).

Besides spin-transfer torques emerging from spin currents through interfaces, spin-transfer torques can also occur at local changes of the magnetization. An example of this effect is skyrmions (cf. Section 3.2). The spin-transfer torque can lead to moving skyrmion solutions of the Landau–Lifshitz–Gilbert equation of motion (Li and Melcher 2018).

Graphene belongs to the materials in which interesting effects for possible use in spintronic applications may occur, especially in the case of zigzag graphene nanoribbons. Density functional theory-based calculations have revealed three different magnetic phases, that is, paramagnetic, ferromagnetic, and antiferromagnetic phases, depending on the electric field applied and the temperature as well as the position (Sivasubramani et al. 2018). Understanding these magnetic phase transitions is an important prerequisite for developing spin-transfer torque-based graphene spintronic devices.

In spin-based logic devices, the spin-transfer torque can be used in the form of spin torque majority gates, applied in nonvolatile logic circuits. Especially in combination with the spin–orbit torque, the problems of difficult circuit configuration could be solved, allowing the creation of a logic family with AND/NAND and OR/NOR functions, working reliably and being able to communicate with other logic units by a spin current flow through a nonmagnetic metal wire (Li et al. 2018c).

In the emerging topic of neuromorphic computing, spin-transfer torque can be used to switch magnetic tunnel junctions, in this way creating artificial neurons (Kondo et al. 2018). In addition, by using thin antiferromagnetic films with biaxial anisotropy, the spin-transfer torque can be used to generate ultrashort "spikes," similar to the signals created by real neurons responding to an external impact (Khymyn et al. 2018).

Nevertheless, it must be mentioned that the spin-transfer torque as a way to switch magnetization in a sample has the drawback of limited speed. This is why nowadays other mechanisms like the spin–orbit torque are also evaluated for possible future use in MRAM or similar nonvolatile magnetic storage media (Rahaman et al. 2018; Khang et al. 2018; Wang et al. 2018d; Seo and Kwon 2021; Li et al. 2024).

7 Spintronics devices

7.1 Electronic parameters of spintronic devices

For a quantitative description of spintronic devices, we first have to define the spin polarization, that is, the intrinsic angular momentum of an elementary particle:

$$P_Q = \frac{q\uparrow - q\downarrow}{q\uparrow + q\downarrow} \tag{7.1}$$

with $q\uparrow$ and $q\downarrow$ denoting majority and minority spins, respectively. The spin polarization can thus reach values between 0 (for equally distributed spins) and 1 (for a vanishing number of minority spins).

In spintronics, the spin polarization is usually applied to the magnetic moments of conduction electrons, resulting in spin-polarized currents as combinations of spin and electronic properties. While spin polarization is often investigated for electrons in metals or magnetic semiconductors, spin polarization can also be attributed to nuclei. The easiest way to produce spin polarization is by applying an external magnetic field.

7.2 Spin valve

A spin valve – or spin filter – in general is an element that allows one spin orientation to pass while it reflects the other one. If a nonpolarized electrical current, composed of electrons, reaches an ideal spin valve, all electrons with spin "up" pass, while the electrons with spin "down" are reflected, or vice versa. Such a spin valve would result in perfectly spin-polarized currents. However, real spin filters are not perfect, creating only partly spin-polarized currents.

The easiest way to create spin-polarized currents is by passing the electrons through a permanently magnetized ferromagnet. If the ferromagnet's majority spins are "up," for example, the resistance of electrons with spin "up" in this layer is relatively low, while the opposite spin orientation leads to reflections of the spin "down" electrons or spin flip scattering in order to find empty states in the ferromagnet's conduction band. This results in higher resistance values for the spin orientation opposite to the ferromagnetic majority spins.

A spin valve is a more sophisticated system that is nevertheless based on the same principle of spin-dependent resistance in a ferromagnetic layer. While there are different possibilities to create a spin valve, the principal idea behind this system is always the same – a nonmagnetic layer is embedded between two magnetic layers with different magnetic properties. One of these magnetic layers is very "soft," that is, it has a small coercive field, allowing for small external magnetic fields switching its

https://doi.org/10.1515/9783111383736-007

magnetization orientation. The other magnetic layer is very "hard," necessitating much larger external magnetic fields to be switched. For the latter, either another material or layer thickness is used than in the soft magnetic layer or it is pinned by a neighboring layer. In the easiest case, this pinning is performed by an AFM layer, resulting in an exchange bias that shifts and broadens the hysteresis loop of the hard FM layer. It is also possible to use a so-called synthetic antiferromagnet, such as Co/Ru/Co, pinning the FM quite strongly.

Figure 7.1 depicts the principal layer setup of a spin valve, as described earlier. Typical layer thicknesses are on the order of magnitude of few nanometers, with the AFM mostly being thicker than the FM layers and the nonmagnetic spacer. Usually, additional layers are added due to technical or physical purposes, for example, acting as a diffusion barrier, to promote adhesion, or to increase the spin-valve effect.

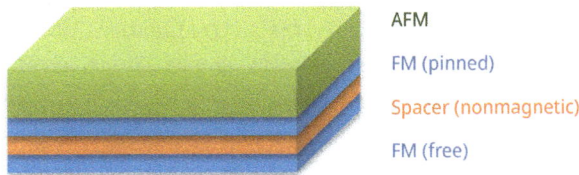

AFM

FM (pinned)

Spacer (nonmagnetic)

FM (free)

Fig. 7.1: Principal layer setup of a spin valve.

Sweeping the external magnetic field now from positive to negative saturation of the whole system, different states occur in which the electrical resistance of the layer system is changed. Figure 7.2 shows these different states that occur during the field sweep. Starting at positive saturation (A) and decreasing the magnetic field, first the free (lower) ferromagnet is switched (B). Sweeping the external field further to negative saturation, the pinned (upper) layer is reversed, too (C). Going back to smaller negative fields again, the pinned layer is switched back first (D).

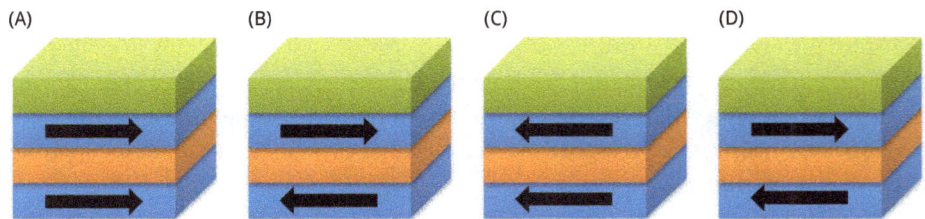

(A)   (B)   (C)   (D)

Fig. 7.2: Magnetic states of a spin valve.

As described above for the case of one ferromagnetic layer, these different magnetization orientations result in different electrical resistances, depending on the spin polarization.

Figure 7.3 depicts a sketch of the typical double-hysteresis loop of a spin-valve system, with the magnetic states A–D as described above. The overall hysteresis loop can be imagined as composed of the different hysteresis loops of the free and the pinned layer.

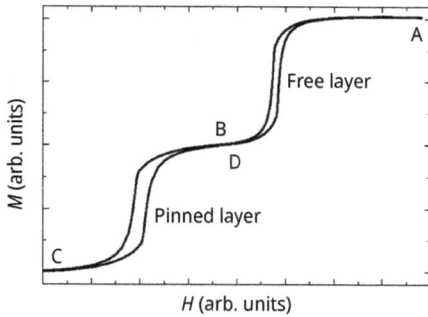

Fig. 7.3: Double-hysteresis loop of a spin-valve system.

In such a spin-valve system, the so-called giant magnetoresistance (GMR) can occur, which was found independently by Peter Grünberg and Albert Fert in 1988 (Baibich et al. 1988, Binasch et al. 1989). The GMR effect is based on the aforementioned ability of electrons to penetrate into a ferromagnetic layer of identical spin orientation, while they are reflected and scattered more strongly in a ferromagnetic layer with the opposite orientation of the magnetic moments. This means that in a spin valve with parallel FM orientation, the resistance is relatively low since electrons with one spin orientation can pass easily. In the states B and D with antiparallel FM orientation, however, electrons with both spin orientations are blocked in one of the layers, resulting in a significantly higher resistance.

Figure 7.4 shows an idealized field-dependent resistance curve. The resistance is much higher for the states B and D with antiparallel orientation of the FM layers than for the completely saturated states A and C.

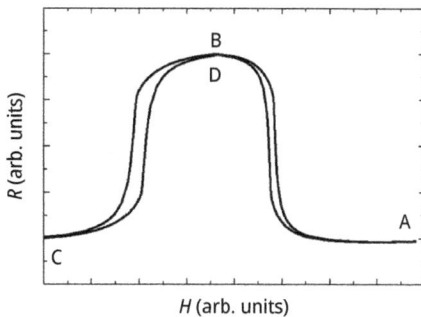

Fig. 7.4: Idealized field-dependent resistance curve of a GMR element.

The size of the GMR effect is defined as

$$\text{GMR} = \frac{\Delta R}{R} = \frac{R_{\uparrow\downarrow} - R_{\uparrow\uparrow}}{R_{\uparrow\uparrow}} \tag{7.2}$$

with the resistances $R_{\uparrow\downarrow}$ and $R_{\uparrow\uparrow}$ for antiparallel and parallel alignment, respectively.

A simple physical explanation is given by the two-current model. Under the assumption that spins are conserved during tunneling, it is possible to evaluate spin currents for spin-up and spin-down electrons independently. This is why the total current can be split into two partial currents, both of which interact differently with the magnetic tunnel junction (MTJ) (Fig. 7.5). In a typical ferromagnet, scattering is the main reason for the electric resistance. The scattering probability depends on the filling of the band that is responsible for the metal's magnetic properties, that is, the 3d band for the usual element ferromagnets such as Fe, Co, or Ni. The density of states in the d band, however, depends on the spin orientation. The density of states – and thus the scattering probability – is significantly higher for minority than for majority spins. It should be mentioned that for some special materials this effect is reversed (Chappert et al. 2007).

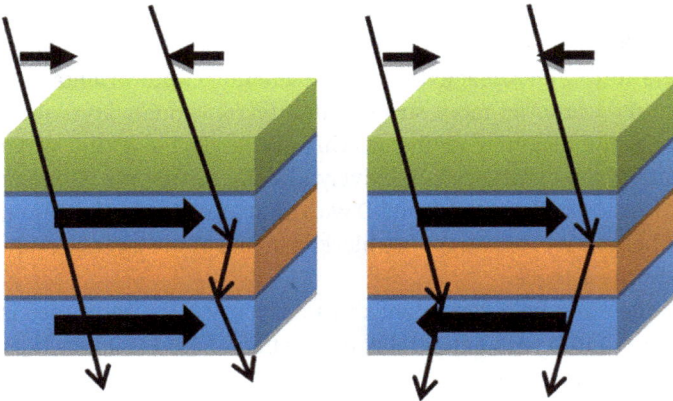

Fig. 7.5: Schematics of the two-current model.

Thus, it is assumed that only the spins that are antiparallel to the magnetization in a ferromagnet are scattered in the respective layer. This means that for parallel orientation of both ferromagnets, the spin-up electrons are not influenced at all while moving through the system, while the spin-down electrons are scattered twice.

On the other hand, for an antiparallel orientation of the ferromagnets, spin-up and spin-down electrons are scattered at one ferromagnet each, while passing the other one without any influence, respectively. In a naïve approach, one could imagine that both cases should not be different. However, the equivalent circuit shows directly that there is indeed a large difference between both situations (Fig. 7.6).

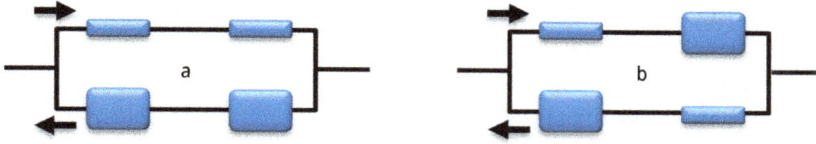

Fig. 7.6: Equivalent circuit for the two-current model.

In the parallel configuration (Fig. 7.6a), the resistance is

$$R_{\uparrow\uparrow} = \frac{2R_\uparrow R_\downarrow}{R_\uparrow + R_\downarrow} \tag{7.3}$$

with the (smaller) resistance $R_\uparrow$ of the majority spins and the (larger) resistance $R_\downarrow$ of the minority spins.

On the other hand, in the antiparallel configuration (Fig. 7.6b), the resistance can be calculated as

$$R_{\uparrow\downarrow} = \frac{1}{2}(R_\uparrow + R_\downarrow). \tag{7.4}$$

This results in a resistance difference of

$$\Delta R = \frac{(R_\uparrow - R_\downarrow)^2}{2(R_\uparrow + R_\downarrow)} \tag{7.5}$$

or in a GMR of

$$\mathrm{GMR} = \frac{\Delta R}{R} = \frac{(R_\uparrow - R_\downarrow)^2}{4R_\uparrow R_\downarrow}. \tag{7.6}$$

This means that the higher the asymmetry between $R_\uparrow$ and $R_\downarrow$, the higher the GMR. Such large asymmetries can be induced by doping with impurities (Fert & Campbell 1976, Dorleijn & Miedema 1977).

This simple model can be expanded to more sophisticated ones, taking also into account the thermal effects, especially in the form of possible spin-flip scattering, that is, spin mixing terms.

While in the first publications of the GMR effect, the resistance differences were in the order of magnitude of some percent at room temperature and up to 50% at low temperatures, depending on the system under examination; afterward even higher values of more than 100% were measured (Zhang et al. 2008). The GMR has not only been detected in epitaxially grown thin film samples (Thangaraj et al. 1994), but also in polycrystalline layers (Modak et al. 1994, Hirohata et al. 2016), microwires (Zhukova et al. 2017), zigzag graphene–graphyne nanoribbon heterojunction devices (Li et al.

2017), microspheres (Xu et al. 2016), or granules of magnetic material dispersed in a nonmagnetic matrix (Dey and Roy 2021).

Technical applications are especially hard disk read/write heads, magnetic field sensors, and the magnetoresistive random access memory (MRAM). In hard disk read heads, the GMR replaced the previously used anisotropic magnetoresistance (AMR) in 1997, resulting in increased areal densities of approximately 1.5 Gbit/in^2 (Bass 2016).

7.3 Magnetic tunnel junctions

MTJs are spintronic elements using the so-called tunnel magnetoresistance (TMR) effect, which will be explained below. In an MTJ, two ferromagnets are separated by a thin isolator (the typical thickness is approximately a few nanometers), one of which is pinned either due to coupling to an antiferromagnet or due to consisting of a material with significantly increased coercive field compared to the free ferromagnetic layer. Electrons injected through contacts on the ferromagnetic layers can tunnel through this isolating barrier if it is thin enough. The general setup of such a MTJ is depicted in Fig. 7.7. In principle, it is quite similar to the setup of a GMR element, besides the different layer between both ferromagnets. Additional layers that are necessary in both spintronic elements due to technical reasons, such as the desired temperature stability, are not discussed here in detail.

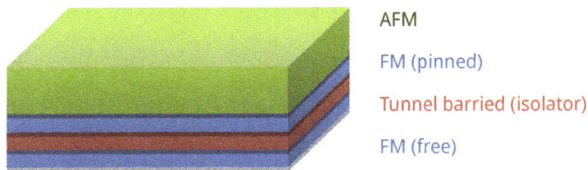

AFM

FM (pinned)

Tunnel barried (isolator)

FM (free)

Fig. 7.7: Principal layer setup of a TMR element.

It should be mentioned that the tunnel barrier must be homogeneous enough to avoid direct electrical contact between both ferromagnets, which means that all layers must be even, with roughnesses significantly below the tunnel barrier thickness. This requirement results in typical thin-film technologies being used for the production of MTJs, such as molecular beam epitaxy (MBE), pulsed laser deposition, electron-beam physical vapor deposition, or magnetron sputtering, the latter mostly being used for industrial-scale production.

The TMR effect is a quantum effect, indicating the phenomenon that the resistance through a MTJ depends on the relative orientation of both ferromagnetic layers to each other. Discovered by Jullière in 1975 at low temperatures and with only a small difference between the resistance for both relative orientations (Julliere 1975), the TMR was first just of academic interest. This changed when larger effects were measured and the TMR was also observed at room temperature.

In 1991, a TMR effect of nearly 3% at room temperature and 3.5% at 77 K was observed (Miyazaki et al. 1991); a value still quite small, but it gave rise to the idea that the resistance change might be applicable at usual temperatures. In 1995, stable values over 10% at room temperature and values around 20% (24%) at 77 K (4.2 K) were observed by increasing the sample preparation technology of CoFe/Al_2O_3/Co or NiFe MTJs (Moodera et al. 1995). In the same year, these values could be increased to 18% at room temperature and 30% at 4.2 K (Miyazaki & Tezuka 1995).

TMR values have strongly increased since. First, polycrystalline or amorphous tunnel barriers were examined, often Al_2O_3, resulting in TMR values of up to 70% (Wang et al. 2004). For other relatively simple systems like Fe as ferromagnet and MgO as the tunnel barrier, the TMR was predicted to be large enough for industrial applications (Butler et al. 2001, Mathon & Umerski 2001). These theoretical models were soon verified experimentally, showing TMR values in this system of up to 180% (Yuasa et al. 2004) or even 220% at room temperature and 300% at low temperatures (Parkin et al. 2004).

For an MTJ built from half-metallic CrO_2 electrodes and an SiCNT tunnel barrier with MgO adsorbed, approximately 100% TMR was found for different bias voltages, combined with a nearly ideal spin filtration of almost 100%, making this tunnel junction an almost perfect spin valve (Choudhary & Choudhary 2017). Higher TMR values of up to 350% at 5 K were found in a system built with a $SrSnO_3$ barrier between $La_{0.67}Sr_{0.33}MnO_3$ layers (Althammer et al. 2016). Here, however, the TMR vanished completely at temperatures above approximately 200 K, making this system not usable for typical room-temperature applications. TMR values of 600% were found in CoFeB/MgO/CoFeB (Ikeda et al. 2008).

Besides the common thin-film magnetic tunnel junctions, as described earlier, the TMR effect was also found in other geometries. Theoretical investigations of graphyne-based molecular MTJs found a TMR above 100%, combined with pure spin currents, depending on the sign of the bias voltage (Yang et al. 2017). Interestingly, when light with wavelengths between infrared and ultraviolet was used to irradiate the system, in some cases, photocurrents with different spin orientations were generated with respect to different electrodes. In this way, different spin currents could be created at the same time.

In an MTJ built by placing a DNA molecule between 3D ferromagnetic electrodes, the next-nearest-neighbor binding has been shown to strongly influence the TMR effect and other properties of the system (Fouladi 2017).

A pure graphene MTJ was simulated by placing armchair graphene between two semi-infinite ferromagnetic armchair graphene nanoribbons, showing that the applied uniaxial strain can result in increasing the TMR by nearly 100% (Fouladi 2016).

The size of the TMR effect is defined as

$$\text{TMR} = \frac{\Delta R}{R} = \frac{R_{\uparrow\downarrow} - R_{\uparrow\uparrow}}{R_{\uparrow\uparrow}} \qquad (7.7)$$

with the resistances $R_{\uparrow\downarrow}$ and $R_{\uparrow\uparrow}$ for antiparallel and parallel alignment, respectively. This equation is identical to the calculation of the GMR. The physical explanation for this effect is also similar to that of the GMR. According to Jullière (1975), the TMR results from the spin polarization of the ferromagnetic electrodes in the MTJ. The spin polarization, defined in Chapter 6 by the numbers of majority and minority spins $q_\uparrow$ and $q_\downarrow$, can also be written using the spin-dependent densities of states $D_\uparrow$ and $D_\downarrow$, respectively:

$$P = \frac{D_\uparrow(E_F) - D_\downarrow(E_F)}{D_\uparrow(E_F) + D_\downarrow(E_F)}. \qquad (7.8)$$

Here, the spin-up electrons have spins parallel to the magnetization, while the spin orientation of the spin-down electrons is antiparallel to it. The value of the TMR effect can now be calculated depending on the spin polarizations of both ferromagnets, P_{free} and P_{pinned}:

$$\text{TMR} = \frac{2P_{\text{free}}P_{\text{pinned}}}{1 - P_{\text{free}}P_{\text{pinned}}}. \qquad (7.9)$$

A slightly enhanced model was suggested by Slonczewski (1989) using an effective spin polarization instead. Other models take into account the effects like a voltage-driven pseudo-torque (Slonczewski 2005).

If no voltage is applied along the junction, the tunnel rates in both directions are statistically identical. A voltage results in a preferred tunneling direction toward the positive electrode. For high spin polarizations $P \sim 1$, the TMR becomes infinitely high, that is, the MTJs become a switch, changing between low and infinitely high resistance. Such high spin polarizations are theoretically predicted, for example, for Heusler alloys. Values near 100% were indeed measured, for example, at the interface between $La_{0.7}Sr_{0.3}MnO_3$ and $SrTiO_3$ (Bowen et al. 2005). For $CoFeTiSi/Fe_2TiSi/CoFeTiSi$, Feng et al. (2022) even reported TMR ratios up to about $3.30 \times 10^8\%$.

Although the principal function of MTJs has been understood for a long time, theoretical explanations are still not completely sufficient to describe all effects correctly that are found in experiments. Even in the relatively simple system Fe/MgO/Fe, theory and experiment are partly incongruent, for example, with respect to the MgO layer thickness depending on the TMR. While experiments with real TMR systems have shown only a weak dependence of the TMR on the tunnel barrier thickness (Parkin et al. 2004, Yuasa et al. 2004), calculations based on density functional theory suggested a severe change in the TMR effect with increasing MgO thickness (Butler et al. 2001, Belashchenko et al. 2005). Recently, Faleev et al. (2016) have published a new model in which they attribute this controversy to the suppression of interfacial resonance states due to interface roughness.

It should be mentioned that the TMR is usually significantly reduced with increasing temperatures or increasing bias voltages. Excitation of magnons and interactions of the electrons with them are one explanation for these effects described in the literature (Schleicher et al. 2014).

TMR elements are nowadays used in hard disk read heads or Hall sensors, but can also be used in MRAM devices as nonvolatile storage elements. In hard disk read heads, they replaced GMR elements in 2004, which were less sensitive, resulting in possible areal data densities of approximately 100 Gbit/in^2. Replacing the previously used Al_2O_3 barrier layer with MgO in 2007 increased the possible areal data density to approximately 700 Gbit/in^2 (Bass 2016). Nowadays, areal densities around 1 Tbit/in^2 are approached and trying to be exceeded. This value correlates with a sensor width of 26 nm. Calculations suggest that higher data densities cannot be reached with TMR read heads due to the physical limits of typical material combinations (Nakamoto et al. 2007, Takagishi et al. 2010).

7.4 Magnetic diode

In electrical circuits, a diode shows asymmetric conductance – in one direction, the resistance is very low, while in the other direction, it is very high. Electrical diodes are usually p–n junctions prepared not only from different semiconductor materials, mostly silicon, but also from gallium arsenide or germanium (Hook and Hall 2001; Kittel 1976). To create a p–n junction, one region of the semiconductor is doped by adding impurities containing electrons, that is, negative charge carriers, resulting in the so-called n-type semiconductor. A neighboring region is doped oppositely with holes (positive charge carriers), producing a p-type semiconductor. This results in an electron flow from n- to p-type semiconductors and the creation of the so-called depletion layer between both regions. Applying an electrical potential to the p side (anode) results in an electron flow through the depletion layer and thus in an electrical current.

Besides these solid-state diodes, vacuum tubes with an anode plate and a heated cathode can also serve as diodes (Guarnieri 2011; Guthrie 1873; Richardson 1929). Alternatively, Schottky diodes are produced combining a metal with a semiconductor, increasing the switching speed (Hasting 2005).

Because of the importance of diodes in electrical circuits, for example, for voltage rectification (transferring AC into DC voltages), signal isolation, voltage reference and amplitude control, mixing and detecting signals, and so on, the question arises whether a similar component can be created for spintronics applications. Adhikari et al. (2011) have investigated the temperature-dependent rectifying properties in ferrite/semiconductor heterostructures. They studied the electrotransport and magnetotransport properties of heterostructures composed of the ferrites Fe_3O_4 or $NiFe_2O_4$ and the semiconductor p-Si.

The systems had a GMR between 200% and 2,000% at low temperatures, respectively. Both systems showed rectifying properties and could be used as spin valves at temperatures below 50 K, while the temperature did not seem to be correlated with the Curie temperatures of the magnetic films. Rectifying here means linear (Ohmic) $I–V$ characteristics above a certain temperature and nonlinear (diode-like) $I–V$ characteristics below: A forward bias works similar to positive voltage at the p-Si substrate. At low temperatures, the $I–V$ characteristics were perfectly diode-like for small magnetic fields, while this behavior is suppressed already for relatively small magnetic fields around 1 T, giving rise to a spin-valve-like action.

In a magnetic p–n junction with an Fe/GaAs Schottky barrier, the p region is realized by a magnetic GaMnAs layer, resulting in spin-polarized electrons being injected from Fe into the n-GaAs region and through this depletion layer into the p-GaMnAs layer. In this way, diode-like $I–V$ characteristics were achieved up to room temperature, which were correlated with only a small magnetoresistance effect around 1% (Chen et al. 2006).

Even higher magnetoresistance effects of up to 100,000% for a magnetic field of 0.8 T were found in Schottky diodes prepared from gold and semi-insulating GaAs (Sun et al. 2004). In this system, an avalanche breakdown of the current was observed at a threshold voltage that could be shifted to higher voltages by applying an external magnetic field, opposite to similar experiments in n-GaAs (Aoki 1991). The large GMR effect is equivalent to a strong current depression under magnetic fields.

In Fe_3O_4/MgO/n-Si(001) heterostructures, giant positive junction magnetoresistance of more than 2,000% was found as well as strong rectifying properties in a broad temperature range up to room temperature (Panda et al. 2014), with a maximum around 60 K. The origin of the GMR was attributed to spin injection/extraction and spin accumulation in n-Si. Because of the relatively low saturation of the external magnetic field, this system was claimed to be advantageous in comparison with other heterostructures for the choice of a magnetic diode.

Another possible material system used as a magnetic diode is a p-$Ge_{1-x}Mn_x$/n-Ge heterostructure (Majumdar et al. 2009). At high temperatures, it behaves like a common diode under forward or reverse bias, respectively, while at low temperatures, it is transferred into a spin valve at forward bias, again combining these two interesting properties.

Tunneling diodes were produced, for example, using the material system (Zn,Mn, Be)Se with dilute magnetic material in a quantum well (Slobodskyy et al. 2003). Here, an external magnetic field resulted in spin splitting of the quantum well levels and thus in splitting of the transmission resonance into two separate peaks, corresponding to tunneling transport through spin polarized levels and thus to a voltage controlled spin filter. It should be mentioned that here the $I–V$ characteristics were less smooth than in a usual diode, but showed resonance peaks in the positive voltage regime.

MTJs prepared from CoFeB/MgO/CoFeB also showed diode characteristics (Ishibashi et al. 2010). Here, ferromagnetic resonance was excited using a radiofrequency

(RF) current. Triggered by tailoring the external magnetic field, the free layer precession was increased. The RF signal could be used to create a large direct-current output voltage. The rectifying properties were equivalent to a typical Schottky diode at room temperature.

A theoretical explanation of this asymmetry was given, for example, by Chshiev et al. (2002) who described the spin-dependent tunneling through asymmetric magnetic double-barrier junctions, introducing ballistic as well as diffuse tunneling into one model. They suggested the transition between the different tunneling regimes to depend on the electron scattering and showed that the strongly asymmetry $I–V$ characteristics and the TMR were based on the quantum well states in the middle metallic layer, with an applied voltage strongly shifting resonances in current and TMR.

Besides the common applications of diodes, magnetic diodes were, for example, suggested for a potential utilization in photovoltaics (Sakr and Yahia 2010; Yahia et al. 2010; Yahia et al. 2013). With a CdTe/CdMnTe heterojunction, a magnetic diode was produced, resulting in an open circuit voltage of around 0.5 V and an increasing reverse current with increasing illumination intensity, showing that this diode can be used as a photodiode in photovoltaics and as a photodetector. Similarly, a diluted magnetic diode prepared from n-CdTe/p-CdMnTe/p-GaAs was used in impedance spectroscopy (Yahia et al. 2015), showing a strong frequency dependence of the density of states at the interface as well as the series resistance.

A completely different application could be possible by applying a special diode-like effect (Tulapurkar et al. 2005, Matsumoto et al. 2011). The so-called spin-torque diode effect can occur in MTJs and is based on the interplay between spin dynamics and spin-dependent transport. If a small alternating current in the RF range is applied to a MTJ, a direct-current voltage can be generated across the device if the applied frequency is in resonance with the spin oscillations arising from the spin-torque effect. This resonance can be tuned using an external magnetic field and results in different resistances for different current directions. While this effect is severely different from the common diode effect in semiconductors, it can be used to create a nanoscale RF detector for telecommunication.

Similarly, different conductive, magnetic, and penetration depth material properties were measured at very high magnetic fields, using a tunnel diode oscillator in an RF-pulsed magnetic field (Coffey et al. 2000). For this, the sample under investigation is placed in an RF tank circuit. The tunnel diode oscillator is stabilized during a magnetic pulse by compensated coils in the tank circuit, enabling, for example, measuring superconducting transition and Shubnikov–de Haas oscillations (oscillations of the conductivity due to strong magnetic fields). Tunnel diode oscillators can generally be used to measure conductivity in highly conductive samples (Ohmichi et al. 2004). While alternatives exist for the extreme conductions in pulsed magnetic fields, tunnel diode oscillators have nevertheless shown higher sensitivity than other detectors for pulsed magnetic field measurements (Altarawneh et al. 2009).

Recently, even higher sensitivities have been reached without the necessity to apply an external bias field, based on an injection-locking effect that occurs for a certain current regime, meaning that the microwave emission frequency is locked to the frequency of the input microwave current (Fang et al. 2016).

As these examples show, there is a broad range of possible applications for magnetic diodes. A special magnetic diode will be described in the next section.

7.5 Magnetic electroluminescent diode

The idea to combine spin injection with light-emitting materials arose several years ago. For example, $Be_xMn_yZn_{1-x-y}Se$ was used to align the spins that were injected into the nonmagnetic semiconductor GaAs/AlGaAs, which serves as a light-emitting diode, resulting in circular polarization of the emitted light (Fiederling et al. 1999).

Generally, an external magnetic field can influence the injection current in organic light-emitting diodes (OLEDs) if non-spin-polarized charge is injected (Kalinowski et al. 2003; Kalinowski et al. 2004). The so-called organic magnetoresistance shows that magnetic field-dependent parameters influence the electrical injection current in organic semiconductive materials (Francis et al. 2004; Sheng et al. 2006). Further research has revealed that the organic magnetoresistance was related to excited states (Wu and Hu 2006; Desai et al. 2007) and that modifying the spin–orbital coupling enabled tailoring the organic magnetoresistance (Wu et al. 2007).

In OLEDs, the organic magnetoresistance was investigated in the dark and under illumination (Desai et al. 2007a). It could be shown that the field dependence of the light-induced organic magnetoresistance was similar to OLEDs with very thin layers. Relatively large values of the organic magnetoresistance of up to 300% were attributed to the interaction of free carriers with triplet excitons inside the system. On the other hand, this means that small magnetic fields should increase the efficiency of organic photovoltaic cells.

The same research group has also investigated the magnetoresistance and efficiency of aluminum tris(8-hydroxyquinoline)-based OLEDs and found positive or negative magnetoresistance, depending on the aluminum tris(8-hydroxyquinoline) layer thickness as well as the external magnetic field (Desai et al. 2007b). In the same system, other groups found the light output also depending on the magnetic field and the applied voltage, shedding light on the formation of the singlet excitons that are necessary for light emission (Kalinowski et al. 2003; Gärditz and Mückl 2005; Lei et al. 2009). By a chemical and thickness modification of the aluminum tris(8-hydroxyquinoline), it was possible to gradually vary the magnetoresistance between small negative and small positive values (Hu and Wu 2007).

Possibilities to tune the magnetoresistance in organic semiconductors have also been investigated by other groups in more detail in theory. Besides the formation of excitons (Prigodin et al. 2006; Desai et al. 2007a) or bipolarons (Bobbert et al. 2007),

there is especially evidence for the importance of the hyperfine coupling for spin mixing (Nguyen et al. 2010). This could be proven by comparing protonated and deuterated polymers, thus modifying the hyperfine interaction, in which the deuterated
polymers showed significantly smaller magnetoelectroluminescence and magnetic
resonance in combination with a significantly increased magnetoresistance.

The magnetic field dependence was attributed to a competition between exciton
formation and spin mixing, a process that could explain the low-field structures
found experimentally in magnetoelectroluminescence measurements (Kersten et al.
2011). In this model, spin mixing was increased by polaron hopping to other sites
within a multisite model before recombination occurred.

In spite of the interesting effects occurring in magnetic electroluminescent diodes,
only little research has been performed recently in this area (Kitahara et al. 2023;
Imai et al. 2024). Nowadays, research is concentrated more on three-pin devices, such
as different forms of transistors, rather than the two-pin diodes. Additionally, large
interest has arisen in the possibility to implement logic operations into spintronics.
Both these spintronics elements will be described in detail in the next sections.

7.6 Single-electron transistor

The transistor is an important part of recent electronic devices, belonging to the IEEE
list of milestones in electronics (IEEE 2009). It can be used to amplify a signal since
the output power can be higher than the controlling power.

The first field-effect transistor was patented in 1926 (Lilienfeld 1926) but could not
be produced at that time due to the lack of sufficiently high-quality semiconductor
material. A similar device was patented by Heil in 1934 (Heil 1934). The first real transistor was created by Bardeen, Brattain, and Shockley in 1947 using high-purity germanium (Shockley et al. 1948; Bardeen and Brattain 1948; Bardeen and Brattain 1949;
Shockley 1976). Nowadays, transistors are most often produced from silicon or germanium, while other semiconductors can also be used.

Two different variations of transistors are often used nowadays: the bipolar transistor in which a small current between the base and emitter controls a larger current
between the collector and emitter; and the field-effect transistor in which a small current at the gate controls a larger current between the source and drain. A well-known
version of the latter is the MOSFET (metal-oxide-semiconductor field-effect transistor),
necessitating only a very low input current to control the output current, as compared
to bipolar transistors (Baskhi and Godse 2007). Besides this possible use as an amplifier, a transistor can not only be used as a switch, for example, in logic gates, but also
in high-power applications. For this, transistors are optimized to switch very fast between an "on" state with very low resistance and an "off" state with very small currents flowing (Kaplan 2003).

Thinking about spintronics, it is more interesting to deal with single bits of information than with large currents. This means that instead of switching or amplifying large amounts of electrons, the aim of a transistor in this research area should be related to few electrons, ideally single ones.

The motion of single electrons can be controlled, for example, by the effect of correlated single-electron tunneling in solid-state structures consisting of conductive areas that are only connected by tunnel barriers (Grabert and Devoret 1992; Averin and Likharev 1991). To control this effect, two approaches have been suggested. In the so-called single-electron logic, single electrons are caught on these islands, working as single bits (Likharev 1988; Averin and Likharev 1992). Alternatively, the single-electron transistor (SET) can be used in digital circuits, switching between "0" and "1" in the form of two different voltage levels (Averin and Likharev 1986; Likharev 1987; Tucker 1992; Lutwyche and Wada 1994). The SET is depicted in Fig. 7.8. Its function is based on the Coulomb blockade (Chapter 6).

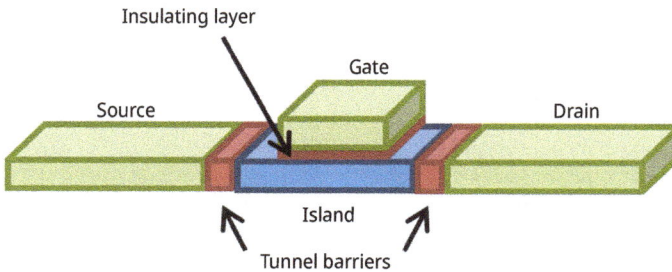

Fig. 7.8: Sketch of an SET.

Such an SET is switched on and off again whenever a single electron is added to its gate; opposite to the common transistor that is only switched on when a current is conducted (Fulton and Dolan 1987; Meirav and Foxman 1995). This unexpected behavior is based on the quantization of charge and energy due to the confinement of electrons in a small area (Goldhaber-Gordon et al. 1998). Quite a similar quantization also occurs when electrons are confined in an atom, leading to the idea that SETs can be regarded as artificial atoms (Kastner 1993; Ashoori 1996), behaving similarly to impurity atoms interacting with the electron gas in a metal (Meir et al. 1993).

Opposite to real atoms, in an SET, the number of electrons can be changed and thus varied between odd and even values, and the coupling to the leads as well as the energy gaps between localized states and Fermi energy can be tailored (Wingreen and Meir 1994). In this way, single localized electron states can be investigated, while their "environment" can be modified to take into account all possible parameters that may affect the electron's state.

One interesting effect that has been investigated in this way is the Kondo effect (for a broader overview, cf. Chapter 6). This effect describes an unusual change in the electrical resistance with the temperature, based on the scattering of conduction electrons at impurities in a metal. The temperature-dependent resistance of electrons in a metal can mathematically be described by the following terms: a constant residual resistance, a term proportional to T^2 (with the absolute temperature T) due to the so-called Fermi-liquid properties, and a term proportional to T^5 due to lattice vibrations. By introducing a term proportional to $\ln(1/T)$, Kondo explained an unexpected increase in the resistance for very small temperatures. This term was attributed to s-d electron scattering (Kondo 1964) and considered as an example of asymptotic freedom, leading to a strong coupling between localized magnetic impurities and itinerant electrons for low temperatures and low energies.

Interestingly, the Kondo effect was also found in quantum dots in which at least one unpaired electron behaves like a magnetic impurity. If such a quantum dot is coupled to a metal, the metal conduction electrons can scatter at the quantum dot, analogously to the formerly described effect of the scattering of conduction electrons at magnetic impurities (Cronenwett et al. 1998; Kouwenhoven and Glazmann 2001).

To investigate the Kondo effect in an SET, using the possibility to tailor the parameters influencing this effect, it is necessary to make the SET very small so that the quantized energy differences between electronic states become important and the temperatures at which the Kondo effect can be expected are not too low (Goldhaber-Gordon et al. 1998). Additionally, the number of electrons in the "electron droplet" in the SET must be odd so that there is one unpaired electron, working as a "magnetic impurity" (Goldhaber-Gordon et al. 1998, Cronenwett et al. 1998). Using approximately 50 electrons, the Kondo effect could also be verified in an SET by transport measurements (Goldhaber-Gordon et al. 1998a).

First SETs were limited to very low temperatures due to the problem that they showed a capacitance of the order of 100 aF. The charging energy – being antiproportional to the capacitance – is required to be significantly higher than the thermal energy, a condition that could only be fulfilled at temperatures below 4 K, which made the investigation of this effect in a typical liquid helium cryostat challenging (Ali and Ahmed 1994). Later, SETs with significantly smaller capacitances were created, allowing for investigating their behavior even at room temperature (Takahashi et al. 1995). For this, it was necessary to create very small elements with low tunnel resistances. This was achieved, for example, by converting a 1D Si wire into a Si island, separated from the leads by tunnel barriers at both ends that were produced by a special pattern-dependent oxidation technique (Takahashi et al. 1995). Alternatively, small islands were formed by self-organization of a fine polysilicon layer, a method that makes the size and structure of the island hard to control (Yano et al. 1993). Other researchers used a scanning tunneling microscope tip to create a point-like contact by oxidizing a fine line in a Ti layer to TiO_x and thus creating a small barrier in a so-called nano-oxidation process (Matsumoto et al. 1995; Matsumoto et al. 1996).

Nevertheless it should be mentioned that SETs are not completely compatible with the working principles of common transistors and thus have to be implemented as logic gates in slightly different ways than in usual CMOS (complementary metal-oxide semiconductor) architecture (Korotkov et al. 1995). Calculations of possible SET architectures are given, for example, by Chen et al. (1996). While they use the fact that SETs can show negative transconductance, opposite to common transistors, they also underline the problem that CMOS gates cannot be reproduced exactly since SETs cannot open in a similarly wide range of gate voltages. This leads to the effect that the SET resistance is always near the asymptotic high-voltage value after optimization of circuit parameters (Chen et al. 1996).

Storing a bit of information in an SET by storing one electron was shown, for example, by Guo et al. (1997). Led by the idea to reduce the size of each single bit more and more, they developed the principle of storing information in a common transistor further. A MOSFET can be used as a nonvolatile semiconductor memory since different numbers of charge carriers on the floating gate influence the transistor's threshold voltage. Decreasing the dimensions of such a nonvolatile memory, the logical limit is reached if one bit is represented by one electron on the gate, that is, the SET. Different technological approaches were used to create such minimized bits, for example, by a device in a fine polysilicon strip in which an electron percolation path is formed as the channel, while nearby polysilicon grains work as the gate (Yano et al. 1994). Since his structure is not very precisely controllable, nanocrystal grains were also used as the gate (Tiwari et al. 1996), still not solving the problem of the broad distribution of channel and gate dimensions due to production parameters being hard to control. These problems have been overcome by more sophisticated production processes, allowing control of the dimensions of all relevant parts much better (Guo et al. 1997).

It should be mentioned that SETs have been realized in diverse forms and for diverse applications recently. An SET was used, for example, to measure the local electrostatic potential of graphene flakes with microvolt sensitivity (Martin et al. 2008), using the SET as a microscope (Yoo et al. 1997; Yacoby et al. 1999). Because of its small tip size, a spatial resolution of the order of 100 nm could be realized in this way, enabling imaging the carrier density landscape of graphene and in this way proving the existence of electron–hole puddles, which have been suggested theoretically before (Hwang et al. 2007).

An SET was also used as a very sensitive electrometer. Similar to a SQUID (superconducting quantum interference device) that is able to measure magnetization very sensitively by "counting" flux quants (Clarke 1989), an SET can "count" electron's charge, as shown by many research groups (Averin and Likharev 1986; Fulton and Dolan 1987; Likharev 1987). Usual SETs, however, are limited to low operation frequencies in the order of magnitude of 1 kHz or even less. The invention of a fast RF SET with significantly reduced $1/f$ noise that was realized by an integrated cryogenic voltage amplifier to minimize the capacitive load (Petersson et al. 1996; Visscher et al. 1996) or by integrating the SET into a fast-oscillating resonant circuit

(Schoelkopf et al. 1998) paved the way for faster measurements with higher sensitivity. Such sensitive electrometers could also be used to define an electron-counting standard for a current, making this recent SI base unit measurable more accurately (Likharev 1999) or allowing to produce better photodetectors (Cleland et al. 1992).

Another potential application is equivalent to the respective application in a "macroscopic" transistor – amplifying signals. This is necessary, for example, to read-out solid-state qubits in quantum computers, that is, amplify the respective single-quantum signals (Devoret and Schoelkopf 2000, Veldhorst et al. 2015). The sensitivity of an SET is theoretically only limited by the quantum shot noise, while it is still higher experimentally. Nevertheless, ideas like using Cooper pair tunneling are assumed to support approaching this theoretical limit.

SETs are also interesting for distance measurements. In several applications, it is necessary to measure distances – or displacements – on the nanometer scale. Thinking about an oscillator in thermal equilibrium with its environment, the displacement amplitude can in principle be reduced by reducing the temperature. Nevertheless, the zero-point motion must also be taken into account, avoiding the displacement amplitude approaching zero. Since an SET works as a very sensitive electrometer, the motion of a nanoresonator can be measured if the SET gate is capacitively coupled to a metal electrode on the resonator and a constant bias voltage is applied at the electrode. In this way, the quantum effect of the zero-point motion was shown to be measurable theoretically (Blencowe and Wybourne 2000; Zhang and Blencowe 2002) and experimentally (Knobel and Cleland 2003). Another research group used an RF SET to measure mechanical oscillations of a quartz disk resonator with a resonance frequency around 9 MHz, finding a measurement sensitivity of the mechanical displacement below 10^{-13} m, reaching the quantum limit of the vibrations (Li et al. 2018a).

Finally, besides the aforementioned SETs based on semiconductors and typical solid-state electronic materials, a completely different way to produce an SET was described by Kubatkin et al. (2003). They built an SET in which the electronic levels of one molecule in differently charged states control the transport properties. The electronic levels in the single molecule are significantly more perturbed than those in the common molecule in a solution, resulting in a strong reduction of the gap between HOMO (highest occupied molecular orbital) and LUMO (lowest unoccupied molecular orbital). This effect is attributed to image charges in the source and drain molecules that strongly localize the charges on the molecule.

After this short overview of the function, applications and possible manifestations of SETs, the next section treats another transistor with different properties.

7.7 Bipolar magnetic transistor

One of the first electronic components that were produced when MBE (Cho and Arthur 1975; Ploog 1980) and metal-organic chemical vapor deposition (Dupuis et al. 1979) had been developed was the bipolar transistor (Kroemer 1982). The idea of this special transistor is using energy gap variation in addition to the electric fields to tailor the forces that work on electrons and holes separately. The forces on electrons and holes may even have the same direction, while in a uniform-gap semiconductor both forces are equal and opposite to each other. In double-heterostructure transistors with wide-gap emitters and collectors, changing the biasing conditions can be used to exchange emitter and collector, making the transistor bipolar. Such bipolar transistors can be realized as n–p–n or p–n–p junctions with a wide-bandgap emitter.

Bipolar transistors with two collectors were found to be influenced by an external magnetic field (Vinal and Nasnari 1984). This means that under the influence of a magnetic field, the currents to both collectors become unequal. This can be explained by an emitter-injection modulation – due to the magnetic field, the injections from different areas of the emitter are increased or suppressed.

As Flatté et al. (2003) have shown, the functionality of bipolar transistors can be significantly increased by introducing a magnetic semiconductor into the base of a bipolar transistor, while both the emitter and collector are still nonmagnetic. They report results on n–p–n transistors (with the magnetic semiconductor being used as p), but mention that the opposite orientation can be used in an equivalent way, that is, a p–n–p transistor (with n being represented by the magnetic semiconductor). In this structure, spin filtering can occur in different ways: for carriers passing from the emitter to the magnetic base; by spin-selective spin-flipping inside the magnetic base; and by spin-selective recombination within the magnetic base. When combined, these three mechanisms lead to a highly spin-polarized electron current into the nonmagnetic collector. Interestingly, this spin-polarized current can be modified by changing the orientation or the absolute value of the magnetization in the base that is possible by common magnetic domain-writing technologies. Additionally, the base magnetization could also be changed by changing the temperature in the device (Ohno et al. 2000) or by optical injection of carriers in the base (Koshihara et al. 1997; Oiwa et al. 2002) to enable a nonmagnetic modification of the spin-filtering properties of the base. Besides ferromagnetic semiconductors, paramagnetic ones can also be used.

Such magnetic bipolar transistors can also be used for current amplification. The possibility to control it by controlling the applied magnetic field gives rise to the new effect of magneto-amplification (Fabian et al. 2004; Fabian and Žutić 2004). Additionally, the magnetic bipolar transistor can work as a spin switch – the current direction is changed when the spin is flipped (Fabian and Žutić 2004a). These new functionalities that are not available in common bipolar transistors can be used for possible applications in logic circuits, nonvolatile memories, signal processing, and so on (Fabian and Žutić 2005). To make these functions better understandable in terms of common electronic components, the so-called Ebers–Moll model has been proposed (Fabian and Žutić 2005) and studied in detail. Recently, modifications of this theoretical model were suggested to increase compatibility with practical characteristics (Masud et al. 2015). In the experiment, magneto-amplification at room temperature could be found, for example, in (In,Mn)As-based magnetic bipolar transistors (Rangaraju et al. 2010).

Finally, magnetic bipolar spin transistors were suggested as parts of a spintronics logic family with cascadable logic gates (Friedman et al. 2015). In this so-called emitter-coupled spin-transistor logic, the emitter-coupled logic was extended to include spintronics features. Here, the magneto-amplification of the magnetic bipolar transistors is modulated by the magnetic field that is created by the current through a differential amplifier. In this way, additional inputs can be added to each logic stage so that each gate shows enhanced functionality, resulting in more compact circuits due to minimized logic components. Since spintronic switching does not necessitate additional currents, the new logic family can be smaller and faster than common emitter-coupled logic, enabling high-performance spintronics. This principle has also been patented (Friedman et al. 2017). If the wire at the output of an emitter-coupled logic element is positioned near a spin transistor in another area of the circuit, more basic logic functions are available (Friedman et al. 2012).

Besides the magnetic bipolar transistor, several other components can be used as magnetic logic gates. Some of them will be introduced in more detail in the following sections.

7.8 Magnetic NOT logic gate

In logic circuits, the NOT gate is identical to an inverter. If the input is true, then the output is false, and vice versa. This logic gate is electronically created using components resulting in a low output signal for a high input signal, and vice versa. Such a function can be realized by combining special transistors (NMOS, i.e., an n-type metal-oxide semiconductor, or PMOS, a p-type metal-oxide semiconductor) with a resistor, or by using two complementary transistors in a CMOS (a metal-oxide semiconductor with oppositely symmetrical pairs of p- and n-type semiconductors) setup so that one of the two transistors is always switched off. Similarly, bipolar junction transistors, as described above, can be used to create a NOT gate. In digital electronics, typically both states are realized by the voltage levels between 0 and 5 V, respectively. In industry, however, NOT gates are often implemented in emergency stop-buttons of machines, switching off the machine if the button is on.

Besides these common NOT gates that are well known and long established, spintronics would also be enriched by possibilities to create circuits in which magnetic bits of data could be used for calculations between writing and detecting them again. This means that several logic operations, among others the NOT gate, should be implemented in spintronics. The first suggestions used sequential conversions between magnetic and electronic states of logic levels in a circuit with successive logic elements (Black Jr. and Das 2000; You and Bader 2000).

One idea to reach this aim with purely magnetic states was suggested by Allwood et al. (2002). Their approach is based on the effect that magnetization in very thin objects, such as ferromagnetic wires, tends to be oriented along these objects due to the so-called shape anisotropy. Between two oppositely oriented magnetic areas in a fine wire, usually a domain wall is formed and shifted by an external magnetic field parallel to the wire (Ono et al. 1999). Additionally, a rotating magnetic field can define the chirality (handedness) of domain walls that can only propagate around corners with identical chirality. This means that by changing the rotational direction of an external magnetic field, the possibility of domain walls traveling around a defined corner can be switched on and off, respectively. In this way, the authors produced an all-metallic ferromagnetic NOT gate in which the output magnetization was inverted, with a short delay in comparison with the input magnetization switching time. It could be shown that bits of information could directly be transferred from one NOT gate to the next one, enabling even the creation of a purely magnetic shift register. Such a shift register is a series of components in which each output is connected with the input of the next component. If new information, in the form of a bit, is added to the input of the first component, the whole chain of information is successively shifted one component further, finally shifting the information on the last component out of the shift register.

Other research approaches are based, for example, on the control of domain wall movement (Cowburn and Welland 2000), magnetoresistance (Reiss and Meyners 2006), and the magnetostatic field created by an array of nanomagnets (Imre et al. 2006). After theoretical approaches dealing with a spin-wave interferometer (Hertel 2004; Vasiliev et al. 2007), this idea was experimentally verified by creating a NOT gate (Kostylev et al. 2005).

Using a chain of soft ferromagnetic nanowires, establishing NOT gates, can be used to build a shift register (Allwood et al. 2002; O'Brian et al. 2009). Since the original symmetric form (i.e., a bidirectional flow of data was possible, dependent on the rotational direction of an external magnetic field) of these NOT gate shift registers was found to need large areas and thus to be unsuitable for high density data storage, an asymmetric form was developed, using a square mesh geometry on which nanowires were created by tessellating an area into the desired nanowire geometry. In this way, significantly smaller areas are necessitated to perform a logic NOT operation based on the control of domain wall movement (Zeng et al. 2010).

Spin waves instead of domain walls are examined, for example, by Khitun et al. (2008). By using the phase of a spin wave to store information (with 0° and 180° being defined as the logic states 1 and 0), it is possible to transfer this information without the necessity of using electrical current. On the other hand, converters must be created to transfer information between voltage and spin and vice versa. In such spin-wave logic devices, it is possible to create diverse logic gates, such as AND, OR, NAND,

NOR, and any other logic function by combining a NOT gate with a majority gate that can be reconfigured using modulators. The possibility to reconfigure different elements in a 2D array of interconnected logic gates is the main feature of a so-called nanofabric (Jacome et al. 2004), which is thus stable against possible defects. Such nanofabrics can be created using the above-described domain-wall logic (Allwood et al. 2005), nanotubes (DeHon and Wilson 2004), or molecular approaches (Likharev and Strukov 2005) as well as with spin-wave propagation (Wang et al. 2005; Khitun and Wang 2006; Kostylev et al. 2005). Using spin waves, their superposition can be used to create logic devices. Depending on the length of a ferromagnetic wire, it can be used as a buffer or as an inverter. A majority logic can be built by evaluating the sum signal of diverse spin waves, which is larger or smaller if more incoming spin waves are in phase or antiphased, respectively.

Finally, it is possible to use MTJs as logic devices. This approach allows using an MTJ at the same time to store information and to process logic operations, realizing a so-called "stateful logic" (Borghetti et al. 2010; Cai et al. 2022; Hoffer & Kvatinsky 2022). Here, one of the main building blocks is the implication IMP, meaning "if a, then b," combined with the opposite operation NIMP. The logic states 1 and 0 can be defined by the high and low resistive states, respectively. It should be mentioned that the errors made in these logic devices do not only depend on the TMR, but also on the topology of the implemented logic (Mahmoudi et al. 2013).

7.9 Other magnetic logic gates and devices

Besides the aforementioned NOT gate, other logic gates have also been implemented in purely magnetically working circuits.

The most interesting logic gates, being referred to as "universal" gates, are NAND and NOR that have two inputs each. From these gates, all other logic operations can be created. While the NOR gate only has "true" as output if both inputs are "false," the "NAND" gate only results in the output "false" if both inputs are "true."

The NOT gates described above could be realized by NOR gates or NAND gates, respectively, by just connecting both their inputs. Other gates, such as OR, AND, XOR, XNOR, and so on, not only need more sophisticated constructions but can also be created unambiguously from NOR or NAND gates.

The NAND gate thus belongs to the spin-wave logic gates that have been developed by Schneider et al. (2008). Similar to Kostylev et al. (2005), they used a Mach–Zehnder interferometer with identical signal and reference interferometer arms and controlled the spin-wave phases by a current in the signal arm to achieve an XNOR (logical equality) process, based on interference patterns – destructive interference is defined as "false" and constructive interference, correspondingly, as "true." Similarly, a NAND (NOT AND) gate was created by placing switches in both interferometer arms that are controlled by currents that may suppress the spin-wave pulse transmission.

A Mach–Zehnder interferometer was also used to create a NOR gate in addition to NOT and NAND gates (Lee and Kim 2008). Interestingly, here the "universal" gates NOR and NAND were built from two Mach–Zehnder interferometers arranged in a serial or parallel configuration, respectively. Nevertheless, they also mention the challenges of controlling and monitoring the constructive/destructive interference as well as creating strong signals from circuits with low power consumption.

Another approach is the domain wall movement already described for the NOT gate. Using sophisticated nanostructures, diverse gates can be realized, often even easier than in common CMOS technology (Allwood et al. 2005). The NAND function, for example, can be combined from a NOT and an AND element in domain-wall logic operations, while it necessitates four transistors in CMOS technology. Similarly, the AND gate can be created by connecting two magnetic nanowires, while six CMOS transistors are needed in the respective technology. Domain wall cross-overs can be realized in a single plane without the necessity to create paths on separated levels. Additionally, as suggested by the authors, the domain-wall logic includes the possibility to extend circuits into 3D neural networks, enabling not only very dense nonvolatile memories but also supplying power at the same time as input signals and even a clock signal to support logic operations.

In a completely different way, AND and OR gates were realized by Zhang et al. (2015). They used skyrmions for these logic operations (cf. Chapter 3). Skyrmions are particle-like magnetic excitations in ferromagnets that have been examined intensively in the last years (Tchoe and Han 2012; Zhou and Ezawa 2014; Boulle et al. 2014). They can be found in thin magnetic films that show the so-called Dyaloshinskii–Moriya interaction (Romming et al. 2013). Such skyrmions are topologically stable since their topological number is quantized (Brown and Rho 2010). Nevertheless, they can be created or destroyed in different ways (Iwasaki et al. 2013; Finazzi et al. 2013). To create logic devices based on skyrmions, it is necessary to move them along linear paths (Fert et al. 2013; Zhang et al. 2015; Zhang et al. 2015a) which can be produced by lithographic techniques to create magnetic strips on materials with large spin–orbit coupling (Moreau-Luchaire et al. 2016; Woo et al. 2016). It is also possible to align them by spatially modulated magnetic anisotropies (Wiesendanger 2016).

Creating logic gates using skyrmions can, for example, be done by doubling them simply by leading them from one nanowire into two separate ones in the right geometry, while the inverse setup can be used to create AND or OR gates, depending on the exact geometry, especially the width of the single nanowire: In a thin nanowire used as output, only one skyrmion needs to be introduced from both inputs to create a new skyrmion (OR), while in the broader nanowire, two skyrmions from both supplying nanowires are necessary to create a (larger) skyrmion in the single nanowire output (AND). Similar to skyrmions, even three-dimensional hopfions (cf. Chapter 3) have been investigated regarding their use in logic devices (Zhang et al. 2023a).

Finally, investigations of graphene nanoflakes have shown that depending on the spin structure in such nanoflakes, the universal NAND and NOR gates can be produced. Theoretically, such graphene-based logic devices should work at room temperature and have only very low error rates (Wang et al. 2009). After the finding that bulk proton irradiated graphite can show magnetic properties (Esquinazi et al. 2003; Ohldag et al. 2007), especially in the case of zigzag-edged nanoribbons (Son et al. 2006) or triangular nanoflakes (Ezawa 2007; Fernández-Rossier and Palacios 2007; Wang et al. 2008), graphene is clearly a candidate of large interest for possible spintronics applications. Theoretical calculations have shown that NOR and NAND gates can be created using a special graphite nanoflake structure and that this spin logic can even be reconfigured by a programming bit to switch between both these logic gates (Wang et al. 2009). This approach is similar to earlier findings of Derycke et al. (2001) who used combinations of p-type and n-type carbon nanotubes field effect transistors.

After exploring these possibilities to perform logic operations on data, the next sections will concentrate on conventional methods to write and read data, using some of the effects described before, and on novel approaches to store data.

7.10 From magnetic tape storage systems to hard disk read/write heads and beyond

This section is partly based on the book "Examination and simulation of new magnetic materials for the possible application in memory cells" (Ehrmann 2014).

Hard disks are the most common storage devices used nowadays. However, several other ideas have been examined in earlier times, some of which are based on similar principles, while others utilized different physical effects. Before describing in detail recent hard disk technology with the corresponding read and write heads, a short overview is given about earlier attempts to store data since the basic ideas behind some approaches may become interesting again for new spintronics storage devices.

One of the first attempts to store data on ferromagnetic wires and tapes worked with analog audio signals (Poulsen and Pedersen 1907). Here, a steel wire or a magnetically coated tape was drawn along a recording head at high speeds (more than 60 cm/s) so that short sections of the medium were magnetized depending on the intensity and polarity of the electrical audio signal that reached the recording head at a certain time. The signals could be read again by using the same process without an external electrical field so that an electrical signal was induced corresponding to the magnetization in the medium. After first tests with transverse magnetization, longitudinal magnetization was recognized as more stable afterward against erroneous twists in the medium. This principle has not only been used for recording audio or, later on, also video signals on magnetic tapes; it has also been utilized for recording (digital) data due to the relatively low costs per information unit, as compared with hard disks, and the expected long lifetimes around 10–30 years, which makes it especially useful for backups (Eleftheriou et al. 2010).

Such magnetic tape storage systems are most often based on the linear tape-open (LTO) standard, which was initiated by IBM, Hewlett-Packard, and Seagate. Typical cartridge speeds of media using this standard are nowadays 750 MB/s (LTO 2018). The linear density of approximately 20 KB/mm has not significantly changed between the last two generations of these media (Quantum 2018), but increased by about one-third during the last 9 years (Quantum 2009). Areal densities of approximately 30 Gb/in^2 have already been proven to be possible corresponding to cartridge capacities of 35 TB (Cherubini et al. 2011). Theoretically, values of up to 10 Gb/in^2 could be reached, which would result in an overall storage capacity of more than 100 TB per tape cartridge (Argumendo et al. 2008).

Magnetic tape storage systems can also work with other principles and standards. In the linear serpentine technique, there are more tracks than heads, while the transverse scan is used to write short data tracks in the direction perpendicular to the tape length. If both methods are mixed, helical or arc-like tracks are written. Finally, other standards like digital linear tape (DLT), developed by Digital Equipment Corporation (DEC) and digital data storage (DDS) show lower capacities and data transfer speed at recent state of research.

A recent alternative to the above-described magnetic tape storage systems are magnetic stripe cards that can be used to store small amounts of data that are never (or only seldom) changed. Information that should be carried, for example, by credit cards, ID cards, or similar objects can be stored in flat bands that contain magnetic iron particles or, especially in the case of important data such as in bank cards, in materials with higher coercivity (IBM 2018).

Magnetic ink that was developed in the 1950s is mostly used for processing checks. Automation of this process can be performed with a special read head that can read the magnetic ink, the latter being used to write down the information on the check translated into a series of numbers, which can be decoded again by the computer (TROY 2007).

While the above-described magnetic tape storage systems have the large disadvantage of long read times, making them only suitable for backups, the drum memory worked by rotation, similar to hard disks, but had a row of read/write heads along the drum axis so that each head was correlated with one track. The system was invented already in 1932, could store a few 10 kilobytes, and was replaced by the semiconductor memory in the 1970s (IBM 1955).

Random access memories (RAM) are direct access memories in which unpredictable positions can be reached for reading or writing. One of the first possibilities to create a RAM is the magnetic-core memory, using small ferromagnetic rings as cores each of which represents a bit. They can be magnetized clockwise or counterclockwise, similar to recent ideas to store information in skyrmions (Wright 1951). In both the cases, the main advantage of the round shape is that there are nearly no stray fields, enabling relatively dense packing of the cores. In the magnetic-core memory, neighboring cores were additionally oriented perpendicular to each other to further minimize undesired interactions due to stray fields (Seleznev et al. 1977).

Opposite to modern memory systems, the magnetic-core memory was not based on lithographic techniques – which did not exist at that time – but on fine wires building a square mesh, with the cores being placed on the crossing points. Besides the wires in the x- and y-direction, a third set of parallel wires was oriented along $\pm45°$, passing also through the cores and working as inhibit/sense lines (Olsen and Best 1959). Writing 1 was performed by a current – the magnetic polarity (i.e., the rotational orientation of the magnetization) was changed by a current applied to the x- and y-wires crossing at the desired core. The idea that only the combined currents in both wires are sufficient to switch the magnetization at the crossing point, while all other cores along the respective wires stay unaltered, is typical for such writing mechanisms. The inhibit lines are used to reduce the net current by applying an identical current as in the x- and y-wires, thus inhibiting a change of the magnetic polarity.

For reading, currents in the opposite direction are used. While nothing happens if a core is in the state 0, the change in the magnetization orientation induces a voltage pulse in the sense line if the state 1 was valid before. This is an example of systems in which reading influences the state of a bit so that the original state has to be restored after reading by rewriting the state 1 (Jones 1976).

Developing this system further, small permalloy dots and printed circuits were used instead of the ferrite cores on wires, resulting in significantly faster access times in the order of 670 ns in combination with a reduction of the dimension from 3D (or actually 2.5D) to 2D (Naval Education and Training Command 1978).

Going back to a real 3D memory, the twister memory was developed in the 1960s. In this memory, a magnetic tape was wrapped around the crossing points, adding a third dimension in the form of solenoids. Using this memory, it was possible to process a complete row of data in one step (Meinken and Stammerjohn 1965). In spite of this advantage, the production was too time-consuming and expensive to be competitive with newer memory systems.

The idea of exchanging data between people processing them on equal systems resulted in the development of external memory cards. The first of them, the so-called CRAM cards, were produced in the early 1960s. They could be handled using vacuum and blowers to carry them from a stack of cards and back again after reading and/or writing them. In this way, they could also be used to store an increased amount of data by handling these cards like books in a library (NCR 1962).

After the CRAM, the floppy disk was invented in the late 1960s. Early floppy disks had one or more index holes in the disk, which were used to measure the starting angle and the rotational speed. Depending on the dimensions of the floppy disk, different slot solutions in their housings were used to allow or impede writing by closing the slot with a tape or leaving it open, respectively. Different materials were used to produce floppy disks, often iron oxide, cobalt, or barium ferrite. The disks were coated on both sides with a nonmagnetic material to reduce the friction in the housing, for example, with Teflon (Brown 2018).

Reading and writing floppy disks work in principle similarly to the first read/write heads for hard disk drives (HDDs). A current flows through a coil in the writing head, moving over the rotating disk, which is sufficient to magnetize a defined area of the disk. The small voltage induced by the magnetic particles is amplified and measured, if a defined area is planned to be read. Opposite to some earlier storage systems, floppy disks necessitate an initialization step, the formatting process, to align the originally chaotic magnetic particles to tracks that are separated by unused tracks in between so that small variations in the rotation speed do not cause errors. Capacities of floppy disks started at 80 kB for the first read-only 8 inch disks (IBM 2018a) to 360 kB (5 ¼ inch DD) to 1.2 MB (5 ¼ inch HD) to finally 1.44 MB (3.5 inch HD) (Brown 2018). Special floppy disks with other capacities were also available, but not common.

The first idea to use special areas in a magnetic material, slightly similar to the above-described approaches to utilize skyrmions for information storage and processing, came up in the early 1970s in the form of the Twistor memory (Rostky 2007). The main idea here was to store and process data by moving the data, not the whole storage medium as it is done in floppy disks or HDDs – an idea similar to the approaches to use moving domain walls or skyrmions for data storage, as described earlier, and which is especially also used in the Racetrack memory that will be described below in more detail.

In the Twistor memory, the first orthoferrite used has an easy magnetic out-of-plane axis. By applying a magnetic field perpendicular to the plane, small magnetic "islands" can be created. These so-called magnetic bubbles are always circular and can be moved if a second magnetic field is applied in-plane. Small ferromagnetic structures can be used to define an exact path along which the bubbles should be moved. The in-plane magnetic field is steadily rotating, in this way defining the moving direction of the bubbles and the empty spaces between them. Smaller bubbles could be created in garnet that allowed for a storage density of approx. 4 kbits/in^2. The bubble memory was configured as a serial memory, similar to disk or drum memories, instead of competing with different sorts of RAM.

The main technical problems were related to material impurities that became especially important when the bubble density was increased. The idea of moving data only, instead of the whole storage medium, made skipping bad sectors difficult. Similar problems are reported in recent scientific literature about Racetrack memory. Finally, the idea of bubble memory was simply too late – it could never clear the initial backlog in comparison with the earlier developed HDD.

The first HDD was developed in 1956 (IBM 1956). An HDD contains one or several rotating disks of nowadays 3.5 inches, 2.5 inches, or smaller diameters, while the first ones had diameters of 14 inches. These disks are basically nonmagnetic, but coated with a thin ferromagnetic layer, typically of 10–20 nm thickness, and an additional protection layer over the ferromagnetic one (Hitachi 2010). Usually, only one read/

write head is used on either side of each disk, while all read/write heads are moved by one common arm (Escotal 2014).

While HDDs nowadays have capacities of several terabytes, the first models were limited to some megabytes (IBM 1956). Similarly, the access times have significantly decreased to a few milliseconds, starting with more than 100 ms in the first HDDs. Rotational speeds nowadays are often 7,200 rounds per minute, sometimes even more (e.g., WD 2017). Data transfer rates are recently in the order of 125–250 MB/s (Seagate 2017).

Similar to Moore's law describing the doubling of the number of transistors on a chip in every 12–24 months (Moore 1965), the areal densities could first be doubled every 18 months. In both cases, however, this steady increase has reduced during the last few years due to technical and partly even physical limits. This is why new ideas to store data have been under investigation for several years now, some of which will be described at the end of this section.

Together with the HDD disks, the read/write heads had to be developed further to enable increasing capacities. The first heads were constructed similar to those used in tape recorders: A ferrite ring with a small gap was wrapped by a fine coil. If a current flowed through the coil, a large magnetic field was induced in the gap that could be used to magnetize the magnetic disk surface directly below the head. Oppositely, the magnetized disk surface induced a current in the coil that enabled reading data. Introducing a small metal piece in the gap resulted in a field concentration and enabled reading and writing smaller areas. Furthermore, the 3D head was exchanged for a lithographically produced thin film head, allowing for the production of HDDs with some GB storage capacity in the mid-1990s (PC Guide 2001).

In the next steps, read heads were developed further by using different physical effects. The AMR effect describes the change in the electrical resistance of a material in a magnetic field. Utilizing the AMR effect, smaller magnetic areas on the disks could be read (IBM 1996).

The next large step was enabled by the GMR, which was first introduced in 1997 into HDD read heads (IBM 2018b, Hitachi 2010). With this effect, the resistance can change by several 10%, depending on the relative magnetization orientation of two or more ferromagnetic layers separated by a nonmagnetic metal, allowing for even smaller structures to be measured.

A further step was the use of the TMR in which the tunnel current between neighboring ferromagnetic layers, separated by nonmagnetic isolators, depends strongly on the relative orientation of the magnetizations in these ferromagnetic layers. The first TMR heads developed by Seagate introduced small heating coils to support the reading and writing processes (Chen et al. 2006a).

Partly parallel to these changes in terms of the physical effects utilized for reading and writing data on HDDs, the magnetic material on the disk has started to be changed. For the typical in-plane recording geometry, the structure size is limited by the superparamagnetic limit – in too small structures, the thermal energy is able to introduce arbitrary changes in the magnetization, resulting in loss of data. Larger structures are more stable since they have larger coercivities, that is, larger external influences are necessary to switch the magnetization. This is why for the next generation of hard disks, perpendicular recording was investigated (Piramanayagam 2007). The difference between both principles is depicted in Fig. 7.9.

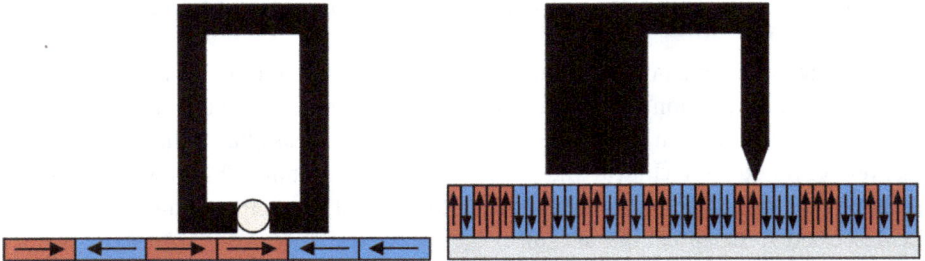

Fig. 7.9: Principle of the difference between the former standard in-plane (longitudinal) recording head (left panel) and the perpendicular recording mechanism (right panel) (adapted from Ehrmann (2014)).

Besides the already mentioned advantage of using a smaller area to store a single bit, another effect is also advantageous in the perpendicular geometry. The magnetic flux created during writing flows from the fine tip at a well-defined position through the recording layer into the soft magnetic bottom layer (marked light grey) and back into the writing head along its broad part, making the flow through the single bit under the writing tip much larger, which enables using materials with higher coercive fields without increasing the currents in the write head. The first HDDs based on the perpendicular recording technology were produced in 2004 (Toshiba 2004) and 2005 (Smith 2005).

While the possible data density reached in HDDs seems to approach technical and even physical limits, how may the future of data storage look like? Several approaches are already in use or under investigation in the moment. A short overview of some of these principles is given now, while the MRAM as a promising candidate for future memory solutions is described in the next section, and a completely different approach, leaving the recent von Neumann architecture behind, is introduced in the following section.

On the one hand, HDDs are being developed further to avoid being outperformed by the solid-state drives (SSDs) that use integrated circuits (ICs) to store data. One problem when a bit shall be written on a smaller area is given by the fields necessary for writing. To allow for using reduced fields and to decrease the influence on the neighboring bits, it is possible to heat up the magnetic material only at the desired spot, resulting in a punctually decreased coercivity. This can be done using a laser spot, using near-field technology to produce a very small near-field laser spot below the laser aperture of the writing head (TDK13), or a microwave pulse in the so-called microwave-assisted magnetic recording (Coughlin and Grochowski 2012; Zhu 2008).

A different way to store information on an HDD was first announced by Toshiba. With bit-patterned media (Richter et al. 2006), approximately 2.5 Tbits/in^2 could be stored (Toshiba 2010). Instead of the usual thin-film technology used for recent HDD drives, self-assembled magnetic dot arrays are used to avoid erroneous influences between the poles of neighboring bits.

Patterned structures are also used in nano-ionic memories (Asamitsu et al. 1997, Kozicki et al. 1999, Beck et al. 2000, Waser and Aono 2007, Borghetti et al. 2010) or especially nanowire-based resistive switching memories (Ielmini et al. 2011, Ielmini et al. 2013). Here, the x–y-wire architecture from the early magnetic-core memories has a revival. The storage nodes are, like in the magnetic-core memory, located at the intersections of the column and row wires. This technique enables bit dimensions below 1 nm. Creating such devices is possible top-down (using optical lithography) or bottom-up (using self-assembled nanowires or nanodots) (Huang et al. 2001, Huang et al. 2001a, Zhong et al. 2003). The wires mostly consist of a metal core, surrounded by a very thin nonconductive oxide shell (Li and Chen 2009, Cagli et al. 2011). When an electrical field is applied across the structure, ion diffusion starts, creating a conductive path between the crossed wires.

Besides the above-described TMR heads, magnetic recording can also be performed using GMR heads in the current perpendicular to the plane geometry. With such GMR heads, areal densities of approximately 400 Gbits/in^2 were found in the first experiments (Carey et al. 2008).

Another technique that may be implemented easily is called shingled writing (Amer et al. 2010). Here, the writing tracks overlap, allowing for smaller grain sizes, which leads to partially destroying data in subsequent tracks during writing and thus necessitates restoring these data in the next step. Theoretically, maximum data densities of 3 Tbits/in^2 can be achieved, which is higher than the paramagnetic limit for conventional recording of 1 Tbit/in^2 (Greaves et al. 2009).

Other approaches try to decrease the number of atoms necessary to store one bit. While theoretically, a bit can be created from just 12 atoms (IBM 2012), the recent borders of lithography have to be taken into account for commercial applications based on bit-patterned media. Typical limits of single features are of the order of 22 nm (Kim et al. 2010) or even 14 nm (Kon et al. 2012, Maruyama et al. 2012, Ogino et al. 2012), with smaller feature sizes becoming technologically problematic due to restrictions of the laser wavelengths used for lithography when working in air. More recent techniques, such as extreme UV lithography or multiple electron beam direct write, are necessary to shift these limits further toward smaller feature sizes (Sharma et al. 2022).

A completely different idea, similar to the earlier developed bubble memory, is the Racetrack memory (or domain-wall memory) which was invented by Stuart Parkin and coworkers at IBM (Parkin et al. 2008). The Racetrack memory uses an array of magnetic nanowires of approximately 100 nm diameter, arranged horizontally or vertically on a silicon chip. In such nanowires, short spin-polarized current pulses can be used to move domain walls, similar to the "normal" currents used to drive magnetic bubbles through the bubble memory substrate. Opposite to common HDD technology, complete "trains" of about 10–100 domain walls can be written or read simultaneously, so that a single nanowire can store several bits of data. Theoretically, the Racetrack memory would enable combining a high storage density with a high read/write speed, as it is nowadays given in flash memory.

The first 3-bit Racetrack memory was experimentally verified early (Hayashi et al. 2008); however, several challenges still prohibit this technology from being used commercially. The first is the velocity with which the domain walls could be moved. First experiments revealed necessary current pulse durations of microseconds, that is, three orders of magnitude larger than expected theoretically. This problem was attributed to imperfections in the nanowires which worked as "traps" for the domain walls (Meier et al. 2007). By eliminating these imperfections, pulse lengths of some nanoseconds were sufficient, as theoretically expected. Another problem is the necessary voltage to drive the domain walls, which is proportional to the nanowire length and thus restricts the possible length, that is, the maximum number of bits per nanowire. Recently, 3D racetrack memories and skyrmion-based racetrack memories belong to the novel approaches to further improve racetrack memories (Rial et al. 2020, Gu et al. 2022, He et al. 2023).

Similar to the Racetrack memory, alternative bit-patterned media can be used to store more than one bit per cell. One approach to realize this is by using a stack of MTJs with different resistance levels, allowing for storing one bit in each tunnel junction. A two-step read/write process was developed for a cell with two bits (Kawahara et al. 2012). First, the smaller resistance bit has to be read with a middle-level reference. Afterward, the higher resistance tunnel junction is read, with a lower or higher reference level, depending on the state of the first bit, to influence the first bit in the desired way (so that it is not changed). Writing two bits with identical values can be achieved simultaneously; otherwise the bit on the MTJ with the larger resistance has to be written first, followed by the other bit, which is opposite to the reading order. Larger numbers of stacked MTJs will necessitate more sophisticated solutions for reading and writing and thus longer process times.

Theoretical and experimental examinations were performed on penta-layers including two pinned layers in antiparallel orientation with a free layer between them, separated on both sides by spacer layers. For such systems, a decrease in the critical current and the switching time could be shown (Fuchs et al. 2005, Makarov et al. 2011, Makarov et al. 2011a).

The switching currents – and thus the minimum bit dimensions – can also be decreased by thermally assisted switching, heating up the desired area for a short time, or by changing to a perpendicular geometry (Ohno 2011, Sbiaa et al. 2011a). This technique, however, still necessitates decreased damping and increase thermal stability (Makarov et al. 2012).

In another approach to store more than one bit per storage position, researchers investigate nanostructured patterns that can be produced by recent lithography techniques and that can principally be read/written by the read/write heads used nowadays. The advantage of such an approach using a technique compatible with the recent one would be that the market barriers that have prevented bubble memories, for example, from entering the market would be much smaller. This necessitates, however, finding magnetic nanoparticles with novel anisotropies, that is, with shape anisotropies that result in more than the usual two stable states at remanence. During the last few years, several publications have discussed magnetic systems with three, four, six, or even eight different state of the magnetization at remanence (Zhang and Haas 2010; Wang et al. 2009a; Thevenard et al. 2010; Huang et al. 2010; Moritz et al. 2011). In most cases, however, it is hard to distinguish between these different states since magnetization has to be measured at different spatial positions of the sample. Alternatively, the symmetry of such particles has to be broken to allow for measuring the macroscopic magnetization, for example, by coating parts of the particles to change their magnetic properties (Bowden and Gibson 2009). This will, on the other hand, make the lithography process more complicated and thus more cost-intensive.

Our group has thus concentrated on developing wire-based fourfold systems or "square rings" with special shapes, which can easily be produced lithographically and allow for measuring four different magnetic states unambiguously (Blachowicz and Ehrmann 2011; Blachowicz and Ehrmann 2013; Blachowicz et al. 2013; Blachowicz and Ehrmann 2013a; Ehrmann et al. 2015; Ehrmann and Blachowicz 2015; Blachowicz and Ehrmann 2015; Blachowicz and Ehrmann 2016; Ehrmann and Blachowicz 2017; Ehrmann and Blachowicz 2017a; Ehrmann and Blachowicz 2017b; Blachowicz and Ehrmann 2017). Figure 7.10 depicts the four magnetic states in such a nanowire systems. These so-called onion-states – in which the magnetization is oriented diagonally from one corner to the opposite one – are typical for such systems and also for lithographically produced samples, that is, samples in which there are no wires with round cross-section but the four "arms" show rectangular cross-sections. Opposite to rounder shapes, such as doughnuts or half-balls, no flux-closed vortex states occur in which the magnetization follows the circle without any domain walls or other inconsistencies.

Fig. 7.10: Possible magnetic states at remanence in a fourfold nanowire system.

Other data storage technologies, such as ferroelectric RAM, utilize no magnetic effects but the polarization of a ferroelectric thin film (e.g., bismuth ferrite, $BeFeO_3$, or other structures of the perovskite type) (Fujitsu 2006). A second storage system without magnetic effects is the phase-change RAM in which two different phases of the storage material are used to define the state of a bit (Müller et al. 2003).

Besides reading magnetic information by a read/write head based on the GMR or TMR effect, it is also possible to read magnetic information optically. The so-called magneto-optical disks (MO disks) looked similar to floppy disk and stored information in magnetic layers, similar to HDDs. Writing, however, worked in a heat-assisted process in which the desired position was heated by a laser, and the local magnetization was changed with an electromagnet. For reading, only a laser with reduced power was necessary that worked on the basis of the magneto-optic Kerr effect (Becker 1998). Nowadays, a similar technique is used in heat-assisted magnetic recording media, which are said to reach about 100 TB, allowing for rewriting data in the magnetic layer on a metal plate instead of the plastic plate of the MO media.

These and other techniques to produce nonvolatile RAM result in different properties, making them suitable for different applications. In the PCRAM, for example, write latency times are significantly higher than in common DRAM, and read latency times are twice as long. Writing additionally needs approximately 50 times more energy than in DRAM, and the writing endurance is approximately 10^6–10^8 lower than in DRAM (Li et al. 2012a).

Finally, the so-called SSDs should be mentioned. In spite of the name containing the word "drive," no moving parts are included here, but ICs to store data permanently. Thus, they are more shock-resistant, more silent during working, and they have shorter access times and lower latencies (Kasavajhala 2011). Most SSDs are based on the so-called NAND-based flash memory, a nonvolatile storage device. SSDs based on RAM are even faster, but volatile, and thus need an additional integrated power source to avoid undesired loss of data (Ekker et al. 2009).

Besides these solutions that are either focused on speed or on nonvolatility, another possibility to store data should be mentioned, which may be able to combine both of these desires.

7.11 Magnetic memory cell: MRAM

The examples described earlier were either nonvolatile, such as the HDD, or fast, such as RAM. An interesting candidate for a possible future "universal" memory, bringing together both paradigms, is the MRAM that has been under investigation since the mid-1990s (Akerman 2005). While the RAM used nowadays stores data in the – volatile – form of electric charges, the MRAM is a nonvolatile magnetic storage device.

The easiest way to create an MRAM is by using spin valves for each single bit. As described earlier, spin valves include two ferromagnetic layers, one of which can be changed more easily, sandwiching a thin insulating layer. This TMR system shows different resistances, depending on the relative orientations between both ferromagnetic layers, which can be used to define a bit corresponding to the low-resistance state and the high-resistance state, respectively (Khvalkovskiy et al. 2013).

Writing the values of an MRAM can be performed using a square grit of write lines, similar to the magnetic-core memory described earlier. Here, however, one of the sets of write lines is located on top of the spin vales, while the perpendicular set of write lines is located below. Parasitic currents can be reduced by an access transistor that is connected in series to the MTJs (Müller et al. 2003).

On the other hand, the minimal size of the single spin valves and thus the bits is physically limited by undesired induction fields, which may reach the neighboring bits and change their values. One possibility to overcome this problem is the utilization of the so-called spin-transfer torque that has been investigated for nearly 20 years now (Slonczewski 1996, Berger 1996, Slonczewski 2002, Sbiaa 2011, Kawahara et al. 2012, Thiyagarajah et al. 2012). This means that spin-polarized electrons can directly rotate domains, which reduces the currents that are necessary for writing single bits, while now spin coherence is necessary (Victora 2012, Khvalkovskiy et al. 2013). This technology is developed further by different companies.

Another approach to reduce the necessary power from initial values of approximately 10^6 A/cm^2 (Wang and Chien 2013) by two orders of magnitude is based on the magnetoelectric effects resulting from electric fields – instead of currents – used for magnetization manipulation (Maruyama et al. 2009, Wu et al. 2011), for example, by changing the interfacial perpendicular magnetic anisotropy at the interface of a tunnel junction (Wang et al. 2012, Wang and Chien 2013) or by driving pure spin currents through MTJs (Liu et al. 2012).

Recent research still concentrates on diverse methods to increase thermal stability and decrease cell size (Nishioka et al. 2020; Garzón et al. 2022; Lack et al. 2024). Using MTJs with perpendicular anisotropy, for example, allows for using lower switching currents and a circular cell shape that makes production on small scales easier. On the other hand, for high polarization combined with low damping and low resistance, new materials have to be found. As in many other applications, graphene is one possibility to create small MRAM cells (for the 8 nm technology node that is not yet used in industry, but is being investigated in research) for spin-transfer torque switching in combination with cobalt/nickel multilayers as magnetic part of the cell, necessitating low write voltage and low power consumption (Varghani et al. 2018).

A significant reduction of the damping in MgO/FeCoB/Ta systems could be achieved by increasing the boron content and by extending the system to a symmetric MgO/FeCoB/Ta/FeCoB/MgO stack, pointing out that Ta impurities in the FeCoB layer are the most important source of damping (Devolder et al. 2018).

The interaction between neighboring MTJs becomes stronger with decreasing distances between adjacent tunnel junctions. Nevertheless, only a few works deal with the impact on the static and dynamic properties of MTJs yet. In a recent model of high-density MTJs, three different MTJ stacks were investigated with respect to the magnetic field-induced coupling and its influence on the static and dynamic properties of the array, depending on undesired deviations from the optimal MTJ characteristics due to the preparation of the MTJ array (Yoon and Raychowdhury 2018).

Another approach to reach these goals is nanocomposites from a $La_{2/3}Sr_{1/3}MnO_3$ matrix and ZnO nanopillars that can be produced by self-assembly. In this composite, an electrical current induced a hysteresis in magnetoresistance measurements, depending on sample temperature and current level, which may give rise to a new spin-valve mechanism that could be used in MRAM applications (Pan et al. 2018).

Besides new materials, new read/write strategies must be investigated to overcome the problem that with smaller devices, their resistance increases and the distribution of values broadens, resulting in higher error rates and thus degradation of read performance. As a possible solution for this problem, a new dual-data line read scheme was suggested in which the current through the memory is recycled during reading to create an additional voltage in a second data line. Combining both signals helps reduce sensing time and energy by nearly 50% (Lee et al. 2018).

A new effect becoming important in smaller and smaller MRAMs is the bias temperature instability, which is well known from MOSFETs where it increases the threshold voltage and thus decreases the drain current and transconductance with time. First simulations propose a different sensing technique to reduce this circuit degradation (Lin et al. 2018).

If the MRAM works someday in the desired way, it has the potential to combine the nonvolatility of HDD with the density of DRAM and the velocity of SRAM (Muneeb et al. 2006). However, in the recent two decades, it has not yet been shown that it will be possible to produce MRAM in an industrial scale with properties similar to those necessary in the simulations and calculations to reach the promised properties.

Quite a different approach – giving up the strict separation between computing and storing data and thus avoiding the so-called von Neumann bottleneck, arising from data transfer between both these parts of recent computers – is described in the next section.

7.12 Beyond von Neumann architecture: neuromorphic computing

Recent computers are based on the so-called von Neumann architecture. This means that they have a processing unit, including an arithmetic logic unit as well as processor registers, and a memory in which data and instructions are stored. Besides, there is an external nonvolatile memory, input and output mechanisms, and finally a control unit with an instruction register (von Neumann 1982; Godfrey and Hendry 1993). This means, on the other hand, that program memory and data memory share one bus, which leads to the von Neumann bottleneck, limiting the performance of a recent computer by the limited data transfer rate between the central processing unit (CPU) and memory (Emma 1997). To be more precise, the CPU could work faster if it got the desired data from the memory faster. Both CPU and memory are nowadays faster than the data transfer between them.

Different possibilities were suggested to overcome the von Neumann bottleneck. One of them is the modified Harvard architecture, providing separate buses for data and instructions. Alternatively, parallel computing (as used in supercomputers) can be used. Nevertheless, both of them are not really different from the principal idea of the original von Neumann architecture. As already postulated by Backus in 1978, an alternative programming style should be found, coupled with new state transition rules (Backus 1978). This, on the other hand, may necessitate new hardware to support this new software.

One such approach on which several research groups are working recently is the so-called neuromorphic computing, also called cognitive computing. The main idea behind this approach is creating biologically inspired computers, working similarly to the human brain, often related to neuronal networks and artificial intelligence. Besides solving special mathematical problem faster, possible applications are autonomous robots, machine hearing, machine vision, and so on. In this research area, brain research and computer research specialists are necessary to support understanding the brain on the one hand and to artificially reproduce the essential functionalities on the other hand.

One of the large projects working in this area was FACETS (Fast Analog Computing with Emergent Transient States), strongly focused on the continuous exchange between biological experiments, software and hardware development (FACETS 2017). Within this project, a new simulator-independent language for neuronal network models was created. The BrainScaleS project followed FACETS (BrainScaleS 2015). Currently, some of the project aims are investigated further within the Human Brain Project, a Flagship Project of the program Horizon 2020 of the European Union (Human Brain Project 2017). The Blue Brain project, aiming at creating a large virtual model of the brain of rats, is another predecessor of the Human Brain Project (Markram 2006). However, in spite of the successful predecessors, the Human Brain Project was not uncontroversial, not only due to the style of it management, but also due to its probably overambitious goals (Theil 2015, Frégnac 2023).

Computer companies also work on this idea. IBM has studied possibilities to create a brain-inspired chip for several years, first working on the SyNAPSE chip that was introduced in 2014 (IBM 2014), before the TrueNorth chip was created, combining one million individually programmable neurons with 256 million individually programmable synapses and more than 4,000 parallel and distributed cores (Modha 2018). Several other smaller or larger projects as well as computer companies also work on this topic. This section introduces some of the main ideas and their possible technological implementation in more detail.

The first mention of the term "neuromorphic computing" was back in 1990, describing very large-scale integration of analog components to mimic biological neural systems (Mead 1990). More generally, the term is used nowadays to non-von Neumann architectures. Usually, neuromorphic architectures are highly parallel, strongly connected, combining processing with storage and also using much lower power than common computers. Another important feature is that they should be learning machines (Schuman et al. 2017), similar to the human brain. While machine-learning algorithms are already being developed, it seems to be most promising to combine them with the approach of neuromorphic computing.

The parallel computation that is known from the human brain has inspired some of the first approaches to create a suitable hardware for neuromorphic computing. The main idea was to create a lot of simple processing units, the "neurons," which were strongly connected by "synapses" (Murray and Smith 1988, Bibyk et al. 1990).

Besides the problem of the von Neumann bottleneck limiting the speed of all calculations, neuromorphic architectures were also assumed to be able to make especially neural network computations faster (Burr 1991, Chiang et al. 1991), which is desired, for example, in the control of autonomous robots (Tarassenko et al. 1991), machine learning, or digital image reconstruction (Lee and Sheu 1990). And last but not least, neuromorphic computing is assumed to desire much lower power than common computers, which is also a driving force toward neuromorphic systems (Leong and Jabri 1992, Schemmel et al. 2010).

Correlated with these reasons to create neuromorphic systems, several technological approaches exist that can be used to implement neuromorphic computing into a technical system. Schuman et al. (2017) split them in five categories:

1. biologically plausible, that is, explicitly modeling realistic neural behavior;
2. biologically inspired, that is, replicating the behavior of biological neural systems;
3. neuron models including other biomimetic components;
4. simple spiking neuron models;
5. a special sort of neuron models, the so-called McCulloch–Pitts neurons, working according a defined equation.

Depending on the desired application, the degrees of complexity and biological inspiration may differ.

Similar to the artificial neurons, the artificial synapses have to be chosen with respect to the planned applications, and they must also fit the chosen neurons. Schuman et al. (2017) differentiate between biologically inspired artificial synapses and others that are typically used in common artificial neural networks. It should be mentioned that the behavior of synapses is correlated with the idea of learning. Thus, ideal artificial synapses should be able to change with time, depending on whether they are often or only scarcely used, similar to plasticity changes in real biological brains. Such "learning" synapses are especially used in biologically inspired neuromorphic systems (Choi et al. 2011; Ramakrishnan et al. 2011). It should be mentioned that "learning," at first glance sounding like a simple term that is well known to all of us, must indeed be defined for neuromorphic systems. Possible rules include, for example, the least-mean-square (Chiang et al. 1991), Gaussian (Lau and Lee 1997), or so-called Hebbian learning rules (Card et al. 1991).

Finally, the created networks should not be forgotten, defined by the rules of how neurons and synapses interact. Here, it should be mentioned that technological restrictions must be taken into account, especially when thinking about 2D structures. Nevertheless, the network should be chosen based on the necessities of the chosen learning system, and it should also fit with the other components, that is, neurons and synapses and the way these parts work. An especially interesting way is the use of self-organizing artificial neural networks that are correlated with unsupervised learning methods (Andreou and Boahen 1989; Mann and Gilbert 1989; Fang et al. 1992; Rodriguez et al. 2015) – this may be the most biologically inspired way to create a neuromorphic system.

Generally, when planning to create a neuromorphic system, one of the first questions is whether it should work like a common digital computer or like an analog circuit – or mix both approaches. While the first ideas of neuromorphic computing were related to analog computing (Mead 1990), computer companies starting to work in this area typically choose the digital path (Modha 2018). The mixed approach is, for example, investigated by research groups working on this subject (BrainScaleS 2015).

After this decision, technical implementations of neurons and synapses have to be created. One of the most important components used in neuromorphic systems is the so-called memristor. This word combines "memory" and "resistor," which actually describes its function completely – the component has a "memory," changing its resistance depending on its history. Similar to normal resistances, capacitors, and coils, the memristor is a passive component. It was suggested to exist in 1971 (Chua 1971), but was found in reality only in 2007 (Wang et al. 2007). The basic principle of a memristor is that due to a current flow, chemical modifications occur in the component that also change the resistance, an effect that can be completely reversed by a current flow in the opposite direction (Strukov et al. 2008). Interestingly, memristors need significantly less power than common DRAMs, can be densely packed, can be processed with recent semiconductor technology, and are nonvolatile (Gutmann 2008).

Because of this "analogue" behavior, enabling learning by the described resistance change, combined with their energy efficiency, it is obvious that memristors are interesting components for the technical implementation in neuromorphic systems. Thus, a large number of publications exist on using memristors as artificial synapses, applying diverse materials and architectures (e.g., Hu et al. 2013; Ziegler and Kohlstedt 2013; Wen et al. 2013; He et al. 2014; Cai et al. 2015; Covi et al. 2016; Hu et al. 2016). MRAMs have also been investigated as multistate cells for neuromorphic computing (Rzeszut et al. 2022).

Another approach to implementing "learning" in neuromorphic systems is using phase-change materials as synapses, sometimes also neurons (Suri et al. 2012; Jackson et al. 2013; Lee et al. 2014; Zhong et al. 2015; Kang et al. 2015). Optical and photonic materials were used as synapses (Rietman et al. 1989; Gholipour et al. 2015; Maier et al. 2016) as well as neurons (Romeira et al. 2016) and some other common electronic components.

Most interesting, within the scope of this book, is the utilization of spin-based devices in neuromorphic systems. This idea was first introduced in 2013, followed by a broad range of scientific literature investigating different spintronic elements (Roy et al. 2013; Roy et al. 2013a; Roy et al. 2014; Zhou and Chen 2021).

This approach is also of high interest since spintronics elements allow for being used as neurons, synapses, and even complete networks. Typically, either magnetic domain walls are utilized, or spin waves, or diverse spin-related components, such as the spin-transfer torque (Roy et al. 2015; Sengupta et al. 2016; Fong et al. 2016; Sengupta and Roy 2018). Interestingly, it is even possible to utilize spintronic devices as basic building blocks of neuromorphic systems, showing synapse as well as neuron functionalities (Sengupta et al. 2016a).

An interesting approach is the use of antiferromagnetic materials that are mostly only used as partners of ferromagnets in exchange bias systems. Recently, possibilities were discovered to switch and detect antiferromagnetic Néel order electrically. Antiferromagnets are of special interest for neuromorphic systems because they have not only the typical two levels of a ferromagnet, but a multitude of levels, which may enable inserting the idea of a learning system. Additionally, they can switch very fast, as compared to ferromagnets, which suggest using them for memory technologies (Zelezny et al. 2018).

Another material with tunable properties is graphene. Since it shows high mobility and low spin–orbit coupling, graphene belongs to the materials often investigated for spintronic applications. For neuromorphic computing, its tunable spin resistance is of high interest, again enabling "learning" by changing weighting factors. An important factor that has only recently been demonstrated practically is the possibility to control the areas of spin injection and spin detection independently. In a recent work, this was achieved by separate gate electrodes for both areas, resulting in locally different spin transport parameters that can be used as a position-dependent resistance network (Anugrah et al. 2018).

In Ta/CoFeB/MgO heterostructures, probabilistic switching occurs in the presence of a spin–orbit torque and thermal noise. While such behavior is unsuitable for common computer technology, probabilistic output in dependence on the input amplitude is a typical element of neuromorphic systems. This spintronic device allows for direct mapping of the function of such a controllable stochastic switch (Shim et al. 2017).

MTJs were shown to be usable similarly to the multilevel weighting in some classes of neuromorphic architectures if written probabilistically. Special MTJs could even show the behavior of "stochastic resonance," which is used in some biological systems for the detection of weak signals (Locatelli et al. 2015).

Another approach is based on the oscillator-like behavior of real neurons in the brain. Thus, it may be useful to model a neuromorphic system with nanoscale nonlinear oscillators as neurons. This approach is complicated due to the noise and the resulting lack of reliability of data processing in nano-oscillators. Recently, MTJs have been shown to be usable as spintronics oscillators, allowing for spoken-digit recognition with high accuracy. In combination with the interaction possibilities of spintronics oscillators, low-energy consumption, and high durability, MTJs may be one way to establish neuromorphic systems based on large networks of such oscillators (Torrejon et al. 2017; Vatajelu and Anghel 2017). MTJs also have the significant advantage of providing nonvolatile memory, which means that memory and processors can be combined (Grollier et al. 2016). Using multiple vertically stacked or serially connected MTJs, compound spintronic synapses were built that showed multiple stable, designable resistance states. Combined with a similarly built compound spintronic neuron, a complete neural network could be created (Zhang et al. 2016, Rzeszut et al. 2019).

Alternatively, MTJs were used in combination with a heavy metal to gain a stochastic binary synapse. Learning was enabled by stochastic switching of the different states of the MTJ that was triggered by the activities in the connected neurons. In this configuration, it was also shown that a combination of long- and short-term stochastic approaches could be used to support the learning process, which was demonstrated by the classification of hand-written digits (Srinivasan et al. 2016).

Digital recognition of handwriting was also used to underline the suitability of a similar approach. Here, multiple MTJs were connected in parallel, operating in a stochastic way and thus jointly behaving like a single synapse, opposite to one binary MTJ. A circuit based on such MTJs as well as stochastic neurons showed high accuracy and reliability under modifications of the device (Zhang et al. 2016a).

As a practical example of a new material eventually suited for neuro-inspired computing, Fig. 7.11 depicts a ferromagnetic nanofiber network with functionalities resulting from its internal physical properties. The inspiration comes from the possibility of creating such networks by electrospinning magnetic polymers or magnetic nanoparticles in nonmagnetic polymers.

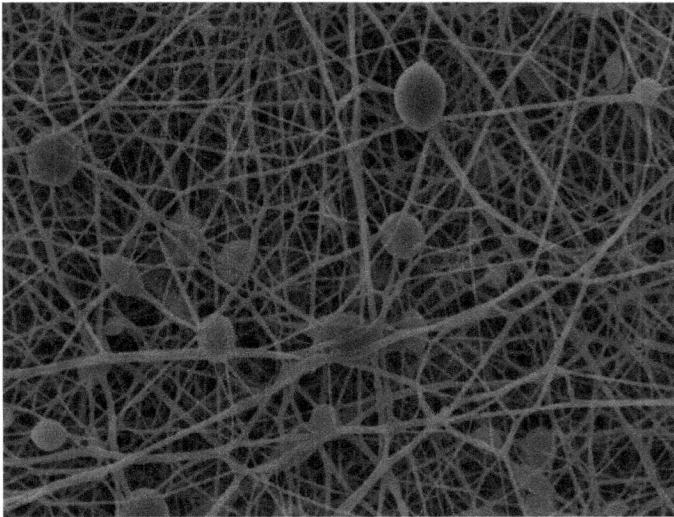

Fig. 7.11: Scanning electron microscopy image of an electrospun polymer-based magnetic nanofiber mat as a potential material for a bioinspired signal-processing environment (modified from Döpke et al. (2019)).

On the basis of the recently accessible knowledge, the following fundamental bio-physical facts that may be important for future research inspirations can be mentioned:

1) there are two basic types of neuro-signals (which is energetically logical): short range ones, chemical in nature, acting between cells, and long range ones, electrical in nature, propagating between different "layers" of the overall neuro system;

2) an elementary bio-cell, the local processing unit or neuron, has the ability to modify the number of local connections with the synapse to enhance or reduce the signal transfer efficiency;

3) the axon connection between two neurons is structurally asymmetrical; there is a single solid coupling with the one neuron (the cause), while the connection with the other one (the effect) is split and the connection efficiency is adaptable in this way;

4) a flow of information is managed by a sorting system covering everything and determining roads for information transfer. These system operations are nonlocal like the human brain;

5) in the neuro system, a given state can be triggered by a series of impulses or spikes instead of a large enough amplitude of a single-step signal.

All the above can be a reason to formulate the following neuro-inspired signal processing postulates in spintronic materials (Fig. 7.12):

1) the elementary unit of information, the local spatial physical property, can be a given domain state possessing its unique spatiotemporal dynamics;

2) the neuro-inspired signal processing can be based on spatial wave-like propagations of local properties in magnetic fibers, the transport of magnetic states in "neurons";

3) the notion of information can be coded in frequency and phase domains;

4) the output should be based on several inputs – logics should be based on multivalued threshold events, similar to the synapse functionality;

5) the input should know about the output – the feedback property;

6) the processing unit, the local magnetoelectronic device, should be adaptable to situations, thus, to the inputs and the outputs;

7) the processing units should be doubled, tripled, or even more multiplied for the same task, corresponding to the redundancy postulate.

As an example of the practical realization of the aforementioned postulates, the propagation of magnetic domains, treated as a signal in the GHz range of frequencies, traveling at the speed of the order of 10^2 m/s, is presented below.

The set of images was created using micromagnetic simulations in permalloy fibers of a rectangular cross section (10 nm × 60 nm) as shown in Figs. 7.13 and 7.14. The source of the in-plane (x–y is the plane of the book page) rotating field was located on the left-hand side of the fibers within a region of 60 nm × 50 nm. The field amplitude equals 1 T (the induction) and the frequency of rotation is 5 GHz.

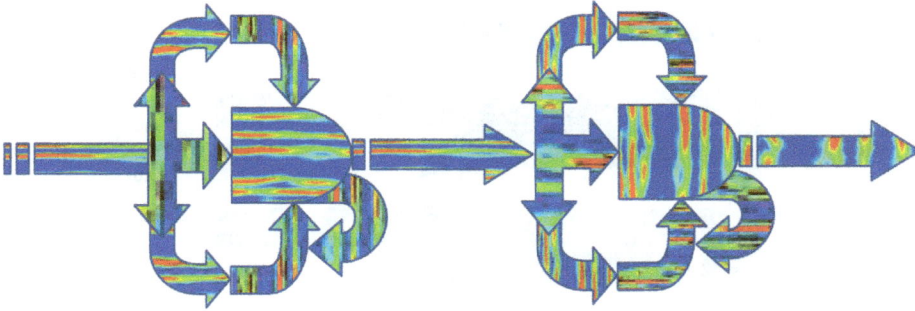

Fig. 7.12: The conception of neuro-inspired signal processing on the basis of magnetic state propagation (cf. the postulates below).

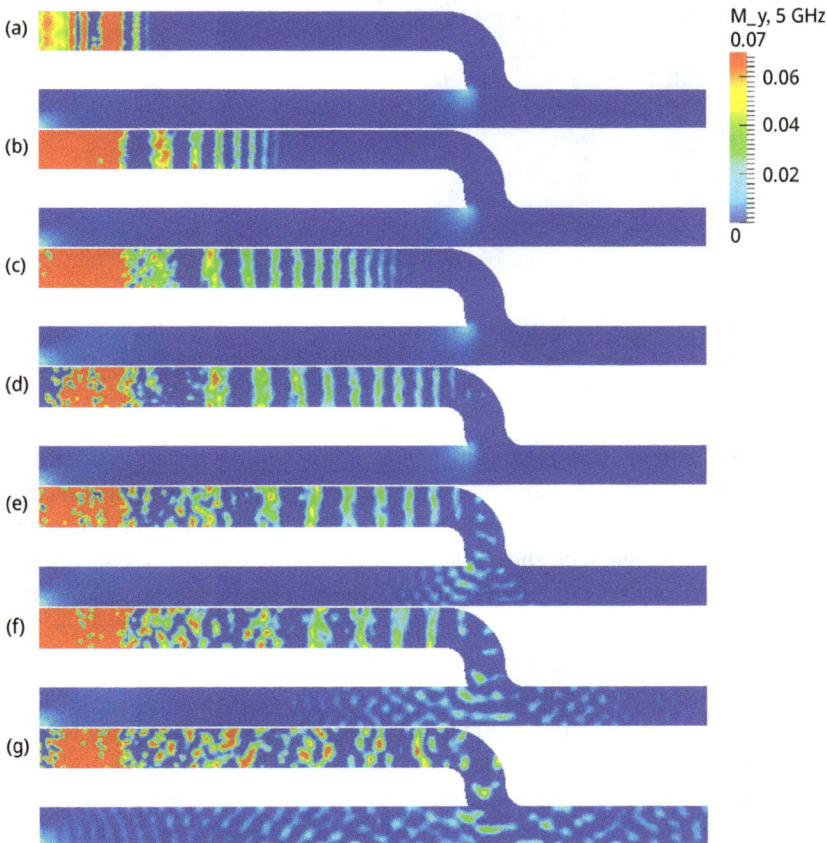

Fig. 7.13: The distribution of magnetization (M_y component) in the permalloy coupler of fibers with source being a rotating magnetic field located on the left-hand side of the upper arm. The periodical signal is split at the region of connection. The width of the single fiber equals 60 nm.

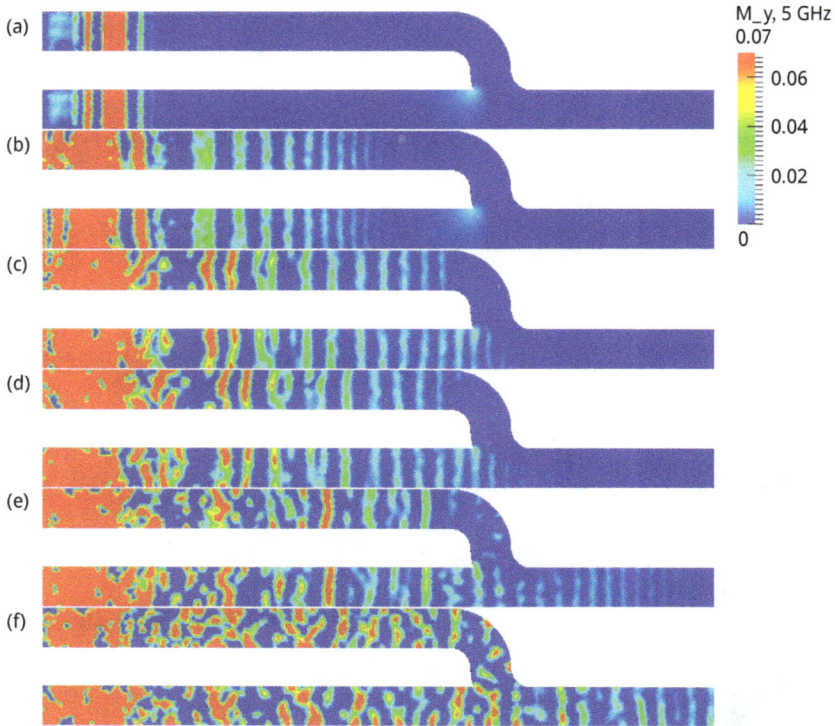

Fig. 7.14: The distribution of magnetization (M_y component) in the permalloy coupler of fibers with two sources of in-phase rotating magnetic fields located on the left-hand sides of the arms. The periodic signals interfere at the region of connection. The width of each fiber equals 60 nm.

Another example of a more complex system that mimics the synapse structure is shown in Figs. 7.15–7.17. Here, the obtained distribution of magnetization also results from the existence of local sources of signals, while some parts of the structure are passive and others are active, with the evident propagation of signals. In Fig. 7.15 the distribution of magnetization is in passive, nonpropagating mode. In Fig. 7.16, for the presented M_z component of magnetization, three skyrmions are seen. Figure 7.17 presents regions of active propagation of magnetic signals. The dimensions of the structures are the same as those for Figs. 7.13 and 7.14.

As shown in these few examples, spintronics is an interesting technology for the area of neuromorphic computing that may pave the way to new hardware and software architecture. Bioinspired computing based on magnetoelectronic realizations is still at the preliminary step of development; however, it seems the most challenging topic for future research.

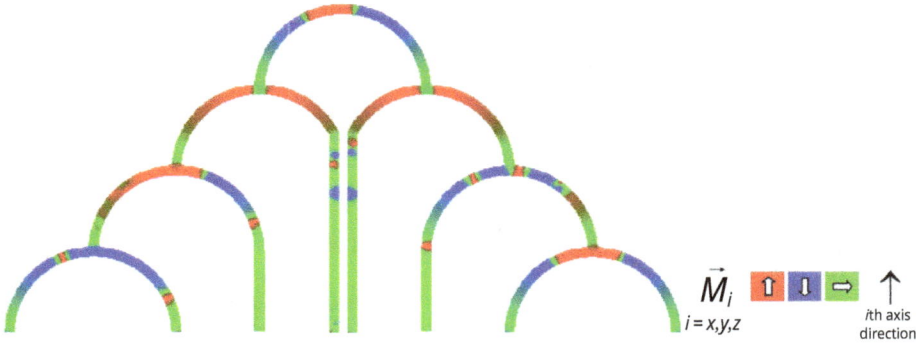

Fig. 7.15: The distribution of magnetization (M_x component) in a ferromagnetic synapse. The system is in the passive mode with no propagation of signals. The meaning of magnetization component is shown on the right-hand side. For example, if the M_x component is parallel to the direction of x-axis, then the color is red.

Fig. 7.16: The distribution of magnetization (M_z component) in a ferromagnetic synapse. The system is in the passive mode with no propagation of signals. The meaning of the magnetization component is shown on the right-hand side. For example, if the M_z component is parallel to the direction of the z-axis, then the color is red. In the figure three skyrmion states are visible.

7.13 CMOS-compatible spintronics and hybrid solutions

One of the main problems of using spintronics elements in computers is based on re-cent producers of computer parts having the equipment necessary for producing con-ventional computer parts. Changing the materials in use or the technologies for production of new elements always means changing a machine – which creates a large market barrier for new spintronics solutions.

The easiest and most logical way to overcome this barrier is by finding spin-tronics solutions that are compatible with recent charge-based technologies or that

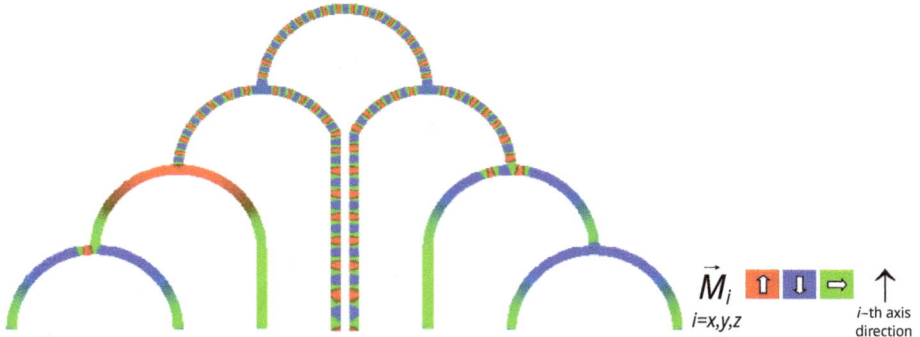

Fig. 7.17: The distribution of magnetization (M_x component) in a ferromagnetic synapse. The system is partially in the dynamic mode with evident propagation of signals in the central region of the structure. The meaning of magnetization components is shown on the right-hand side. For example, if the M_x component is parallel to the direction of x-axis, then the color is red; if it is antiparallel, then the color is blue. In the case of perpendicular orientation the color is green.

can unambiguously be blended with them. Thus, searching for CMOS-compatible spintronics elements or hybrid solutions is one of the large challenges in recent spintronics research.

In general, silicon can be used for spintronics applications due to the long spin lifetimes possible in this material, which can still be increased by applying uniaxial stress. On the other hand, the spin injection efficiency is low, and electrical manipulation of the spins is complicated (Sverdlov and Selberherr 2015). These problems must be overcome to enable CMOS-compatible spintronics devices. Another issue that should be mentioned is that with ongoing miniaturization of CMOS technology, the critical switching current density for MTJs need careful scaling, too, to allow for switching elements prepared according to the 16 nm technology node or even smaller (Dorrance et al. 2012).

MTJs, for example, can be built on CMOS logic circuits. The MTJ can store computation results as a buffer, while the logic operations are still performed by the CMOS elements, or the MTJs are used as the main computing elements, making the architecture of the logic circuit much easier than in the hybrid solution. With MTJs, MRAM cells can also be integrated to combine logic and nonvolatile memory in one circuit (Makarov et al. 2016). The idea of integrating nonvolatile memory into circuits aims at a reduction of the power nowadays necessary for refreshment cycles and initializing data in recently unused circuit parts (Sverdlov et al. 2017). Additionally, loss of data in the case of an unintentional shutdown can be avoided. For this application, current-induced domain-wall motion can be used as a fast, low-power nonvolatile switching mechanism with high data density. For this, a magnetic flash analog–digital converter was developed that was faster and more efficient than common CMOS solutions and nevertheless compatible with CMOS technology (Upadhyaya et al. 2016).

A hybrid solution combining magnetic and CMOS properties was used to store and process data by magnetic and electrical fields. This nanoscaled cell, containing a spin-transfer torque magnetic MTJ, could be applied in MRAM as well as for reprogrammable computing and showed promising values of read/write speed, energy consumption, and storage density (Jovanovic et al. 2014).

On the other hand, MTJ-based elements can be used for diverse other applications, such as magnetic field sensors (Lu et al. 1997; Lacour et al. 2002; Leitao et al. 2016) or random number generators (Fukushima et al. 2015; Yuasa et al. 2013).

Spintronics oscillators can also be integrated in CMOS circuits. Spin-torque oscillators, for example, are quite small elements compared to a voltage-controlled oscillator prepared in CMOS technology (Makarov et al. 2016).

Another way to deeper integrate CMOS and spintronics is by creating a magnetic shift register (Windbacher et al. 2013; Zhao et al. 2011) or a nonvolatile magnetic flip-flop (Windbacher et al. 2013a; Windbacher et al. 2013b; Windbacher et al. 2015; Windbacher et al. 2015a; Windbacher et al. 2015b, Jung et al. 2014; Ryu et al. 2012).

More recently, Song (2018) suggested a realistic switch compatible with CMOS technology. While previous experiments on colossal magnetoresistance or magnetoresistance, finding values of several orders of magnitude, were based on large magnetic fields and very low temperatures (Jin et al. 1994; Ramirez 1997; Ali et al. 2014; Gao et al. 2017), the new approach reached similar values without magnetic field and at moderately low temperatures (liquid nitrogen or liquid helium, respectively). The respective device uses yttrium iron garnet strips on graphene, applying two gate voltages, which results in full-polarization reversal in the free ferromagnet and thus a switch from metal-like to insulator-like conductivity.

While switching MTJs by spin-transfer–torque or spin-orbit–torque interactions is typically limited to some nanoseconds, all-optical magnetization switching has long been known to be much faster, with achievable switching times below picoseconds (Banerjee et al. 2021). For this purpose, the inverse Faraday effect is applied, using circularly polarized light on the sample. It is possible to use CMOS-compatible plasmonic materials to reverse the magnetization in nanomagnets, that is, materials allowing for the confinement of light on dimensions are smaller than the actual diffraction limit, which is necessary for the manipulation of magnetization on the nanoscale. Using localized surface plasmon resonances, light can be coupled to nanomagnets, resulting in severely larger magneto-optic constants and thus a larger influence of the impinging light on the magnetization reversal than with usual freely propagating light (Dutta et al. 2017).

A CMOS-compatible spin field effect transistor with voltages applied at the gate and drain was theoretically investigated, using a Monte-Carlo simulation especially developed for modeling spintronics devices including spin transport (Thorpe et al. 2017). Here, the spin transport in the gate was simulated. The model showed a nonuniform decay of the overall magnetization between the source and drain. Applying a high electric field between the gate and drain resulted in magnetization recovery.

Thus, the spin polarization of the drain current could be triggered by the gate and source–drain voltages.

Similarly, a CMOS-compatible spin-transfer-torque MOSFET prepared from a Heusler alloy showed high endurance and short writing times. In addition, it could be reconfigured. In this way, field programmable gate arrays could be created with spintronics elements that outperformed conventional CMOS-based components in terms of magnetocurrent and necessary area (Saito et al. 2011; Saito et al. 2011a).

Another approach is based on spin waves that can have very short wavelengths on the nanoscale, allowing for producing correspondingly small devices. The wave-vector-independent detection of spin waves with wavelengths around 150 nm was enabled by the inverse spin-Hall effect in ultrathin waveguides. The material systems used for this technology are compatible with semiconductor circuits and spintronics, allowing for the creation of links between spin waves and CMOS technology (Brächer et al. 2017).

An interesting topic related to spin waves is their parallel parametric amplification, meaning that microwave photons are converted into magnons at half the microwave frequency without changing their phases. In this way, spin waves can be excited in a broad range of wave-vectors, and even pairs of waves with correlated phases can be created. To build magnonic networks, this parallel parametric amplification technology must be applied to CMOS-compatible structures. An application to metallic microstructures was recently reported, making a large step toward the practical application of magnonics (Brächer et al. 2017a).

Different frequencies are applied in RF components that are often used in information and communication technology. CMOS-compatible RF devices can be used for microwave signal generation and detection as well as signal modulation (Ebels et al. 2017).

Besides these typical spintronics devices, some researchers concentrate on increasing the compatibility on the material side. Recent electronics is typically based on silicon (Si), while germanium (Ge), for example, is a possible candidate for the integration of spin quantum bits into Si electronics. Usual Ge/Si heterojunctions, however, have the disadvantages of misalignments in epitaxy and an undesired band alignment. Recently, these problems were solved by growing a 2D electron gas in Ge quantum dots on Si, showing significantly longer spin lifetimes and the possibility to tailor the electron g-factor over a broad range that was impossible in previous Si-based spintronics devices (Giorgioni et al. 2016).

Another approach is based on patterned nickel nanowires, electrodeposited in nanoporous polyimide membranes that are integrated on silicon wafers. Polyimide is often used in microsystem technology, and nanopores can be ion-etched into this material. In this way, nickel nanowires of 20–60 nm diameter were created with lengths around 2 μm. This approach offers a new possibility to create complex nanostructures for miniaturized electromechanical systems (Walewyns et al. 2011).

As shown in these few examples, development of CMOS-compatible spintronic elements or CMOS/spintronics hybrid systems is of great interest for the industrialization of spintronics elements. Besides this direct integration of spintronics elements into existing CMOS devices, however, more possibilities exist to use spintronics devices and their physical effects for special applications, a few of which are described in the next sections.

7.14 Quantum interferometers, Aharonov–Bohm, and Fano effects

Investigating quantum effects in spintronics devices may lead to interesting and sometimes technologically usable features. One of them, for example, is the so-called quantum beat.

Beats always occur when two harmonic oscillations with very similar frequencies are superposed, resulting in a new more or less harmonic oscillation with increasing and decreasing amplitude, the beat. Quantum beats occur in quantum dots, quantum wells, or hetero-structures due to an interference of quantum states, for example, electron spins (Heberle et al. 1994). Spintronics necessitates long coherence lengths, which means that there should be possibilities to observe interfering spin waves.

The Aharonov–Bohm effect is one of the effects that influence electrons and result in an unexpected modification of the interference. Here, electrons are guided along two possible paths in a double-slit experiment and interfere on an observation screen. By switching on a magnetic field between both paths, this interference pattern is modified – however, the magnetic field and the electron's wave function do not considerably overlap (Aharonov and Bohm 1959; Aharanov and Bohm 1961; Batelaan and Tonomura 2009). Similarly, quantum interference in a loop structure could occur due to the Larmor precession of the spin (Tralle 1999). This has been examined for static (Tralle and Paśco 2003) as well as time-dependent magnetic fields (Tralle and Paśco 2003a). In the latter case, the spins are treated as in the ballistic regime, that is, the phase relaxation length is larger than the structure length. Here the transmission coefficient was found to be proportional to the cosine of the phase difference of the electrons, which was attributed to the 4π (720°) symmetry of the spinors. As a possible application of this finding, the authors suggest developing spin-based SQUIDs, allowing for measuring even smaller deviations of magnetic fields than with recent instruments.

Similarly, the Aharonov-Bohm effect was used to develop a spin-flip mechanism. Controlling spins of electrons transported through quantum wires and other 2D geometries by Aharonov–Bohm fluxes allows for influencing the magnetoconductance in the ballistic regime (Frustaglia et al. 2004).

In a theoretical work, Zhang et al. (2005) studied the transmission coefficients as well as spin currents in a triple-arm quantum-ring, including spin–orbit coupling in their calculation. They found that spin–orbit coupling induced an effective flux in the quantum ring and could thus introduce persistent spin currents as well as increase or decrease them and even switch the direction of the spin currents by changing its intensity, in this way allowing for controlling the spin currents.

Another study went one step further and included a quantum dot in a mesoscopic Aharonov–Bohm ring, applying a magnetic field to the quantum dot (Wei et al. 2005). This magnetic field significantly influenced the density of states for the different spins in the quantum dot. In addition, the spin-dependent transmission amplitudes and phases showed unexpected effects, such as a peak in the transmission phase.

With a two-quantum-dot setup, Pan and Lue (2008) investigated the influence of the spin–orbit coupling on the spin polarization in an Aharonov–Bohm interferometer. They found that the amplitude and sign of the spin polarization in the quantum dots could be tailored by the magnetic flux, gate voltage, and bias. It was even possible to create opposite spin polarizations in both quantum dots, showing the broad range of spin manipulations that can be enabled by such a setup.

On the other hand, using an Aharonov–Bohm loop with four leads enabled electrical control of the charge and the spin and thus the possibility to create either pure spin currents or pure charge currents. It should be mentioned that in this special system, negative conductance values were also found (Dolcini 2011).

The Aharonov–Bohm effect can also be used to create pure spin currents. In combination with the Aharonov–Casher effect, this was achieved in a mesoscopic ring with a tunnel barrier on which an Aharonov–Bohm flux along the ring axis was applied. By parametric pumping of spin and charge currents, a pure spin current could be created (Citro and Romeo 2006).

In addition to good spin filtering properties, a large TMR was found for properly adjusted system parameters in a closed Aharonov–Bohm interferometer, making these systems possible candidates for spin valves and other spintronics applications (Wu and Sun 2007). A system of three quantum dots was theoretically examined, resulting in a spin valve that could be tuned by the magnetic field (Tosi and Aligia 2011). A new spintronics device, the so-called topological spin transistor, could be created by an Aharonov–Bohm ring of a material with special physical properties and enabled the control of spin rotation by a magnetic flux (Maciejko et al. 2010). Connecting quantum dots with thin wires in the form of a diamond and using this system in an Aharonov–Bohm ring leads not only to full spin polarization, with the polarization orientation being tunable by electric and magnetic fields but also to the creation of a spin field-effect transistor (Aharony et al. 2011; Matityahu et al. 2013, Zhu et al. 2014).

Another effect that can sometimes be found in the same systems as the Aharonov–Bohm effect is the Fano effect, describing an asymmetry in resonant scattering, for example, of electrons (Fano 1961). The spin-dependent electron transport through parallel coupled quantum dots in an Aharonov–Bohm interferometer with an asymmetrical connection to the leads, for example, shows a spin-dependent Fano effect, which may be used for spin manipulations in spintronics applications (Chi et al. 2008).

In a similar system, Wu (2009) and Wu et al. (2010) found that the magnetic flux and the spin polarization influenced the conduction in the closed Aharonov–Bohm interferometer, while the conductance line shape was mostly affected by the magnetic flux. In this system, not only was a large tunneling magnetoresistance (TMR) was found, but even negative values could be achieved by carefully adjusting the system parameters.

Electron charge and spin can be pumped through an Aharonov–Bohm ring with ferromagnetic leads or one ferromagnetic and one nonmagnetic metal lead by an oscillating electric field along the quantum dot. It was found that in the second situation, charge and spin currents can be pumped in antiparallel directions (Pan et al. 2010).

As these examples have shown, Aharonov–Bohm rings can solve one of the most important problems in spintronics, the creation of ideally spin-polarized currents (Majhi and Maiti 2023). The second point that should be underlined – although it is not as often mentioned here – is the possibility to use a spin filter not only for production but also for the detection of spin polarization.

The following section will introduce some new ideas of a possible application of spintronics in the research and development area of life sciences.

7.15 Magnetic markers and biochips

MTJ sensors can be used, among others, to detect nanoparticles – not only in technical environments, but also in biological and medical applications. Magnetic nanoparticles can be introduced into the body of an animal or a man, allowing for monitoring their path through the organism. In addition, they can be directly attached to proteins or other biomolecules and trace their way. This necessitates highly sensitive detectors that should ideally also be inexpensive and transportable (Piedade et al. 2006; Liang et al. 2024).

To reach these goals, recognition assays can be created using the GMR or the TMR effect (Graham et al. 2004; Wang et al. 2005; Shen et al. 2005), with MTJs enabling larger magnetoresistance values, that is, a better sensitivity for the small magnetic fields (usually below 1 Oe) expected in biological applications. MTJs have the additional advantages of needing two orders of magnitude lower power and being nearly three orders of magnitude less expensive than a SQUID as the typical highly sensitive measurement instrument for magnetic fields. Finally, typical MRT materials can even be applied on flexible substrates which broaden the range of possible applications (Freitas et al. 2016).

Using such an MTJ, it was possible, for example, to trace DNA labeled with magnetite nanoparticles of only 16 nm diameter (Shen et al. 2008). Improving such sensors further necessitates noise reduction and/or an increase of the detection level. This can be reached by a reduction of the sensor area if magnetic fields should be measured that are varied on short distances (Freitas et al. 2016). Miniaturization offers another advantage, that is, increasing the sensor resolution further into the submicron region to combine magnetic sensing with microfluidic devices (Cardoso et al. 2017).

Another attempt to build such biosensors with high sensitivity is using magnetic nanowires, for example, from permalloy, with receptor sites for magnetic nanoparticles in an aqueous solution. For this situation, micromagnetic simulations showed that the changes in AMR occurred near the receptor sites at the bridges between neighboring nanowires (Will et al. 2015).

The shape of the magnetic labels also influenced the magnetoresistive changes, with square nanoparticles seeming to be superior to other shapes (Bowerbank et al. 2012). In another investigation, round nanoparticles were found to be superior to square or triangular ones (Will et al. 2014). The latter shape may also be advantageous to avoid undesired agglomerations of nanoparticles in a fluid since in round nanoparticles, typically magnetic vortices are formed that avoid large stray fields.

Nowadays, such platforms for the detection of biomolecules, based on MTJ, are already commonly available. After showing the first prototype in 2009 (Germano et al. 2009), for example, the plugged-in electronic platform "MR-biochip" was recently used to sense nanoparticles used for tuberculosis detection even quantitatively due to the proportionality between the respective bacteria and the magnetoresistive signal. In this way, a low-cost test could be established, which would be of special importance in poorer nations (Barroso et al. 2018). Similarly, the "magnetic sifter" is another microfluidic chip with a dense magnetic pore array in which tumor cells are captured, labeled with magnetic nanoparticles, and afterward characterized with a GMR nanosensor (Earhart et al. 2014).

Another possibility to integrate spintronics in biochips is using the planar Hall effect of spin-valve structures. This effect occurs in ferromagnetic materials and means that the AMRs change if an external magnetic field is applied in the sensor plane, opposite to the normal Hall effect where magnetic field components perpendicular to the plane are detected. Such planar Hall effect sensors usually show an electronic readout proportional to the applied magnetic field, while the latter scales with the bead concentration (Dalslet et al. 2011). Here, large ratios of the Hall voltage in relation to the measured magnetic field are necessary to gain sufficient sensitivity to use such spin valves as spintronic biochips (Tu et al. 2009).

Using such a planar Hall effect sensor, magnetic nanoparticles with moments smaller than 10^{-5} emu could be detected, a value that was reached, for example, by maghemite nanobeads of less than 0.2 μg (Volmer and Avram 2013).

While this chapter gave an overview of spintronics effects and possible applications of spintronics elements, the next one will investigate the influence of external influences such as temperature, light, or electric fields.

8 Influence of external factors and physical fields onto spintronic devices work

In laboratory experiments, we are typically used to excluding all possibly disturbing external influences. Thinking of spintronics in possible future applications, however, we must also take into account such possible external influences, from temperature changes or electric fields to mechanical impact. On the other hand, writing information is known to be supportable in most cases by increasing the temperature of the bit that is to be changed.

This chapter gives an overview of typical external factors influencing the function of spintronics devices in a desired way or possibly disturbing it.

8.1 Temperature-supported switching of elements

Several approaches exist to support magnetization switching by heating. For MRAMs, the required thermal stability of stored information necessitates a large energy barrier and correspondingly higher anisotropies and higher switching fields, resulting in larger writing power consumption (Prejbeanu et al. 2007).

Beech et al. (2000) tried to overcome this problem by Joule heating a GMR bit, using a current pulse through the memory cell and the word line. In this way, the temperature of the FeMn pinning layer was increased above its ordering temperature, allowing for switching the pinning direction by a magnetic field generated by a current in the word line. Similarly, Daughton and Pohm (2003) suggested overcoming the problem of desired thermal stability and low drive currents.

In a slightly different approach, Abraham and Trouilloud (2000), as well as Dieny and Redon (2001), used the direct current flow through the magnetic tunnel junction for heating it. This process is more efficient than the indirect diffusive heating in the previous method.

Microwave-assisted magnetization switching was investigated by Moriyama et al. (2007). In a magnetic tunnel junction with a coplanar waveguide, they showed that the coercive field of a permalloy layer was significantly decreased by a microwave of defined frequency and power. Microwaves were also used by Hayashi et al. (2012) to inject domain walls into magnetic nanowires.

Several theoretical and experimental approaches aimed at investigating the influence of the current applied to the free layer and the critical current of spin-torque switching on the switching probability of the magnetization in a thermally activated region. Koch et al. (2004) developed a mathematical formula to describe this correlation, which was used for the evaluation of different experimental investigations (Hayakawa et al. 2008; Yakata et al. 2009). Modifications of the original formula were suggested later on by different groups due to inconsistencies with their experimental

https://doi.org/10.1515/9783111383736-008

findings (Heindl et al. 2011; Butler et al. 2012; Taniguchi and Imamura 2012; Taniguchi and Imamura 2012a).

In spite of possible problems with the so-called resistive-open defects in MRAM, that is, imperfect circuit connections (opposite to stuck-open defects, in resistive-open defects the connection is still existent but shows a wrong resistance) (Azevedo et al. 2012), thermally assisted switching is a promising technology to create reliable and strongly miniaturized MRAMs, so-called TAS-MRAMs (thermally assisted switching MRAM) (Zhao et al. 2009, Rao et al. 2021, Ren et al. 2021).

Similarly, heat-assisted magnetic recording (HAMR) belongs to the promising technologies to reach high areal densities in perpendicular magnetic recording (Rott-mayer et al. 2006, Hsu and Victora 2022). This technique is based on laser heating of the magnetic medium to reduce its coercivity. Several challenges have to be addressed in HAMR media, such as the necessity that the materials can withstand high temperatures, that the magnetic grains are highly ordered and thermally anisotropic, that heat flow is controllable on the nanometer scale, and so on (Kief and Victora 2018). Because of the small dimensions of bits, it is additionally necessary to reduce the laser spot to a region below the diffraction limit, typically using a so-called near-field transducer (Wang and Komvopoulos 2018).

Recently, Datta and Xu (2018) discussed possibilities to increase the performance of near-field transducers. They underlined the importance of the antenna design as well as the possibility of exciting such near-field transducers at larger wavelengths, resulting in different possibilities to optimize them. Diverse near-field transducer designs have been examined since (Katayama and Sugiura 2020, Venuthurumilli et al. 2021).

Another challenge is keeping the near-field transducer relatively cool. The surface plasmon polaritons are concentrated at the tip of the magnetic write head, using a plasmonic waveguide with the near-field transmitter at the end, to heat the recording area without heating the near-field transmitter too strongly. Nakagawa et al. (2018) developed protection layers, for example, from carbon in diamond structure, to keep the temperature of the near-field transmitter below 100 °C. Matlak et al. (2018) and Matlak & Komvopoulos (2018) investigated different layered coatings, such as amorphous carbon/SiN/NiCr, that were sequentially deposited on Au or FeCo base layers. These new coatings showed better stability to heating at 350 °C and may thus, after further optimization, offer a solution to the problem of thermal damage of the HAMR heads.

An optimization of such systems, in addition to near-field transducers, was recently suggested by Deng et al. (2018). They designed a nanocavity to enhance the light–matter interaction in a thin FePt film inside an $Ag/SiO_2/FePt$ trilayer. By optimizing the thickness of the middle layer, they could reduce light reflection by more than a factor of 2, thereby strongly increasing the efficiency of the laser heating of FePt.

A possible alternative to near-field transducers that suffer from a complex design and correspondingly high production costs is plasmonic mirrors. Using a planar plasmonic mirror without small or complex geometries, Ogut et al. (2018) could focus surface plasmons to small optical spots, one to two orders of magnitude smaller than the light wavelength. With a parabolic plasmonic mirror, short-wavelength surface plasmons could be excited, depending on the refractive index and the numerical aperture of the parabolic mirror.

Other spintronics systems can also be supported by heat-assisted processes. Narayanapillai et al. (2014) investigated thermally assisted domain wall generation in trilayer nanowires and examined magnetization reversal by the anomalous Hall effect. They managed to generate a defined number of domain walls at defined positions in the nanowires. This method is naturally useful for materials with strongly temperature-dependent coercive fields, enabling the support of magnetization switching by a heat pulse.

Depinning of domain walls from single defects was shown by Attané et al. (2006). They introduced a nanoscale defect in a wire and found that the relaxation times of the domain wall showed a statistical distribution, an effect that cannot be neglected in spintronics applications. Nevertheless, Ravelosona et al. (2005) suggested thermally assisted current-driven domain wall motion as an efficient method to control domain wall motion on the nanometer scale.

Another idea was tested by Wang et al. (2017). They used phase-locked laser pulses from two strong lasers to create a transient optical grating and a spatiotemporal temperature distribution. In this way, they produced elastic as well as heat waves along the surface and inside the sample. They found that the spin current induced by both effects propagated along larger distances than expected for the pure thermal effect, making this procedure usable for detecting spin caloritronics effects (cf. next section).

8.2 Spin caloritronics

While most of the effects described in the last section are based on the idea of supporting magnetization switching by adding thermal energy at a well-defined position, spin caloritronics describes investigations of thermally driven spin fluxes, that is, spin transport and other spin-related phenomena outside the thermodynamic equilibrium.

Generally, spin and heat can be correlated in different ways. Magnetization and heat can be transported commonly by magnons, that is, spin waves, while conduction electrons transport spin, charge, and entropy together. Using thermal excitation, larger spin fluxes can be driven than in other ways, for example, by ferromagnetic resonance or electrical pumping. The first thermodynamic description of spin-dependent transport was given by Johnson and Silsbee (1985), who also investigated thermodynamic processes at interfaces (Johnson and Silsbee 1987, 1988). Some of the effects examined in spin caloritronics are described in this section.

The Seebeck effect and the Peltier effect are often summarized as the thermoelectric effect; the Thomson effect, as an extension of the models of Peltier and Seebeck, is also typically included in this definition. The Seebeck effect describes the conversion of heat into electricity using an appropriate device (Seebeck 1822). If two different conductors or semiconductors form an electrical circuit, and the contact points between the materials have different temperatures, then a thermovoltage is created that is approximately proportional to the temperature difference and to the difference of the Seebeck coefficients that are usually given with platinum as the reference value of 0 µV/K. In this way, thermocouples can measure temperature differences, or by exchanging the voltmeter with a load, a thermocurrent can be generated.

The Peltier effect describes the opposite case – a current flow through the interface between two different conductors can heat or cool the junction (Peltier 1834). This effect is used in Peltier elements, in which one side is cooled by the current flow while the other one is heated.

In spintronics, instead of these classical effects, the spin Seebeck effect and the spin Peltier effect are of interest. The spin Seebeck effect, based on the coupling between phonons and magnons, can be used to efficiently generate a spin current using a longitudinal thermal gradient (Uchida et al. 2010). It was found not only in ferromagnetic structures but also in antiferromagnetic (Prakash et al. 2016), ferrimagnetic (Ramos et al. 2013), and even paramagnetic materials (Wu et al. 2015). Measurements are usually performed by coupling the material in which the spin Seebeck effect occurs to a heavy metal with strong spin–orbit coupling, in this way converting the spin current to a charge current using the inverse spin Hall effect, and measuring the small charge current using a nanovoltmeter (Wu et al. 2015a; Wu et al. 2016).

Interestingly, the origin of the spin Seebeck effect is not absolutely clear yet. First, the spin Seebeck effect was understood as a magnon spin current along the bulk of yttrium iron garnet (YIG) in YIG/Pt heterostructures, created by a temperature gradient (Rezende et al. 2014; Rezende et al. 2016). On the other hand, the thermal gradient was found to rise quicker than the spin Seebeck signal, indicating the importance of bulk magnon transport (Aldosary et al. 2016). Another investigation using time-resolved MOKE, however, indicated that spin accumulation was much too fast to include bulk temperature gradients (Kimling et al. 2017). Next, interface observations revealed a strong dependence of the temperature-dependent spin current on the interface properties (Guo et al. 2016). More recently, a significant difference between the spin Seebeck versus temperature curves for zero field cooling and field cooling was found, indicating the existence of a spin glass state near the YIG/Pt interface and again the importance of the interface structure between both materials (Wang et al. 2018). Other recent research concentrates, for example, on the spin Seebeck effect in magnetic tunnel junctions (Daqiq 2018), on coupling YIG to an ultrathin metallic antiferromagnet (Arana et al. 2018), or on modeling the spin Seebeck effect within a Boltzmann transport theory of coupled magnon/phonon transport (Schmidt et al. 2018).

The spin Peltier effect describes, again, the opposite case in which cause and effect are exchanged. Here, a spin current injected into a magnetic material results in a temperature difference (Flipse et al. 2014; Daimon et al. 2016). As expected, the spin

Peltier effect shows a similar temperature dependence as the spin Seebeck effect (Yagmur et al. 2018).

Following the idea of the Seebeck effect, the next step is thermal spin injection across the interface of a ferromagnet and a nonmagnetic metal. In the ferromagnet, the Seebeck effect is spin-dependent; thus, a heat current results in a spin current. At the interface between the ferromagnet and the nonmagnetic metal, the abrupt change of the spin current leads to a spin accumulation, with its scale depending on the relaxation lengths in the ferromagnet and the nonmagnetic metal (Slachter et al. 2010). Several research groups concentrated on this new possibility to inject spins thermally since electrical currents could be avoided in this way. Using a ferromagnetic CoFeAl electrode for thermal spin injection, a high spin-dependent Seebeck coefficient could be created and thus a very efficient thermally driven spin current (Hu et al. 2014), especially when combined with electrical spin injection (Hu and Kimura 2014). Alternatively, a temperature gradient across a tunnel contact can be used, for example, for spin injection from CoFe into the semiconductor Ge over an MgO tunnel barrier (Jeon et al. 2012).

Another possibility to influence magnetization by heat is the thermal spin torque. Reversing the magnetization orientation in a ferromagnet by a spin-polarized thermoelectric heat current, which is correlated to a spin-transfer torque, was predicted by Hatami et al. (2007). This effect means that a thermally induced torque works on the magnetization at the interface between a ferromagnet and a normal metal and was first discussed for a spin valve consisting of a ferromagnet/normal metal/ferromagnet stack. First theoretical calculations found a higher efficiency of the thermal spin torque for thicker magnetic layers and a higher probability to fulfill the desired prerequisites in spin valves containing magnetic semiconductors.

An interesting point about the thermal spin torque is its possible efficiency. Small temperature differences in the order of 1 K were calculated to result in spin torques in domain walls, which were two orders of magnitude larger than those created by the maximum possible electrical current density (Yuan et al. 2010). Temperature gradients of only 0.2 K/nm were as efficient as large charge current densities (Bauer et al. 2010). Experimentally, it could be shown that asymmetric spin valves are necessary to create a thermal spin torque, and that the effect of the heat current on the magnetization can be interpreted as coupling of heat, charge, and spin currents, corresponding to the transport thermodynamics described by the Onsager relations (Yu et al. 2010). In a metallic spin valve, a picosecond laser pulse was used to create intense, ultrafast heat currents that could be used to manipulate magnetization; the thermal spin torque could be tailored by the composition of the ferromagnetic layers and the heat sink layer's thickness (Choi et al. 2015). On the other hand, spin waves propagating over a single-crystal YIG thin film could be amplified by a transverse temperature gradient, which was attributed to the thermal spin torque counteracting relaxation (Padrón-Hernández et al. 2011).

While these first sections concentrated on the influences of temperature and heat on spintronics systems, the next sections will discuss the impact of mechanical influences, such as pressure or strain.

8.3 Pressure

One of the most interesting effects of pressure on different materials is the possibility to make them superconducting in this way. Simple iron is converted to a nonmagnetic state at pressures above 10 GPa (Cort et al. 1982; Taylor et al. 1991), but becomes superconducting at higher pressures between 15 and 30 GPa (Shimizu et al. 2001). Mazin et al. (2002) showed by first principles calculations of the electron–phonon coupling in this system that the conventional phonon-based mechanism can generally explain the superconductivity, but not the fast disappearance around 30 GPa, resulting in the idea of an unconventional magnetic origin of the effect, that is, spin fluctuations being responsible for the superconductivity.

Lithium, as the lightest metal, was predicted to become superconducting at high pressures with a relatively high transition temperature (Allen and Cohen 1969; Richardson and Ashcroft 1997), which was in the meantime verified experimentally (Lin and Dunn 1986; Shimizu et al. 2002).

At temperatures below 0.14 K, and thus more of academic interest than useful for spintronics applications, ferromagnetic UIr becomes superconducting under a pressure of approximately 2.6–2.7 GPa (Akazawa et al. 2004).

In insulating black phosphorus, even two different crystal phases were found to be superconducting under pressures of 10 GPa (Kawamura et al. 1985) or 262 GPa (Akahama et al. 2000), respectively. These results could be underlined theoretically by density functional theory (Ostanin et al. 2003). Phosphorus is of high interest for spintronics applications since multilayer systems containing phosphorus can be grown by molecular beam epitaxy, that is, epitaxially with high control of layer height and crystal orientation (Pearton et al. 2003). Using suitable templates, the bcc phase of phosphorus can be stabilized, for example, in Fe/P/Fe multilayers (Ostanin et al. 2003). In this system, the total magnetic moments are slightly reduced in comparison with pure Fe. In addition, the proximity effect in ferromagnetic/superconducting heterostructures leads to changes in the superconducting transition temperature depending on the thickness of the ferromagnet. In another type of junction, for example, V/Fe/P including the superconductor vanadium, spontaneous currents parallel to the interfaces can be expected (Ostanin et al. 2003), making such stacks also interesting for spintronics applications.

Another possible influence of pressure on spintronics materials is the change in the Curie temperature of ferromagnetic films with pressure. This effect was, for example, reported for bulk MnAs in the α phase (De Blois and Rodbell 1963; Yamada et al. 2002).

In CoMnCrAl and CoFeCrGe, an induced pressure was shown to result in a transition from the half-metallic to a metallic state (Enamullah et al. 2015; Enamullah et al. 2016). In GaN/CrN superlattices, a crystallographic phase transition from NaCl structure to wurtzite structure was found theoretically at a pressure of 13.5 GPa, also correlated with a phase transition from half-metallic to metallic behavior (Espitia-Rico et al. 2015).

Multilayered MoS_2 was expected to show a semiconductor/metal transition under high pressure (Bhattacharyya and Singh 2012; Johari and Shenoy 2012). After some papers reported a small decrease in electrical resistance (Hromadová et al. 2013; Zhu et al. 2013), an intermediate state starting from approximately 10 GPa and finally a pressure-induced semiconductor/metal transition was shown theoretically and experimentally by Nayak et al. (2014).

An interesting difference is visible between the half-metallic Heusler alloys Co_2VAl and Co_2VGa. While the latter shows a magnetic moment at low temperatures that is independent of the applied pressure, indicating the half-metallic electronic structure, the former shows a decrease of the Curie temperature with increasing pressure, opposite to other Heusler alloys with Mn atoms included (Kanomata et al. 2010).

While several materials show a semiconductor/metal transition under high pressure, BiTeI oppositely performs a phase transition from a semiconductor into a topological insulator (Bahramy et al. 2012).

8.4 Strain

Similar effects can be created by mechanical strain instead of pressure.

A mechanical strain can be used to support heat-assisted magnetic switching, as described in Chapter 8.1.

In ferromagnetic thin films, a strain can change the magnetic properties. Stoffel et al. (1970) have shown that in epitaxial and polycrystalline MnAs films, the Curie temperature was changed by several Kelvin due to the influence of an external strain. A stronger effect was reached by Zhu and Li (2016) in boron nitride nanoribbons, where they managed to switch the material between half-metal and spin-gapless semiconductor by a mechanical strain. In nanoribbons prepared from MoS_2, the ferromagnetic states could be significantly improved due to the increased difference between nonmagnetic and magnetic states by nearly a factor of 5, using a strain of 10% (Pan and Zhang 2012).

In strained crystals, a spin splitting can be induced, resulting in internal magnetic fields (Awschalom and Samarth 2009), which, on the other hand, can be used to manipulate spins by the spin–orbit coupling, for example, for spin precession and electrically driven spin resonance (Kato et al. 2004), for manipulation of spin transport in different semiconductors (Kato et al. 2005; Ghosh et al. 2006) or even to develop a completely current-based form of spintronics, without the necessity to use magnetic materials or external magnetic fields (Engel et al. 2007).

In the multiferroic material $BiFeO_3$, strain can modify the original cycloidal spin modulation into pseudo-collinear antiferromagnetism at high tensile or compressive strain. The spin-angle was found to vary from in-plane to out-of-plane with increasing strain, which can be used to tailor other magnetic properties, such as exchange bias and giant magnetoresistance (Sando et al. 2013).

$NbSe_2$ and NbS_2 single layers have revealed switching from an antiferromagnetic state at low strain, resulting from the superexchange interaction that aligns the next-nearest neighbors of Nb atoms, to a ferromagnetic state based on the double exchange (Xu et al. 2014). In Mn_2Au thin films, the antiferromagnetic domains could be switched into a preferred orientation by either using a large magnetic field of 70 T or by straining the thin films (Sapozhnik et al. 2017).

Similar results, showing that magnetic and other properties are influenced by mechanical strain, have been reported for diverse materials. As an example of the influence of strain in molecular spintronics, switching single molecules should be mentioned. Frisenda et al. (2016) have shown that the spin state of terpyridine ligands within homoleptic Fe-II complexes could be controlled mechanically by pulling them, as revealed by current–voltage measurements at low temperatures. With increasing stretching of the molecules, the conductance increased by 1–2 orders of magnitude, which was explained by better transmission for the high-spin configuration that could be reached by stretching the molecule. This mechanical manipulation was performed by movable nanoelectrodes.

8.5 Electric field

It is long known that electric fields can be used to change the conductivity of semiconductors. Tailoring magnetic properties by an electric field, however, is less usual but is possible (Qin et al. 2019).

Ohno et al. (2000) have shown that the transition temperature of hole-induced ferromagnetism could be varied reversibly by an electric field. For this, they prepared field-effect transistors containing the magnetic semiconductor (In,Mn)As with a transition temperature below 30 K (Dietl et al. 2000). By applying a gate bias voltage and measuring hysteresis loops at a temperature slightly below the transition temperature, they could switch the ferromagnetism off or, oppositely, increase it, depending on the sign of the voltage. This process was fully reversible.

The tunnel magnetoresistance (TMR) measured in spin valves, built from two ferromagnetic electrodes that are separated by a thin tunnel barrier, is in most materials positive; that is, the resistance through this device is smaller for the parallel orientation of both ferromagnets. In some materials, however, the magnetoresistance can be negative (George et al. 1994) for a spin-dependent electron transmission probability

(Žutić et al. 2004). This offers a possibility to tailor the magnetoresistance by a gate field (Schäpers et al. 2000). Sahoo et al. (2005) have demonstrated this idea for a system of carbon nanotubes connected with ferromagnetic electrodes, which were capacitively coupled to a back-gate (Sahoo et al. 2005a) and found regular oscillations in a gate voltage range below 1 V that were attributed to quantum interference.

An interesting class of materials is the multiferroics (cf. Chapter 9). In some of these materials, an electric field can be used to change their magnetic properties (Lebeugle et al. 2008). This correlation between magnetic field and electric polarization, as well as between electric field and magnetization, is called linear magnetoelectric effect. Typically, antiferromagnetic insulators are used for this purpose, which show magnetic and ferroelectric order (Khomskii 2006). In materials such as $YMnO_3$, Cr_2O_3, or $LuMnO_3$, however, low temperatures or an additional external magnetic field are necessary to tailor magnetization reversal processes (Laukhin et al. 2006; Skumryev et al. 2011). Nevertheless, materials like $HoMnO_3$ show magnetoelectric effects, which are by several orders of magnitude larger than in other materials (Lottermoser et al. 2004). At room temperature, $BiFeO_3$ thin films and crystals can be used (Bibes and Barthélémy 2008; Lebeugle et al. 2009). Combined with ferromagnetic $Co_{0.9}Fe_{0.1}$, an electric field can be used to reverse the ferromagnetic magnetization at room temperature (Heron et al. 2011).

Finally, it should be mentioned that electric fields can also be used to switch magnetic skyrmions (Hsu et al. 2017), an effect that could also be used in the planned skyrmion racetrack memories (Fert et al. 2013; Parkin et al. 2008).

8.6 Light

Besides the aforementioned possibilities of using strong light pulses for heating defined positions to support magnetization reversal, some other effects of light on spintronics elements have also been examined.

By shaping a light pulse with circular polarization into an elliptical form and focusing the beam on a magnetic material, Satoh et al. (2012) generated spin waves by the inverse Faraday effect. Because of the orientation of the major axis, they could modify the direction of the energy flow.

Kanda et al. (2011) showed the possibility of controlling 2D oscillations of the magnetization in antiferromagnetic NiO by using two femtosecond lasers with twisted polarization. The phase and amplitude of both degenerate magnetic oscillation modes could be manipulated by the polarization angle of the laser pulses.

Heintze et al. (2012) switched molecular nanowires using light with significantly lower power than in other optically induced magnetization switching methods (Gerrits et al. 2002; Koopmans et al. 2010) by a so-called kickoff process in which a photon is absorbed in a magnetic nanowire, resulting in creation of an exciton that can switch its spin at lower energies than normal spins and thus allows for introducing domain walls into the spin chain. The domain wall, however, can only start propagating after the exciton has decayed.

In the area of molecular spintronics, Einati et al. (2015), for example, recently demonstrated the possibility of changing the spin-filtering properties of bacteriorhodopsin. Such spin-filtering properties can be found in several chiral molecular systems (Naaman and Waldeck 2012). While the wild type of bacteriorhodopsin was also formerly shown to have spin-filtering properties (Mishra et al. 2013), a mutant of this protein can be used as a light-controlled spin switch (Einati et al. 2015).

8.7 Electric current and switching of spintronic devices

Magnetization switching in spintronics devices by an electric current belongs to the fields that are of large interest in recent research attempts since high magnetic fields would be necessary to switch bits in recent storage devices with large data density. Reversing the magnetization in magnetic nanostructures can be performed by a torque (Tsoi et al. 1998; Albert et al. 2000); however, there are technical limits restricting the possible current densities. Lower current densities are necessary to achieve magnetization reversal in ferromagnetic semiconductors (Yamanouchi et al. 2004).

The orientation of the magnetization of a vortex core may be used to create a vortex-based memory. Triggering the resonant dynamics of the vortex by an electrical current without a magnetic field was shown in magnetic nanodisks (Yamada et al. 2007).

Interestingly, Miron et al. (2011) managed to switch a perpendicularly magnetized cobalt dot by an in-plane current at room temperature. This effect was attributed to the magnetic anisotropy of the cobalt layer and a Rashba interaction due to asymmetric platinum and aluminum oxide interface layers (Rodmacq et al. 2009; Miron et al. 2010) and to the torque induced in the platinum layer by the spin Hall effect (Liu et al. 2011).

Finally, similar to light and other stimuli, an electric current can also be used to manipulate molecular spintronics systems. For example, Komeda et al. (2011) showed that it was possible to switch the spin state of a molecule, in this case a bis(phthalocyaninato)terbium(III) complex ($TbPc_2$) adsorbed on an Au(111) surface, by an electric current that was applied by a scanning tunneling microscope. In this way, one of the Pc ligands in the $TbPc_2$ could be rotated reversibly, allowing writing information into single molecules.

After this chapter about the general possibilities to manipulate magnetization, that is, spins, in diverse materials, the next chapter will give a short overview of materials recently used in commercial applications, as well as promising materials on which recent research approaches concentrate.

9 Materials in commercial applications

While research is being performed on a broad range of materials that may be used in spintronic applications, there are spintronic applications that have been used for several years now, starting from hard disk write/read heads to the first magnetoresistive random access memories (MRAMs). This chapter aims at giving an overview of some of the typical materials used in commercial spintronic applications.

9.1 Metals and half-metals

Different metals occur in spintronic devices. Spin valves, for example, consist of two ferromagnetic layers separated by a thin nonmagnetic metal. While copper is a typical material used for the nonmagnetic spacer, the ferromagnetic layers often consist of alloys of iron, nickel, and cobalt (Dieny et al. 1991).

Permalloy, that is, $Ni_{80}Fe_{20}$ or similar blends, is also often used in different spintronic components. The spin Seebeck effect, for example, can be used as a thermal spin current injection method, necessitating less energy than electrical spin injection. In permalloy/p-Si bilayers, the spin Seebeck effect could be used for thermal spin generation, while a large inverse spin Hall effect could be applied to detect the spin Seebeck effect (Bhardwaj et al. 2018). The first MRAMs were based on the anisotropic magnetoresistance (AMR) in NiFe and similar alloys (Baugh et al. 1982; Schwee et al. 1982; Pohm et al. 1988); however, since the AMR was not very large, other materials showing giant magnetoresistance (GMR) or tunnel magnetoresistance (TMR) were investigated later on for this purpose.

A typical material often used in MRAMs as free layers is CoFeB which is typically sputter-deposited and enables high read distribution separation (Tehrani et al. 2006). NiFe and CoFe are also typical materials for the ferromagnetic layers in MRAMs (Slaughter 2009). Materials with perpendicular magnetocrystalline anisotropy, which can be used in MRAMs, are, for example, $L1_0$-FePt (i.e., $L1_0$-ordered FePt with perpendicular magnetic anisotropy) (Seki et al. 2006; Yoshikawa et al. 2008) or TbCoFe (Nakayama et al. 2008). In this way, other shapes of the nanoparticles used as bits could be enabled, and the necessary current for spin-torque switching could be reduced.

Half-metals often show ideal spin polarization, which is necessary for large magnetoresistance. Possible materials used in spintronic devices, such as spin valves or MRAMs, are Cr_2O_5, MoO_2, $Mo_xCr_{1-x}O_y$, Fe_3O_4/CrO_2, or Fe_3O_4/Ag (Kimishima 2009).

https://doi.org/10.1515/9783111383736-009

9.2 Diluted magnetic semiconductors

While common electronic elements use doped semiconductors to influence their electronic properties, spintronics necessitates special magnetic properties. To enable spin generation and manipulation in a single material, ferromagnetism and semiconductivity should be combined (Ohno 1998; Jedema et al. 2001). Ideally, such diluted magnetic semiconductors should have Curie temperatures above room temperature to allow for using them in spintronic devices without additional cooling (Gupta et al. 2020).

An often-used diluted magnetic semiconductor is Ga(Mn)As. Unfortunately, this material has a Curie temperature significantly below room temperature, with the exact value varying with sample preparation (Wang et al. 2005b).

An interesting diluted magnetic semiconductor predicted to have a high Curie temperature is zirconia, stabilized by manganese (Lajavardi et al. 2000). While this material is typically used as "synthetic diamond" or recently also as a catalyst (Dravid et al. 1994), its magnetic properties have been studied less often before $Zr_{1-x}Mn_xO_2$ was found to exhibit Curie temperatures up to several 100 °C, using different calculation methods (Ostanin et al. 2007).

The ferrimagnetic material Ti_2MnAl can be grown by magnetron sputtering on a silicon substrate, making it possibly useful for commercial spintronic applications. This ferrimagnet is a spin-gapless semiconductor at room temperature with a high Curie temperature well above room temperature, a small coercive field, and small magnetization at saturation (Feng et al. 2015). Interestingly, in this ferrimagnetic semiconductor, the magnetoresistance shows negative field dependence at room temperature and positive field dependence at lower temperatures, similar to Mn_2CoAl, which exhibited comparable effects (Ouardi et al. 2013).

9.3 Multiferroics

Multiferroics belong to the materials of high interest for spintronic applications (Muneeswaran et al. 2020; Yakout 2021); however, commercialization is not yet predictable. They can be used to control the magnetic properties of a system by an electric field, resulting in significantly reduced energy consumption for magnetization reversal, as compared to current-driven processes (Meisenheimer et al. 2018). This magnetoelectric coupling can be strain-mediated. One of the multiferroic heterostructures under research is $Pb(Mg_{1/3}Nb_{2/3})_{0.7}Ti_{0.3}O_3$, showing a ferroelectric domain structure, which influences the possible electrical field control of the magnetic properties, such as a 90° rotation of the easy magnetic axis (Chen and Zhao 2018).

While perovskites belong to the most important material classes to exhibit multiferroic properties, these cannot be expected in highly symmetrical cubic crystals. Nevertheless, spin-induced ferroelectric polarization was found in the multiordered cubic perovskite $LaMn_3Cr_4O_{12}$. A similar multiordered perovskite, $BiMn_3Cr_4O_{12}$, showed two different multiferroic phases, enabling creation of four distinguishable states of the polarization. In addition, large ferroelectric polarization and strong magnetoelectric coupling could be reached at the same time (Zhou et al. 2018).

9.4 Antiferromagnets

Antiferromagnetic (AFM) materials are necessary in spin valves, used in GMR heads. For these applications, often synthetic antiferromagnets (two ferromagnetic layers separated by a coupling layer) are used (Tsang et al. 1994; Wang et al. 2023).

Besides ferromagnetic materials, AFM materials can also be used to create systems exhibiting an AMR. To be more exact, this is possible with the so-called compensated antiferromagnets with vanishing net magnetization (Shick et al. 2010; MacDonald and Tsoi 2011; Gomonay and Loktev 2014). Typical bistable memories work for AFM metals, such as FeRh (Marti et al. 2014; Moriyama et al. 2015) or IrMn (Park et al. 2011), by switching between an AFM and a ferromagnetic phase or by coupling the AFM to a ferromagnet, resulting in the usual two different magnetic states at remanence. In MnTe, however, more stable states at vanishing external field are available, based on the AMR which changes with the angle between the current and the external magnetic field, making them viable for applications in neuromorphic computing (Kriegner et al. 2016). In addition, these stable states are not deleted by strong magnetic perturbation fields, as long as the AFM material is held well below the Néel temperature; that is, such magnetic memories would be quite long-term stable.

9.5 Organic spintronics

In organic spintronics, organic semiconductors and other organic materials are used to create spintronic devices, exhibiting magnetoresistance and other promising features (Privitera et al. 2021; Pandey et al. 2023). The advantages of organic materials are, among others, the relatively inexpensive production and the typical large flexibility, making them useful for commercial applications (Sanvito 2011; Boehme and Lupton 2013; Zhang et al. 2017a; Majumdar et al. 2006; Szulczewski et al. 2009). Other special properties typical for organic semiconductors are weak hyperfine interaction and spin–orbit coupling, suggesting the possibility to reach long spin lifetimes and correspondingly large spin diffusion lengths (Dediu et al. 2009; Wagemans and Koopmans 2011; Zheng and Wudl 2014; Sun et al. 2014). Measurements have revealed spin lifetimes up to the order of seconds (Pramanik et al. 2007; Szulczewski et al. 2009a).

Organic spin valves consist of two ferromagnetic electrodes and an intermediate organic nonmagnetic layer (Yao et al. 2018). Such devices were built, for example, by $La_{0.7}Sr_{0.3}MnO_3$/sexithienyl/$La_{0.7}Sr_{0.3}MnO_3$ (Dediu et al. 2002), $La_{0.7}Sr_{0.3}MnO_3$/Alq$_3$/ $La_{0.7}Sr_{0.3}MnO_3$ (Xiong et al. 2004), or $La_{0.67}Sr_{0.33}MnO_3$/pentacene/ $La_{0.67}Sr_{0.33}MnO_3$

(Ikegami et al. 2008). Other organic molecules used in spin valves are C_{60} (Gobbi et al. 2011), C_{70} (Liang et al. 2016), or rubrene (Shim et al. 2008). Even tests with purely organic spin valves were performed (Li et al. 2011); however, until now without satisfying results. Nevertheless, several attempts are being made to drive this approach further, for example, by using a prefabricated tunnel junction with exposed sides as a test bench for molecular junctions bridging the metal electrodes (Tyagi et al. 2014, 2015).

9.6 Other materials in layered systems

Besides the aforementioned materials, several others are necessary, which often play a less prominent role but are nevertheless important for the reliable function of a spintronic component.

The substrates on which spintronic elements are built usually consist of silicon, possibly followed by a silicon nitride or silicon oxide buffer layer (Lee et al. 2017). While silicon as a base material was first chosen due to its long-term use in wafers and electronic components, recent research has pointed out another interesting feature, making it possibly feasible as a real base material for spintronics. Generally, Si shows only small spin–orbit coupling and spin–lattice interactions, which are the main reasons for spin relaxations. Si thus enables large spin diffusion lengths, for example, up to 6 µm in n-Si (Ishikawa et al. 2017; Dash et al. 2009). Lou and Kumar (2018) created thin films on n-Si as only 2 µm thin substrates, consisting of Pd/Ni$_{80}$Fe$_{20}$/MgO, and showed that this system revealed a field-dependent thermal transport as well as spin Hall magnetoresistance which were attributed to the Rashba spin–orbit coupling at the interface. A second-order AFM phase transition was observed near 0 °C, resulting from a spin accumulation. These findings could enable building Si-based spintronic elements.

Top and bottom electrodes of MRAMs typically consist of Ta or TaN, while spacer layers in MRAMs can be prepared, for example, by Ru (Slaughter 2009), 2D materials (Zhang et al. 2021) or even organic materials (Li and Yu 2021).

Magnetic tunnel junctions, for example, contain a tunnel barrier that is typically built from aluminum oxide (Simmons 1963; Brinkman et al. 1970; Moodera et al. 1995; Miyazaki and Tezuka 1995a; Parkin et al. 1999; Akerman et al. 2004). Combined with usual ferromagnetic materials or alloys, aluminum oxide tunnel barriers result in high magnetoresistance values around 50%. For MgO, even up to 200% magnetoresistance and higher values were gained (Parkin et al. 2004; Yuasa et al. 2004; Yuasa et al. 2006; Lee et al. 2007). This material was used, for example, in the Everspin MRAM (Dave et al. 2006). Nowadays, MXenes such as Ti_2CO_2 and other 2D materials are also investigated regarding their use in tunnel barriers, ferromagnetic electrodes, etc. (Zhang et al. 2021; Das et al. 2022).

Bibliography

Abdallah, F. S., Cheri, S. M., Bouamama, K., Roussigne, Y., & Hsu, J. H. 2018. Effect of deposition temperature on morphological, magnetic and elastic properties of ultrathin $Co_{49}Pt_{51}$ films. Appl. Surf. Sci. 433: 647–652.

Abraham, D. W., & Trouilloud, P. L. 2000. Thermally-assisted magnetic random access memory (MRAM). US patent US6385082B1.

Ackerman, P. J., van de Lagemaat, J., & Smalyukh, I. I. 2015. Self-assembly and electrostriction of arrays and chains of hopfion particles in chiral liquid crystals. Nature Comm. 6: 6012.

Adhikari, R., Sarkar, A., Patta, G. R., & Das, A. K. 2011. Magnetic diode exploiting giant positive magnetoresistance in ferrite semiconductor heterostructures. Appl. Phys. Lett. 98: 183504.

Aharonov, Y., & Bohm, D. 1959. Significance of electromagnetic potentials in quantum theory. Phys. Ref. 115: 485–491.

Aharonov, Y., & Bohm, D. 1961. Further considerations on electromagnetic potentials in the quantum theory. Phys. Rev. 123: 1511–1524.

Aharony, A, Tokura, Y., Cohen, G. Z., Entin-Wohlman, O., & Katsumoto, S. 2011. Filtering and analyzing mobile qubit information via Rashba-Dresselhaus-Aharonov-Bohm interferometers. Phys. Rev. B 84: 035323.

Ahrens, J., Geveci, B., & Law, Ch. 2005. ParaView: An End-User Tool for Large Data Visualization. In: (Ch. D. Hansen and Ch. R. Johnson, eds.) Visualization Handbook. Elsevier, Amsterdam, Boston, Heidelberg, London, New York, Oxford, Paris, San Diego, San Francisco, Singapore, Sydney, Tokyo, pp. 717–731.

Akahama, Y., Kawamura, H., Carlson, S., Le Bihan, T., & Häusermann, D. 2000. Structural stability and equation of state of simple-hexagonal phosphorus to 280 GPa: phase transition at 262 GPa. Phys. Rev. B 61: 3139.

Akansel, S., Venugopal, V. A., Kumar, A., Gupta, R., Brucas, R., George, S., Neagu, A., Tai, C. W., Gubbins, M., Andersson, G., & Svedlindh, P. 2018. Effect of seed layers on dynamic and static magnetic properties of $Fe_{65}Co_{35}$ thin films. J. Phys. D. 51: 305001.

Akansel, S., Kumar, A., Behera, N., Husain, S., Brucas, R., Chaudhary, S., & Svedlindh, P. 2018a. Thickness-dependent enhancement of damping in Co_2FeAl/β-Ta thin films. Phys. Rev. B. 97: 134421.

Akazawa, T., Hidaka, H., Fujiwara, T., Kobayashi, T. C., Yamamoto, E., Haga, Y., Settai, R., & Onuki, Y. 2004. Pressure-induced superconductivity in ferromagnetic UIr without inversion symmetry. J. Phys. Cond. Matter. 16: L29.

Åkerman, J., DeHerrera, M., Durlam, M., Engel, B., Janesky, J., Mancoff, F., Slaughter, J., & Tehrani, S. 2004. Magnetic tunnel junction based magnetoresistive random access memory. In: Magnetoelectronics, ed. M Johnson, 5: 231–272. Elsevier, Amsterdam.

Akerman, J. 2005. Towards a universal memory. Science. 308: 508–510.

Alam, J., Bran, C., Chiriac, H., Lupu, N., Óvári, T. A.; Panina, L. V.; Rodionova, V., Varga, R.; Vazquez, M., & Zhukov, A. 2020. Cylindrical micro and nanowires: fabrication, properties and applications. J. Magn. Magn. Mater. 513: 167074.

Albert, F. J., Katine, J. A., Buhrman, R. A., & Ralph, D. C. 2000. Spin-polarized current switching of a Co thin film nanomagnet. Appl. Phys. Lett. 77: 3809–3811.

Albrecht, J. D., & Smith, D. L. 2002. Electron spin injection at a Schottky contact. Phys. Rev. B. 66: 113303.

Aldosary, M., Li, J., Tang, C., Xu, Y., Zheng, J.-G., Bozhilov, K. N., & Shi, J. 2016. Platinum/yttrium iron garnet inverted structures for spin current transport. Appl. Phys. Lett. 108: 242401.

Ali, D., & Ahmed, H. 1994. Coulomb blockade in a silicon tunnel junction device. Appl. Phys. Lett. 64: 2119.

Ali, M. N., Siong, J., Flynn, S., Tao, J., Gibson, Q. D., Schoop, L. M., Liang, T., Haldolaarachchige, N., Hirschberger, M., Ong, N. P., & Cava, R. J. 2014. Large, non-saturating magnetoresistance in WTe_2. Nature. 514: 205–208.

https://doi.org/10.1515/9783111383736-010

Ali, M., Bello, F., Abadía, N., Huang, F. M., & Donegan, J. 2022. Elliptical plasmonic near-field transducer and V-shape waveguide designs for heat assisted magnetic recording. Optics Continuum 1: 1529–1541.

Allen, P. B., & Cohen, M. L. 1969. Pseudopotential calculation of the mass enhancement and superconducting transition temperature of simple metals. Phys. Rev. 187: 525–538.

Allwood, D. A., Xiong, G., Cooke, M. D., Faulkner, C. C., Atkinson, D., Vernier, N., & Cowburn, R. P. 2002. Submicrometer ferromagnetic NOT gate and shift register. Science. 296: 2003–2006.

Allwood, D. A., Xiong, G., Faulkner, C. C., Atkinson, D., Petit, D., & Cowburn, R. P. 2005. Magnetic domain-wall logic. Science. 309: 1688–1692.

Altarawneh, M. M., Mielke, C. H., & Brooks, J. S. 2009. Proximity detector circuits: an alternative to tunnel diode oscillators for contactless measurements in pulsed magnetic field environments. Rev. Sci. Instr. 80: 066104.

Althammer, M., Singh, A. V., Keshavarz, S., Yurtisigi, M. K., Mishra, R., Borisevich, A. Y., LeClair, P., & Gupta, A. 2016. Investigation of the tunnel magnetoresistance in junctions with a strontium stannate barrier. J. Appl. Phys. 120: 233903.

Alvarado, S. F., & Renaud, P. 1992. Observation of spin-polarized-electron tunneling from a ferromagnet into GaAs. Phys. Rev. Lett. 68: 1387.

Ambrose, T., & Chien, C. L. 1998. Dependence of exchange field and coercivity on cooling field in NiFe/CoO bilayers. J. Appl. Phys. 83: 7222.

Amer, A., Long, D. D. E., Miller, E. L., Paris, J.-F., & Schwarz, T. 2010. Design Issues for a Shingled Write Disk System. Proc. of 26th IEEE (MSST2010) Symposium on Massive Storage Systems and Technologies, Incline Village, NV / USA, May 3–7, 2010.

Anderson, P. W. 1950. Antiferromagnetism. Theory of superexchange interaction. Phys. Rev. 79: 350.

Anderson, P.W. 1961. Localized magnetic states in metals. Phys. Rev. 124: 41–53.

Andrade, J. A., & Cornaglia, P. S. 2016. Spin filtering and thermopower in star-coupled quantum dot devices. Phys. Rev. B 94: 235112.

Andreev, A. F. 1964. Thermal conductivity of the intermediate state of superconductors. Sov. Phys. JETP. 19: 1228.

Andreou, A. G., & Boahen, K. A. 1989. Synthetic neural circuits using current-domain signal representations. Neural Comput. 1: 489–501.

Anugrah, Y., Hu, J. X., Stecklein, G., Crowell, P. A., & Koester, S. J. 2018. Independent gate control of injected and detected spin currents in CVD graphene nonlocal spin valves. AIP Adv. 8: 015129.

Aoki, K. 1991. Cross-over instability during impurity avalanche breakdown in semiconductors under longitudinal magnetic field. Solid State Commun. 77: 87.

Appelbaum, I., Huang, B. Q., & Monsma, D. J. 2007. Electronic measurement and control of spin transport in silicon. Nature. 447: 295–298.

Apostolov, A. T., Apostolova, I. N., & Wesselinowa, J. M. 2018. $La_{1-x}Sr_xMnO_3$ nanoparticles for magnetic hyperthermia. Phys. Stat. Sol. B. 255: 1700587.

Arava, H. & Phatak, C. M. 2023. Magnetic hopfion rings in new era for topology. Nature 623: 702–703.

Arana, M., Gamino, M., Silva, E. F., Barthem, V. M. T. S., Givord, D., Azevedo, A., & Rezende, S. M. 2018. Spin to charge current conversion by the inverse spin hall effect in the metallic antiferromagnet Mn_2Au at room temperature. Phys. Rev. B. 98: 144431.

Aranda, G. R., Gonzales, J. M., del Val, J. J., & Guslienko, K. Y. 2010. Limits for the vortex state spin torque oscillator in magnetic nanopillars: micromagnetic simulations for a thin free layer. J. Appl. Phys. 108: 123914.

Argumendo, A. J., Berman, D., Biskeborn, R. G., Cherubini, G., Cideciyan, R. D., Eleftheriou, E., Häberle, W., Hellman, D. J., Hutchins, R., Imaino, W., Jelitto, J., Judd, K., Jubert, P.-O., Lantz, M. A., McClelland, G. M., Mittelholzer, T., Narayan, C., Ölçer, S., & Seger, P. J. 2008. Scaling tape-recording areal densities to 100 Gb/in^2. IBM J. Res. Develop. 52: 513–527.

Aronov, A. G., Pikus, G. E., & Titkov, A. N. 1983. Spin relaxation of conduction electrons in p-type Ill-V compounds. Sov. Phys. JETP. 57: 680.

Arrayás, M. & Trueba, J. L. 2017. Collision of two hopfions. J. Phys. A. Math. Theor. 50: 085203.

Asam, N., Yamanoi, K., & Kimura, T. 2018. Modification of the magnetization dynamics of a NiFe nanodot due to thermal spin injection. J. Phys. D. 51: 224004.

Asamitsu, A., Tomioka, Y., Kuwahara, H., & Tokura, Y. 1997. Current-switching of resistive states in colossal magnetoresistive oxides. Nature. 388: 50–52.

Ashoori, R. C. 1996. Electrons in artificial atoms. Nature. 379: 413.

Attané, J. P., Ravelosona, D., Mary, A., Samson, Y., & Chappert, C. 2006. Thermally activated depinning of a narrow domain wall from a single defect. Phys. Rev. Lett. 96: 147204.

Averin, D. V., & Likharev, K. K. 1985. Coulomb blockade of single-electron tunneling, and coherent oscillations in small tunnel junctions. J. Low Temp. Phys. 62: 345–373.

Averin, D. V., & Likharev, K. K. 1986. Coulomb blockade of single-electron tunneling, and coherent oscillations in small tunnel junctions. J. Low Temp. Phys. 62: 345.

Averin, D. V., & Likharev, K. K. 1991. Single Electronics: A Correlated Transfer of Single Electrons and Cooper Pairs in Systems of Small Tunnel Junctions. In: Mesoscopic Phenomena in Solids, edited by B. Altshuler, P. Lee, and R. Webb. Elsevier, Amsterdam, p. 173.

Averin, D. V., & Likharev, K. K. 1992. Possible Applications of the Single Charge Tunneling. In: Single Charge Tunneling, Plenum, New York, p. 311.

Avishai, Y., & Band, Y. B. 2022. Graphene bilayer and trilayer moiré lattice with Rashba spin-orbit coupling. Phys. Rev. B 106: L041406.

Avsar, A., Yang, T.-Y., Bae, S. K., Balakrishnan, J., Volmer, F., Jaiswal, M., Yi, Z., Ali, S. R., Güntherodt, G., Hong, B. H., Beschoten, B., & Özyilmaz, B. 2011. Toward wafer scale fabrication of graphene based spin valve devices. Nano Lett. 11: 2362–2368.

Avsar, A., Tan, J. Y., Taychatanapat, T., Balakrishnan, J., Koon, G. K. W., Yeo, Y., Lahiri, J., Carvalkoh, A., Rodin, A. S., O'Farrell, E. C. T., Eda, G., Gastro Neta, A. H., & Özyilmaz, B. 2014. Spin-orbit proximity effect in graphene. Nat. Commun. 5: 4875.

Awschalom, D., & Samarth, N. 2009. Trend: spintronics without magnetism. Physics. 2: 50.

Azadian, S., Tehranchi, M. M., Mohseni, S. M., & Mohseni, S. M. 2017. Reduction and control of permalloy thin film damping factor under microwave irradiation. J. Alloys Comp. 723: 960–966.

Azevedo, J., Virazel, A., Bosio, A., Dilillo, L., Girard, P., Todri, A., Prenat, G., Alvarez-Herault, J., & Mackay, K. 2012. Impact of resistive-open defects on the heat current of TAS-MRAM architectures. Des. Autom. Test in Europe. 2012.: 532–537.

Backus, J. 1978. Can programming be liberated from the Von Neumann style? A functional style and its algebra of programs. Commun. ACM. 21: 613–641.

Bahramy, M. S., Yang, B.-J., Arita, R., & Nagaosa, N. 2012. Emergence of non-centrosymmetric topological insulating phase in BiTeI under pressure. Nat. Comm. 3: 679.

Bai, X. F., Chi, F., Zheng, J., & Li, Y. N. 2012. Generating and reversing spin accumulation by temperature gradient in a quantum dot attached to ferromagnetic leads. Chinese Phys. B. 21: 077301.

Baibich, M. N., Broto, J. M., Fert, A., Nguyen Van Dau, F., Petroff, F., Etienne, P., Creuzet, G., Friederich, A., & Chazelas, J. 1988. Giant magnetoresistance of (001)Fe/(001)Cr magnetic superlattices. Phys. Rev. Lett. 46: 2472–2475.

Bainsla, L., Yilgin, R., Tsujikawa, M., Suzuki, K. Z., Shirai, M., & Mizukami, S. 2018. Low magnetic damping for equiatomic CoFeMnSi Heusler alloy. J. Phys. D. Appl. Phys. 51: 495001.

Bakshi, U. A., & Godse, A. P. 2007. MOSFETs. In: Electronic Circuits. Technical Publications. Pune, pp. 8–1 to 8–22. ISBN 978-81-8431-284-3.

Balakrishnan, R., Dandoloff, R., & Saxena, A. 2023. Exact hopfion vortices in a 3D Heisenberg ferromagnet. Phys. Lett. A 480, 128975.

Banerjee, C., Rode, K., Atcheson, G., Lenne, S., Stamenov, P., Coey, J. M., D., & Besbas, J. 2021. Ultrafast Double Pulse All-Optical Reswitching of a Ferrimagnet. Phys. Rev. LEtt. 126: 17202.

Bardeen, J., & Brattain, W. H. 1948. The transistor, a semi-conductor triode. Phys. Rev. 74: 230.

Bardeen, J., & Brattain, W. H. 1949. Physical principles involved in transistor action. Phys. Rev. 75: 1208.

Barman, J., & Ravi, S. 2018. Magnetization reversal and tunable exchange bias behavior in Mn-substituted NiCr2O4. J. Mater. Sci. 53: 7187–7198.

Barnaś, J., & Weymann, I. 2008. Spin effects in single-electron tunnelling. J. Phys. Cond. Matter. 20: 423202.

Barroso, T. G., Martins, R. C., Fernandes, E., Cardoso, S., Rivas, J., & Freitas, P. P. 2018. Detection of BCG bacteria using a magnetoresistive biosensor: a step towards a fully electronic platform for tuberculosis point-of-care detection. Biosens. Bioelectron. 100: 259–265.

Bass, J. 2016. CPP magnetoresistance of magnetic multilayers: a critical review. J. Magn. Magn. Mater. 408: 244–320.

Batelaan, H., & Tonomura, A. 2009. The Aharonov-Bohm effects: variations on a subtle theme. Phys. Today. 62: 38–43.

Bauer, G. E. W., Bretzel, S., Brataas, A., & Tserkovnyak, Y. 2010. Nanoscale magnetic heat pumps and engines. Phys. Rev. B. 81: 024427.

Baugh, C. W., Cullom, J. H., Hubbard, E. A., Mentzer, M. A., & Fedorak, R. 1982. Fabrication and characterization of a crosstie random access memory. IEEE Trans. Mag. 18: 1782–1784.

Beck, A., Bednorz, J. G., Gerber, C., Rossel, C., & Widmer, D. 2007. Reproducible switching effect in thin oxide films for memory applications. Appl. Phys. Lett. 77: 139–141.

Becker, S. 1998. Feingebrannt – höhere speicherdichten bei magnetooptischen wechselplatten. c't. 25/ 1998: 190.

Bednik, G. 2019. Hopfions in a lattice dimer model. Phys. Rev. B 100: 024420.

Beech, R. S., Anderson, J. A., Pohm, A. V., & Daughton, J. M. 2000. Curie point written magnetoresistive memory. J. Appl. Phys. 87: 6403.

Beenakker, C. W. J. 2000. Why does a metal-superconductor junction have a resistance? Quantum mesoscopic phenomena and mesoscopic devices in microelectronics. NATO Sci. Ser. 559: 51–60.

Behera, A. K., Mishra, S. S., Mallick, S., Singh, B. B., & Bedanta, S. 2018. Size and shape of skyrmions for variable Dzyaloshinskii-Moriya interaction and uniaxial anisotropy. J. Phys. D. Appl. Phys. 51: 285001.

Behera, D., Abraham, J. A., Sharma, R., Mukerjee, S. K., & Jain, E. 2022. First Principles Study of New d^0 Half-Metallic Ferromagnetism in CsBaC Ternary Half-Heusler Alloy. J. Supercond. Nov. Magn. 35: 3431–3437.

Beiranvand, R., Hamzehpour, H., & Alidoust, M. 2016. Tunable anomalous Andreev reflection and triplet pairings in spin-orbit-coupled graphene. Phys. Rev. B. 94: 125415.

Belanger, D. P., & Young, A. P. 1991. The random field Ising model. J. Magn. Magn. Mat. 100: 272.

Belashchenko, K. D., Velev, J., & Tsymbal, E. Y. 2005. Effect of interface states on spin-dependent tunneling in Fe/MgO/Fe tunnel junctions. Phys. Rev. B. 72: 140404(R).

Berger, L. 1996. Emission of spin waves by a magnetic multilayer traversed by a current. Phys. Rev. B. 54: 9353–9358.

Berk, C., Ganss, F., Jaris, M., Albrecht, M., & Schmidt, H. 2018. All-optical measurement of interlayer exchange coupling in Fe/Pt/FePt thin films. Appl. Phys. Lett. 112: 052401.

Bhardwaj, R. G., Lou, P. C., & Kumar, S. 2018. Spin seebeck effect and thermal spin galvanic effect in $Ni_{80}Fe_{20}$/p-Si bilayers. Appl. Phys. Lett. 112: 042404.

Bhattacharyya, S., & Singh, A. K. 2012. Semiconductor-metal transition in semiconducting bilayer sheets of transition-metal dichalcogenides. Phys. Rev. B. 86: 075454–075454.

Bhatti, S., Sbiaa, R., Hirohata, A., Ohno, H., Fukami, S., & Piramanayagam, S. N. 2017. Spintronics based random access memory: a review. Mater. Today. 20: 530–548.

Bibes, M., & Barthélémy, A. 2008. Towards a magnetoelectric memory. Nat. Mater. 7: 425–426.

Bibyk, S., Ismail, M., Borgstrom, T., Adkins, K., Kaul, R., Khachab, N., & Dupuie, S. 1990. Current-mode neural network building blocks for analog MOS VLSI. IEEE Int. Symp. Circuits Syst. 4: 3283–3285.

Binasch, G., Grünberg, P., Saurenbach, F., & Zinn, W. 1989. Enhanced magnetoresistance in layered magnetic structures with antiferromagnetic interlayer exchange. Phys. Rev. B. 39: 4828–4830(R).

Bir, G. L., Aronov, A. G., Pikus, G. E. 1975. Spin relaxation of electrons due to scattering by holes. Zh Eksp. Teor. Fiz. 69: 1382–1397.

Biternas, A. G., Nowak, U., & Chantrell, R. W. 2009. Training effect of exchange-bias bilayers within the domain state model. Phys. Rev. B. 80: 134419.

Biternas, A. G., Chantrell, R. W., & Nowak, U. 2010. Behavior of the antiferromagnetic layer during training in exchange-biased bilayers within the domain state model. Phys. Rev. B. 82: 134426.

Biternas, A. G., Chantrell, R. W., & Nowak, U. 2014. Dependence of training effect on the antiferromagnetic structure of exchange-bias bilayers within the domain-state model. Phys. Rev. B. 89: 184405.

Blachowicz, T., & Ehrmann, A. 2011. Fourfold nanosystems for quaternary storage devices. J. Appl. Phys. 110: 073911.

Blachowicz, T., & Ehrmann, A. 2013. Six-state, three-level, six-fold ferromagnetic wire system. J. Magn. Magn. Mat. 331: 21–23.

Blachowicz, T., Ehrmann, A., Steblinski, P., & Palka, J. 2013. Directional-dependent coercivities and magnetization reversal mechanisms in fourfold ferromagnetic systems of varying sizes. J. Appl. Phys. 113: 013901.

Blachowicz, T., & Ehrmann, A. 2013a. Micromagnetic simulations of anisotropies in coupled and uncoupled ferromagnetic nano-wire systems. Sci World J. 2013: 472597.

Blachowicz, T., & Ehrmann, A. 2015. Magnetization reversal modes in fourfold Co nano-wire systems. J. Phys.: Conf. Ser. 633: 012100.

Blachowicz, T., & Ehrmann, A. 2016. Stability of magnetic nano-structures against erroneous shape modifications. AIP Conf. Proc. 1727: 020004.

Blachowicz, T., & Ehrmann, A. 2016a. Magnetization reversal in 3D nano-structures of different shapes. AIP Conf. Proc. 1727: 020003.

Blachowicz, T., & Ehrmann, A., 2017. Square nano-magnets as bit-patterned media with doubled possible data density. Elsevier Mater. Today: Proc. 4: S226–S231.

Blachowicz, T., & Ehrmann, A. 2018. Magnetization reversal in bent nanofibers of different cross-sections. J. Appl. Phys. 124: 152112.

Blachowicz, T. & Ehrmann, A. 2020. Magnetic Elements for Neuromorphic Computing. Molecules 25: 2550.

Blachowicz, T., & Ehrmann, A. 2021. Exchange bias in thin films – an update. Coatings 11: 122.

Błachowicz, T., Ehrmann, A., Stebliński, P., & Palka, J. 2013. Directional-dependent coercivities and magnetization reversal mechanisms in fourfold ferromagnetic systems of varying sizes. J. Appl. Phys. 113: 013901.

Blachowicz, T., Ehrmann, A., & Mahltig, B. 2017. Magneto-optic measurements on uneven magnetic layers on cardboard. AIP Adv. 7: 045306.

Blachowicz, T., Döpke, C., & Ehrmann, A. 2020. Micromagnetic simulations of chaotic ferromagnetic nanofiber networks. Nanomaterials 10: 738.

Blachowicz, T., Grzybowski, J.; Steblinski, P., & Ehrmann, A. 2021. Neuro-inspired signal processing in ferromagnetic nanofibers. Biomimetics 6: 32.

Blachowicz, T.; Ehrmann, A.; & Wortmann, M. 2023. Exchange bias in nanostructures: an update. Nanomaterials 13: 2418.

Black, Jr., W., & Das, C., B. 2000. Programmable logic using giant-magnetoresistance and spin-dependent tunneling devices. J. Appl. Phys. 87: 6674.

Blencoweve, M. P., & Wybourne, M. N. 2000. Sensitivity of a micromechanical displacement detector based on the radio-frequency single-electron transistor. Appl. Phys. Lett. 77: 3845–3847.

Blonder, G. E., Tinkham, M., & Klapwijk, T. M. 1982. Transition from metallic to tunneling regimes in superconducting microconstrictions: excess current, charge imbalance, and supercurrent conversion. Phys. Rev. B. 25: 4515.

Bobbert, P. A., Nguyen, T. D., van Oost, F. W. A., Koopmans, B., & Wohlgenannt, M. 2007. Bipolaron mechanism for organic magnetoresistance. Phys. Rev. Lett. 99: 216801.

Bo, L., Ji, L. Z., Hu, C. L., Zhao, R. Z., Li, Y. X., Zhang, J., & Zhang, X. F. 2021. Spin excitation spectrum of a magnetic hopfion. Appl. Phys. Lett. 119: 212408.

Boehme, C., & Lupton, J. M. 2013. Challenges for organic spintronics. Nat. Nanotech. 8: 612–615.

Bogdanov, A., & Shestakov, A. 1998. Vortex states in antiferromagnetic crystals. Phys. Sol. Stat. 40: 1350–1356.

Bogdanov, A. N., & Yablonskii, D. A. 1989. Thermodynamically stable „vortices" in magnetically ordered crystals. The mixed state of magnets. Sov. Phys. JETP 68: 101–103.

Bogdanov, A. N., & Hubert, A. 1994. Thermodynamically stable magnetic vortex states in magnetic crystals. J. Mag. Magn. Mat. 138: 255–269.

Bonda, A., Uba, S., Zaleski, K., Dubowik, J., & Uba, L. 2018. Ultrafast magnetization dynamics in epitaxial Ni-Mn-Sn heusler alloy film. Act. Phys. Pol. A. 133: 501–504.

Bordoloi, A., Zannier, V., Sorba, L., Schönenberger, C., Baumgartner, A. 2020. A double quantum dot spin valve. Communications Physics 3: 135.

Borghetti, J., Snider, G. S., Kuekes, P. J., Yang, J. J., Stewart, D. R., & Williams, R. S. 2010. 'Memristive' switches enable 'stateful' logic operations via material implication. Nature. 464: 873–876.

Boulle, O., Buda-Prejbeanu, L. D., Jué, E., Miron, I. M., & Gaudin, G. 2014. Current induced domain wall dynamics in the presence of spin orbit torques. J. Appl. Phys. 115: 17D502.

Boutaleb, M., Doumi, B., Mokaddem, A., Sayede, A., & Tadjer, A. 2018. The doping effect on ferromagnetic arrangement and electronic structure of cubic AlAs with low concentration of 3d (V, Cr, and Mn) impurities. J. Superconduc. Novel Magn. 31: 2157–2163.

Bovenzi, N., Breitkreiz, M., Baireuther, P., O'Brian, T. E., Tworzydlo, J., Adagideli, I., & Beenakker, C. W. J. 2017. Chirality blockade of andreev reflection in a magnetic weyl semimetal. Phys. Rev. B. 96: 035437.

Bowden, S. R., & Gibson, U. J. 2009. Optical characterization of all-magnetic NOT gate operation in vortex rings. IEEE Trans. Magn. 45: 5326.

Bowen, M., Barthélémy, A., Bibes, M., Jacquet, E., Contour, J. P., Fert, A., Wortmann, D., & Blügel, S. 2005. Half-metallicity proven using fully spin-polarized tunnelling. J. Phys.: Condens. Matter. 17: L407–L409.

Bowerbank, T., Ding, A., Lu, C., & Xu, Y. B. 2012. Shape effect of magnetic nanoelements as biomolecular labels for magnetic biosensors. IEEE Trans. Magn. 48: 3681–3683.

Brächer, T., Fabre, M., Meyer, T., Fischer, T., Auffret, S., Boulle, O., Ebels, U., Pirro, P., & Gaudin, G. 2017. Detection of short-waved spin waves in individual microscopic spin-wave waveguides using the inverse spin hall effect. Nano Letters. 17: 7234–7241.

Brächer, T., Pirro, P., & Hillebrands, B. 2017a. Parallel pumping for magnon spintronics: amplification and manipulation of magnon spin currents on the micron-scale. Phys. Rep. – Rev. Sec. Phys. Lett. 699: 1–34.

BrainScale, S. 2015. Brain-inspired multiscale computation in neuromorphic hybrid systems. http://brainscales.kip.uni-heidelberg.de, accessed April 2018.

Brinkman, W. F., Dynes, R. C., & Rowell, J. M. 1970. Tunneling conductance of asymmetrical barriers. J. Appl. Phys. 41: 1915.

Broomhall, T. J., & Hayward, T. J. 2017. Suppression of stochastic domain wall pinning through control of gilbert damping. Sci. Rep. 7: 17100.

Brown, G. How Floppy Disk Drives Work. HowStuffWorks (https://computer.howstuffworks.com/floppy-disk-drive.htm, accessed March 2018).

Brown, G. E., & Rho, M. 2010. The Multifaced Skyrmions. World Scientific, Singapore.

Burr, J. B. 1991. Digital neural network implementations. Neural networks, concepts, applications, and implementations. 3: 237–285.

Burzuri, E., & van der Zant, H. S. J. 2014. Single-molecule spintronics. Mol. Magn.: Phys. Appl., 297–318.

Butler, W. H., Zhang, X.-G., Schulthess, T. C., & MacLaren, J. M. 2001. Spin-dependent tunneling conductance of Fe|MgO|Fe sandwiches. Phys. Rev. B. 63: 054416.

Butler, W. H., Mewes, T., Mewes, C. K. A., Visscher, P. B., Rippard, W. H., Russek, S. E., & Heindl, R. 2012. Switching distributions for perpendicular spin-torque devices within the macrospin Approximation. IEEE Trans. Magn. 48: 4684–4700.

Butenko, G. 2013. Phenomenological theory of chiral states in magnets with Dzyaloshinskii-Moriya interactions, PhD dissertation, Technischen Universität Dresden, Dresden: 91–93.

Cagli, C., Nardi, F., Harteneck, B., Tan, Z., Zhang, Y., & Ielmini, D. 2011. Resistive-switching crossbar memory based on Ni-NiO core-shell nanowires. Small. 7: 2899–2905.

Cai, W., Ellinger, F., & Tetzlaff, R. 2015. Neuronal synapse as a memristor: modeling pair-and triplet-based STDP rule. IEEE Trans. Biomed. Circuits Syst. 9: 87–95.

Cai, W. L., Wang, M. X., Cao, K. H., Yang, H. W., Peng, S. Z., Li, H. S., & Zhao, W. S. 2022. Stateful implication logic based on perpendicular magnetic tunnel junctions. Science China Information Sciences 65: 122406.

Camara, I. S., Tacchi, S., Garnier, L. C., Eddrief, M., Fortuna, F., Carlotti, G., & Marangolo, M. 2017. Magnetization dynamics of weak stripe domains in Fe-N thin films: a multitechnique complementary approach. J. Phys. Cond. Mater. 29: 465803.

Card, H., Schneider, C., & Moore, W. 1991. Hebbian plasticity in MOS synapses. IEE Proc. F (Radar and Signal Processing). 138: 13–16.

Cardona, M., Maruschak, V. A. & Titkov, A. N. 1984. Stress-induced splitting of the conduction bands of GaAs and GaSb. Solid State Commun. 50: 701.

Cardona, M., Christensen, N. E., & Fasol, G. 1988. Relativistic band structure and spin-orbit splitting of zinc-blende-type semiconductors. Phys. Rev. B. 38: 1806.

Cardoso, S., Leitao, D. C., Dias, T. M., Valadeiro, J., Silva, M. D., Chcharo, A., Silverio, V., Gaspar, J., & Freitas, P. P. 2017. Challenges and trends in magnetic sensor integration with microfluidics for biomedical applications. J. Phys. D – Appl. Phys. 50: 213001.

Carey, M. J., Maat, S., Smith, N., Fontana, R. E. E. Jr., Druist, D., Carey, K. J., Katine, J. A., Robertson, N., Boone, T. D. Jr., Alex, M., Moore, J. O., & Tsang, C. H. 2008. All-Metal current-perpendicular-to-plane giant magnetoresistance sensors for narrow-track magnetic recording. IEEE Trans. Magn. 44: 90–94.

Castillo-Sepúlveda, S., Cacilhas, R., Carvalho-Santos, V. L., Corona, R. M., & Altbir, D. 2021. Magnetic hopfions in toroidal nanostructures driven by an Oersted magnetic field. Phys. Rev. B 04: 184406.

Chappert, C., Fert, A., & Nguyen Van Dau, F. 2007. The emergence of spin electronics in data storage. Nat. Mater. 6: 813–823.

Cha, J. J., Williams, J. R., Kong, D. S., Meister, S., Peng, H. L., Bestwick, A. J., Gallagher, P., Goldhaber-Gordon, D., & Cui, Y. 2010. Magnetic doping and kondo effect in Bi2Se3 nanoribbons. Nano Lett. 10: 1076–1081.

Chan, Y. J., Huan, C. H., Chang, Y. M., Lu, Y. Y., Wu, S. Y., Wei, D. H., & Kuo, C. C. 2018. Dipolar magnetism in assembled Co nanoparticles on graphene. Phys. Chem. Chem. Phys. 20: 20629–20634.

Chen, R. H., Korotkov, A. N., & Likharev, K. K. 1996. Single-electron transistor logic. Appl. Phys. Lett. 68: 1954–1956.

Chen, P., Moser, J., Kotissek, P., Sadowski, J., Zenger, M., Weiss, D., & Wegscheider, W. 2006. All electrical measurement of spin injection in a magnetic p-n junction diode. Phys. Rev. B. 74: 241302(R).

Chen, Y., Liu, F. X., Chen, X., Xu, B., Lu, P.-L., Patwari, M. S., Xi, H., Chang, C. H., Miller, B., Ménard, D., Pant, B. B., Loven, J., Duxstad, K., Li, S., Zhang, Z., Jonston, S. B., Lamberton, R. W., Gubbins, M. A., McLaughlin, T. K., Gadbois, J. B., Ding, J., Cross, B., Xue, S. S., & Ryan, P. J. 2006a. Commercial TMR heads for hard disk drives: characterization and extendibility at 300 gbit/in^2. IEEE Trans. Magn. 42: 97–102.

Chen, A. P., Gonzalez, J., & Guslienko, K. 2018. Magnetization reversal modes in short nanotubes with chiral vortex domain walls. Materials. 11: 101.

Chen, M.-C., Sengupta, A., & Roy, K. 2018a. Magnetic skyrmion as a spintronic deep learning spiking neuron processor. IEEE Trans. Magn. 54: 1500207.

Chen, J., Liang, J. J., Yu, J. H., Qin, M. H., Fan, Z., Zeng, M., Lu, X. B., Gao, X. S., Dong, S., & Liu, J. M. 2018b. Dynamics of distorted skyrmions in strained chiral magnets. New J. Phys. 20: 063050.

Chen, X., Qin, J. Y., Yu, T., Han, X. F., & Liu, Y. W. 2018c. Micromagnetic simulation of spin torque ferromagnetic resonance in nano-ring-shape confined magnetic tunnel junctions. Appl. Phys. Lett. 113: 142406.

Chen, A. T., & Zhao, Y. G. 2018. Progress of converse magnetoelectric coupling effect in multiferroic heterostructures. Act. Phys. Sin. 67: 157513.

Cherubini, G., Cideciyan, R. D., Dellmann, L., Eleftheriou, E., Häberle, W., Jelitto, J., Kartik, V., Lantz, M. A., Ölcer, S., Pantazi, A., Rothuizen, H. E., Berman, D., Imaino, W., Jubert, P.-O., McClelland, G., Köppe, P. V., Tsuruta, K., Harasawa, T., Murata, Y., Musha, A., Noguchi, H., Ohtsu, H., Shimizu, S., & Suziki, R. 2011. 29.5 Gb/in^2 Recording areal density on barium ferrite tape. IEEE Trans. Magn. 47: 137–144.

Choi, S.-J., Kim, G.-B., Lee, K., Kim, K.-H., Yang, W.-Y., Cho, S., Bae, H.-J., Seo, D.-S., Kim, S.-I., & Lee, K.-J. 2011. Synaptic behaviors of a single metal-oxide-metal resistive device. Appl. Phys. A. 102: 1019–1025.

Choi, G.-M., Moon, C.-H., Min, B.-C., Lee, K.-J., & Cahill, D. G. 2015. Thermal spin-transfer torque driven by the spin-dependent seebeck effect in metallic spin-valves. Nat. Phys. 11: 576–581.

Choudhary, G., & Choudhary, S. 2017. First-Principle study of effects of magnesium oxide adsorption in SiCNT-based magnetic tunnel junction. J. Superconduc. Novel Magn. 30: 2302–2308.

Chi, F., Bai, X.-F., Xiu, X.-M., & Gao, X. J. 2008. Controlling spin-dependent transport via inhomogeneous magnetic flux in double-dot Aharonov-Bohm interferometer. Phys. Lett. A. 372: 1123–1128.

Chi, F., & Yuan, X.-Q. 2009. Triple quantum dot molecule as a spin-splitter. Chinese Phys. Lett. 26: 097301.

Chi, F., Dai, X.-N., & Sun, L.-L. 2010. A quantum dot spin injector with spin bias. Appl. Phys. Lett. 96: 082102.

Chiang, M., Lu, T., & Kuo, J. 1991. Analogue adaptive neural network circuit. IEE Proc. G (Circuits, Devices and Systems). 138: 717–723.

Cho, A. Y., & Arthur, J. R. 1975. Molecular beam epitaxy. Prog. Solid State Chem. 10: 157–191.

Chshiev, M., Stoeffler, D., Vedyayev, A. & Ounadjela, K. 2002. Magnetic diode effect in double-barrier tunnel junctions. Europhys. Lett. 58: 257.

Chua, L. O. 1971. Memristor – the missing circuit element. IEEE Trans. Circuit Theor. 18: 507–519.

Chung, K. H., Kim, S. N., & Lim, S. H. 2018. Magnetic parameters in giant magnetoresistance spin valve and their roles in magnetoresistance sensitivity. Thin Solid Films. 650: 44–50.

Citro, R., & Romeo, F. 2006. Pumping in a mesoscopic ring with Aharonov-Casher effect. Phys. Rev. B. 73: 233304.

Clarke, J. 1989. Principles and applications of SQUIDs. Proc. IEEE. 77: 1208.

Cleland, A. N., Esteve, D., Urbina, C., & Devoret, M. H. 1992. Very low noise photodetector based on the single electron transistor. Appl. Phys. Lett. 61: 2820.

Coffey, T., Bayindir, Z., DeCarolis, J. F., Bennett, M., Esper, G., & Agosta, C. C. 2000. Measuring radio frequency properties of materials in pulsed magnetic fields with a tunnel diode oscillator. Rev. Sci. Instr. 71: 4600–4606.

Corona, R. M., Saavedra, E., Castillo-Sepulveda, S., Escrig, J., Altbir, D., & Carvalho-Santos, V. L. 2023. Curvature-induced stabilization and field-driven dynamics of magnetic hopfions in toroidal nanorings. Nanotechnology 34: 165702.

Cort, G., Taylor, R. D., & Willis, J. O. 1982. Search for magnetism in hcp ε-Fe. J. Appl. Phys. 53: 2064–2065.

Coughlin, T., & Grochowski, E. 2012. Years of destiny: HDD capital spending and technology developments from 2012–2016. IEEE Magnetics Society Meeting.

Covi, E., Brivio, S., Serb, A., Prodromakis, T., Fanciulli, M., & Spiga, S. 2016. Analog memristive synapse in spiking networks implementing unsupervised learning. Front. Neurosci. 10: 482.

Cowburn, R. P., & Welland, M. E. 2000. Room temperature magnetic quantum cellular automata. Science. 287: 1466–1468.

Craco, L., Carara, S. S., da Silva Pereira, T. A., & Milosevic, M. V. 2016. Electronic states in an atomistic carbon quantum dot patterned in graphene. Phys. Rev. B. 93: 155417.

Cronenwett, S. M., Oosterkamp, T. H., & Kouwenhoven, L. P. 1998. A tunable kondo effect in quantum dots. Science. 281: 540–544.

Crooker, S. A., & Smith, D. L. 2005. Imaging spin flows in semiconductors subject to electric, magnetic, and strain fields. Phys. Rev. Lett. 94: 236601.

Cubukcu, M., Martin, M. B., Laczkowski, P., Vergnaud, C., Marty, A., Attane, J. P., Seneor, P., Anane, A., Deranlot, C., Fert, A., Auffret, S., Ducruet, C., Notin, L., Vila, L., & Jamet, M. 2015. Ferromagnetic tunnel contacts to graphene: contact resistance and spin signal. J. Appl. Phys. 117: 083909.

Cuadrado, R., Oroszlany, L., Deak, A., Ostler, T. A., Meo, A., Chepulskii, R. V., Apalkov, D., Evans, R. F. L., Szunyogh, L., & Chantrell, R. W. 2018. Site-Resolved contributions to the magnetic-anisotropy energy and complex spin structure of Fe/MgO sandwiches. Phys. Rev. Appl. 9: 054048.

Cygorek, M., Tamborenea, P. I., & Axt, V. M. 2016. Nonperturbative correlation effects in diluted magnetic semiconductors. Phys. Rev. B. 93: 035206.

Daimon, S., Iguchi, R., Hioki, T., Saitoh, E., & Uchida, K.-I. 2016. Thermal imaging of spin peltier effect. Nat. Commun. 7: 13754.

Dalslet, B. T., Damsgaard, C. D., Donolato, M., Stromme, M., Stromberg, M., Svedlindh, P., & Hansen, M. F. 2011. Bead magnetorelaxometry with an on-chip magnetoresistive sensor. Lab on a Chip. 11: 296–302.

Dankert, A., Kamalakar, M. V., Bergsten, J., & Dash, S. P. 2014. Spin transport and precession in graphene measured by nonlocal and three-terminal methods. Appl. Phys. Lett. 104: 192403.

Daqiq, R. 2018. Thermoelectric effects of resonant magnetic tunnel junctions. J. Magn. Magn. Mater. 465: 237–245.

Das, S., Kabiraj, A., & Mahapatra, S. 2022. Room temperature giant magnetoresistance in half-metallic Cr_2C based two-dimensional tunnel junctions. Nanoscale 14: 9409–9418.

Dash, S. P., Sharma, S., Patel, R. S., de Jong, M. P., & Jansen, R. 2009. Electrical creation of spin polarization in silicon at room temperature. Nature. 462: 491–494.

Datta, S., & Das, B. 1990. Electronic analog of the electro-optic modulator. Appl. Phys. Lett. 56: 665.

Datta, A., & Xu, X. F. 2018. Optical and thermal designs of near field transducer for heat assisted magnetic recording. Jap. J. Appl. Phys. 57: 09TA01.

Daughton, J. M., & Pohm, A. V. 2003. Design of curie point written magnetoresistance random access memory cells. J. Appl. Phys. 93: 7304.

Dave, R. W., Steiner, G., Slaughter, J. M., Sun, J. J., Craigo, B., Pietambaram, S., Smith, K., Grynkewich, G., DeHerrera, M., Akerman, J., & Tehrani, S. 2006. MgO-based tunnel junction material for high-speed toggle MRAM. IEEE Trans. Mag. 42: 1935–1939.

Davidović, D., & Tinkham, M. 1999. Spectroscopy, interactions, and level splitting in Au Nanoparticles. Phys. Rev. Lett. 83: 1644–1647.

De, A., Mondal, S., Banerjee, C., Chaurasiya, A. K., Mandal, R., Otani, Y., Mitra, R. K., & Barman, A. 2017. Investigation of magnetization dynamics in 2D Ni80Fe20 diatomic nanodot arrays. J. Phys. D. 50: 385002.

De Blois, R. W., & Rodbell, D. S. 1963. Magnetic first-Order phase transition in single-crystal MnAs. Phys. Rev. 130: 1347.

de Jong, M. J. M., & Beenakker, C. W. J. 1995. Andreev reflection in ferromagnet-superconductor junctions. Phys. Rev. Lett. 74: 1657–1660.

Dediu, V., Murakami, M., Matacotta, F. C., Taliani, C., & Barbanera, S. 2002. Room temperature spin polarized injection in organic semiconductors. Solid State Commun. 122: 181–184.

Dediu, V. A., Hueso, L. E., Bergenti, I., & Taliani, C. 2009. Spin routes in organic semiconductors. Nat. Mater. 8: 707–716.

Deepa, E., & Therese, H. A. 2018. Hierarchical nickel nanowire synthesis using polysorbate 80 as capping. Appl. Surf. Sci. 449: 48–54.

DeHon, A., & Wilson, M. J. 2004. Nanowire-based sublithographic programmable logic arrays. Proceedings of the 2004 ACM/SIGDA 12th international symposium on Field programmable gate arrays, 123–132.

Demirci, E., Dadashbaba, M., & Kurt, H. 2018. Local long-distance spin transport in single layer graphene spin filter. AIP Adv. 8: 055911.

Dempsey, K. J., Ciudad, D., & Marrows, Ch. 2011. Single electron spintronics. Phil. Trans. Roy. Soc. A, Phys. Eng. Sci. 369: 3150–3174.

Deng, C. H., Song, H. M., Parry, J., Liu, Y. H., He, S. L., Xu, X. H., Gang, Q. Q., & Zeng, H. 2018. Nanocavity induced light concentration for energy efficient heat assisted magnetic recording media. Nano Ener. 50: 750–755.

Derycke, V., Martel, R., Appenzeller, J., & Avouris, Ph. 2001. Carbon nanotube inter- and intramolecular logic gates. Nano Lett. 1: 453–456.

Desai, P., Shakya, P., Kreouzis, T., Gillin, W. P., Morley, N. A., & Gibbs, M. R. J. 2007. Magnetoresistance and efficiency measurements of Alq3-based OLEDs. Phys. Rev. B. 75: 094423.

Desai, P., Shakya, P., Kreouzis, T., & Gillin, W. P. 2007a. Magnetoresistance in organic light-emitting diode structures under illumination. Phys. Rev. B. 76: 235202.

Desai, P., Shakya, P., Kreouzis, T., & Gillin, W. P. 2007b. The role of magnetic fields on the transport and efficiency of aluminum tris(8-hydroxyquinoline) based organic light emitting diodes. J. Appl. Phys. 102: 073710.

Devolder, T., Kim, J. V., Swerts, J., Couet, S., Rao, S., Kim, W., Mertens, S., Kar, G., & Nikitin, V. 2018. Material developments and domain Wall-Based nanosecond-scale switching process in perpendicularly magnetized STT-MRAM cells. IEEE Trans. Magn. 54: 3400109.

Devoret, M. H., Estève, D., Grabert, H., Ingold, G.-L., Pothier, H., & Urbina, C. 1990. Effect of the electromagnetic environment on the Coulomb blockade in ultrasmall tunnel junctions. Phys. Rev. Lett. 64: 1824–1827.

Devoret, M. H., & Schoelkopf, R. J. 2000. Amplifying quantum signals with the single-electron transistor. Nature. 406: 1039–1046.

Dewhurst, J. K., Shallcross, S., Gross, E. K. U., & Sharma, S. 2018. Substrate-controlled ultrafast spin injection and demagnetization. Phys. Rev. Appl. 10: 044065.

Dey, R., Register, L. F., Banerjee, S. K. 2020. Two-dimensional transport model of spin-polarized tunneling in a topological-insulator/tunnel-barrier/ferromagnetic-metal heterostructure. Phys. Rev. B 102: 144414.

Dey, P. & Roy, J. N. 2021. Giant Magnetoresistance (GMR). In Spintronics, pp. 75–101, Springer, Singapore.

Dhiman, A. K., Gieniusz, R., Gruszecki, P., Kisielewski, J., Matczak, M., Kurant, Z., Sveklo, I., Guzowska, U., Tekielak, M., Stobiecki, F., & Maziewski, A. 2022. Magnetization statics and dynamics in $(Ir/Co/Pt)_6$ multilayers with Dzyaloshinskii–Moriya interaction. AIP Advances 12: 045007.

Diao, Z., Panchula, A., Ding, Y. F., Pakala, M., Wang, S. Y., Li, Z. J., Apalkov, D., Nagai, H., Driskill-Smith, A., Wang, L.-C., Chen, E., & Huai, Y. M. 2007. Spin transfer switching in dual MgO magnetic tunnel junctions. Appl. Phys. Lett. 90: 132508.

Dieny, B., Speriosu, V. S., Metin, S., Parkin, S. S. P., Gurney, B. A., Baumgart, P., & Wilhoit, D. R. 1991. Magnetotransport properties of magnetically soft spin-valve structures (invited). J. Appl. Phys. 69: 4774.

Dieny, B., & Redon, O. 2001. Magnetic device with magnetic tunnel junction, memory array and read/write methods using same. European patent EP 1466329B1.

Diep, H. T., El Hog, S., & Bailly-Reyre, A. 2018. Skyrmion crystals: dynamics and phase transition. AIP Adv. 8: 055707.

Dietl, T., Ohno, H., Matsukura, F., Cibert, J., & Ferrand, D. 2000. Zener model description of ferromagnetism in zinc-blende magnetic semiconductors. Science. 287: 1019–1022.

Ding, J. J., Kakazei, G. N., Liu, X. M., Guslienko, K. Y., & Adeyeye, A. O. 2014. Higher order vortex gyrotropic modes in circular ferromagnetic nanodots. Sci. Rep. 4: 4796.

Ding, Y., Zhao, H., Bashir, M. A., Goncharov A., & van der Heijden, P. A. 2022. Sub-Nanosecond Switching of Spin-Transfer-Torque Device for Energy-Assisted Perpendicular Magnetic Recording, IEEE Transactions on Magnetics 58: 3101106.

Dlubak, B., Martin, M. B., Deranlot, C., Bouzehouane, K., Fusil, S., Mattana, R., Petroff, F., Anane, A., Seneor, P., & Fert, A. 2012. Homogeneous pinhole free 1 nm Al_2O_3 tunnel barriers on graphene. Appl. Phys. Lett. 101: 203104.

Dolcini, F. 2011. Full electrical control of charge and spin conductance through interferometry of edge states in topological insulators. Physica E – Low-Dimensional Syst. Nanostruct. 43: 1543–1568.

Donahue, M. J., & Porter, D. G. 1999. OOMMF User's Guide, Version 1.0, Interagency Report NISTIR 6376, National Institute of Standards and Technology, Gaithersburg, MD.

Donnelly, C., Guizar-Sicairos, M., Scagnoli, V., Gliga, S., Holler, M., Raabe, J., & Heyderman, L. J. 2017. Three-dimensional magnetization structures revealed with X-ray vector nanotomography. Nature 547: 328–331.

Donnelly, C. & Scagnoli, V. 2020. Imaging three-dimensional magnetic systems with x-rays. J. Phys. Condens. Matter 32: 213001.

Döpke, C., Grothe, T., Klöcker, M., Steblinski, P., Blachowicz, T., Ehrmann, A. 2019. Electrospun multifunctional nanofiber nonwovens for bio-inspired computers. Technical Textiles, 62: E45–E47.

Dorleijn, J. W. F., & Miedema, A. R. 1977. The residual resistivities of dilute iron-based alloys in the two-current model. J. Phys. F: Met. Phys. 7: L23.

Dorrance, R., Ren, F. B., Toriyama, Y., Hafez, A. A., Yang, C. K. K., & Markovic, D. 2012. Scalability and design-space analysis of a 1T-1MTJ memory cell for STT-RAMs. IEEE Trans. Electron Devices. 59: 878–887.

Dravid, V. P., Ravikumar, V., Notis, M. R., Lyman, C. E., Dhalenne, G., & Revcolevschi, A. 1994. Stabilization of cubic zirconia with manganese oxide. J. Am. Ceram. Soc. 77: 2758–2762.

Dubowik, J., Goscianska, I., Zaleski, K., Glowinski, H., Ehresmann, A., Kakazei, G., & Bunyaev, S. A. 2011. Epitaxial growths and magnetization dynamics of Ni2MnSn heusler alloy films. Act. Phys. Pol. A. 121: 1121–1123.

Dugay, J., Tan, R. P., Meffre, A., Blon, T., Lacroix, L. M., Carrey, J., Fazzini, P. F., Lachaize, S., Chaudret, B., & Respaud, M. 2011. Room-Temperature tunnel magnetoresistance in self-assembled chemically synthesized metallic iron nanoparticles. Nano Lett. 11: 5128–5134.

Duine, R. A., Lee, K. J., Parkin, S. S. P., & Stiles, M. D. 2018. Synthetic antiferromagnetic spintronics. Nature Phys. 14: 217–219.

Dupuis, R. D., Moudy, L. A., & Dapkus, P. D. 1979. Preparation and properties of Ga1-xAlxAs-GaAs heterojunctions grown by metal-organic chemical vapor deposition. IOP Conf. Ser. 45: 1–9.

Dushenko, S., Ago, H., Kawahara, K., Tsuda, T., Kuwabata, S., Takenobu, T., Shinjo, T., Ando, Y., & Shiraishi, M. 2016. Gate-tunable spin-charge conversion and the role of spin-orbit interaction in graphene. Phys. Rev. Lett. 116: 166102.

Dutta, A., Kildishev, A. V., Shalaev, V. M., Boltasseva, A., & Marinero, E. E. 2017. Surface-plasmon opto-magnetic field enhancement for all-optical magnetization switching. Opt. Mater. Express. 7: 4316–4327.

Dyakonov, M. I. 2017. Spin Physics in Semiconductors, Springer International Publishing AG, Basel.

Dyakonov, M. I., & Perel, V. I. 1971. Spin orientation of electrons associated with the interband absorption of light in semiconductors. Sov. J. Exp. Theor. Phys. 33: 1053.

Dyakonov, M. I., & Kachorovskii, V. Yu. 1986. Spin relaxation of two-dimensional electrons in noncentrosymmetric semiconductors. Sov. Phys. Semicond. 20: 110.

Dzialoshinskii, I. E. 1957. Thermodynamic theory of "weak" ferromagnetism in antiferromagnetic substance. Sov. Phys. JETP. 5: 1259–1272.

Dzyaloshinsky, I. 1958. A thermodynamic theory of "weak" ferromagnetism of antiferromagnetics. J. Phys. Chem. Solids. 4: 241–255.

Earhart, C. M., Hughes, C. E., Gaster, R. S., Ooi, C. C., Wilson, R. J., Zhou, L. Y., Humke, E. W., Xu, L. Y., Wong, D. J., Willingham, S. B., Schwartz, E. J., Weissman, I. L., Jeffrey, S. S., Neal, J. W., Rohatgi, R., Wakeleebe, H. A., & Wang, S. X. 2014. Isolation and mutational analysis of circulating tumor cells from lung cancer patients with magnetic sifters and biochips. Lab on a Chip. 14: 78–88.

Ebels, U., Hem, J., Purbawati, A., Calafora, A. R., Murapaka, C., Vila, L., Merazzo, K. J., Jimenez, E., Cyrille, M. C., Ferreira, R., Kreissig, M., Ma, R., Ellinger, F., Lebrun, R., Wittrock, S., Cros, V., & Bortolotti, P. 2017. Spintronic based RF components. Joint Conference of the European Frequency and Time Forum and the IEEE International Frequency Control Symposium, 66–67.

Ehrmann, A., & Blachowicz, T. 2011. Adjusting exchange bias and coercivity of magnetic layered systems with varying anisotropies. J. Appl. Phys. 109: 083923.

Ehrmann, A. 2014. Examination and simulation of new magnetic materials for the possible application in memory cells. Logos Verlag Berlin, ISBN 978-3-8325-3772-2.

Ehrmann, A., Blachowicz, T., Komraus, S., Nees, M.-K., Jakobs, P.-J., Leiste, H., Mathes, M., & Schaarschmidt, M. 2015. Magnetic properties of square Py nanowires: irradiation dose and geometry dependence. J. Appl. Phys. 117: 173903.

Ehrmann, A., & Blachowicz, T. 2015. Influence of shape and dimension on magnetic anisotropies and magnetization reversal of Py, Fe, and Co nano-objects with four-fold symmetry. AIP Adv. 5: 097109.

Ehrmann, A., & Blachowicz, T. 2017. Interaction between magnetic nanoparticles in clusters. AIMS Mater. Sci. 4: 383–390.

Ehrmann, A., & Blachowicz, T. 2017a. Influence of the distance between nanoparticles in clusters on the magnetization reversal process. J. Nanomater. 2017: 5046076.

Ehrmann, A., & Blachowicz, T. 2017b. Magnetization reversal in ferromagnetic nano-rings of fourfold symmetries. Adv. Mater. Sci. Eng. 2017: 3149682.

Ehrmann, A., & Blachowicz, T. 2017d. Angle and rotational direction dependent horizontal loop shift in epitaxial Co/CoO bilayers on MgO(100). AIP Adv. 7: 115223.

Ehrmann, A., & Blachowicz, T. 2018. Systematic study of magnetization reversal in square Fe nanodots of varying dimensions in different orientations. Hyperfine Interact. 239: 8.

Ehrmann, A., & Blachowicz, T. 2019. Vortex and double-vortex nucleation during magnetization reversal in Fe nanodots of different dimensions. J. Magn. Magn. Mater. 475: 727–733.

Einati, H., Mishra, D., Friedman, N., Sheves, M., & Naaman, R. 2015. Light-Controlled spin filtering in bacteriorhodopsin. Nano Lett. 15: 1052–1056.

Ekker, N., Coughlin, T., & Handy, T. 2009. Solid State Storage 101 – An introduction to Solid State Storage. SNIA – Solid State Storage Initiative.

El Hog, S., Bailly-Reyre, A., & Diep, H. T. 2018. Stability and phase transition of skyrmion crystals generated by Dzyaloshinskii-Moriya interaction. J. Magn. Magn. Mater. 455: 32–38.

Eleftheriou, E., Haas, R., Jelitto, J., Lantz, M., & Pozidis, H. 2010. Trends in storage technologies. IEEE Data Eng. Bull. 33: 4–13.

Elliott, R. J. 1954. Theory of the effect of Spin-Orbit coupling on magnetic resonance in some semiconductors. Phy. Rev. 96: 266–279.

Emma, P. G. 1997. Understanding some simple processor-performance limits. IBM J. Res. Dev. 41: 215–232.

Enamullah, D. D., Johnson, K. G., & Suresh, A. Alam. 2016. Half-metallic Co-based quaternary Heusler alloys for spintronics: defect- and pressure-induced transitions and properties. Phys. Rev. B. 94: 184102.

Venkateswara, Enamullah, Y., Gupta, S., Varma, M. R., Singh, P., Suresh, K. G., & Alam, A. 2015. Electronic structure, magnetism, and antisite disorder in CoFeCrGe and CoMnCrAl quaternary Heusler alloys. Phys. Rev. B. 92: 224413.

Engel, H. A., Rashba, E. I., & Halperin, B. I. 2007. Out-of-Plane spin polarization from In-Plane electric and magnetic fields. Phys. Rev. Lett. 98: 036602.

Escotal.com. 2014. Computer Training – Internet Consulting. Hard Drives (http://www.escotal.com/hard drive.html, accessed March 2018).

Espitia-Rico, M. J., Díaz-Forero, J. H., & Castillo-Méndez, L. E. 2015. First-principles calculations of the pressure dependence on the structural and electronic properties of GaN/CrN superlattice. Rev. fac. ing. univ. Antioquia. 76:143–147.

Esquinazi, P., Spemann, D., Höhne, R., Setzer, A., Han, K.-H., & Butz, T. 2003. Induced magnetic ordering by proton irradiation in graphite. Phys. Rev. Lett. 91: 227201.

Evans, R. F. L., Hinzke, D., Atxitia, U., Nowak, U., Chantrell, R. W., & Chubykalo-Fesenko, O. 2012. Stochastic form of the Landau-Lifshitz-Bloch equation. Phys. Rev. B. 85: 014433.

Ezawa, M. 2007. Metallic graphene nanodisks: electronic and magnetic properties. Phys. Rev. B. 76: 245415.

Ezawa, M. 2011. Compact merons and skyrmions in thin chiral magnetic films. Phys. Rev. B 83: 100408(R).

Fabian, J., & Žutić, I. 2004. Spin-polarized current amplification and spin injection in magnetic bipolar transistors. Phys. Rev. B. 69: 115314.

Fabian, J., & Žutić, I. 2004a. Spin switch and spin amplifier: magnetic bipolar transistor in the saturation regime. Acta Phys. Pol. A. 106: 109–118.

Fabian, J., Žutić, I., & Das Sarma, S. 2004. Magnetic bipolar transistor. Appl. Phys. Lett. 84: 85.

Fabian, J., & Žutić, I. 2005. The Ebers-Moll model for magnetic bipolar transistors. Appl. Phys. Lett. 86: 133506.

FACETS. 2017. The FACETS Project. http://facets.kip.uni-heidelberg.de/index.html, accessed April 2018.

Falci, G., Feinberg, D., & Hekking, F. 2001. Correlated tunneling into a superconductor in a multiprobe hybrid structure. Europhys. Lett. 54: 255–261.

Faleev, S. V., Mryasov, O. N., & Parkin, S. S. P. 2016. Strength of the symmetry spin-filtering effect in magnetic tunnel junctions. Phys. Rev. B. 94: 174408.

Fang, W.-C., Sheu, B. J., Chen, O.-C., & Choi, J. 1992. A VLSI neural processor for image data compression using self-organization networks. IEEE Trans. Neural Networks. 3: 506–518.

Fang, B., Carpentieri, M., Hao, X., Jiang, H., Katine, J. A., Krivorotov, I. N., Ocker, B., Langer, J., Wang, K. L., Zhang, B., Azzerboni, B., Amiri, P. K., Finocchi, G., & Zeng, Z. 2016. Giant spin-torque diode sensitivity in the absence of bias magnetic field. Nature Commun. 7: 11259.

Fano, U. 1961. Effects of configuration interaction on intensities and phase shifts, Phys. Rev. 124: 1866–1878.

Fedak, W. A., & Prentis, J. J. 2009. The 1925 Born and Jordan paper "On quantum mechanics". Am. J. Phys. 77: 128–139.

Feng, W. W., Fu, X., Wan, C. H., Yuan, Z. H., Han, X. F., Quang, N. V., & Cho, S. 2015. Spin gapless semiconductor like Ti_2MnAl film as a new candidate for spintronics application. Phys. Stat. Sol. Rapid Res. Lett. 9: 641–645.

Fen, Y., Ding, H. N.; Li, X. H., Wu, B., & Chen, H. 2022. Spin transport properties of highly lattice-matched all-Heusler-alloy magnetic tunnel junction. J. Appl. Phys. 131, 133901.

Fernandez-Corbaton, I. & Vavilin, M. 2023. A Scalar Product for Computing Fundamental Quantities in Matter. Symmetry 15, 1839.

Fernandez-Roldan, J. A. & Chubykalo-Fesenko, O. 2022. Dynamics of chiral domain walls under applied current in cylindrical magnetic nanowires. APL Mater. 10, 111101.

Fernández-Rossier, J., & Palacios, J. J. 2007. Magnetism in graphene nanoislands. Phys. Rev. Lett. 99: 177204.

Ferry, D. K., Ramey, S., Shifren, L., & Akis, R. 2003. The effective potential in device modeling: the good, the bad, and the ugly. J. Comput. Electron. 1: 59–65.

Fert, A., & Campbell, I. A. 1976. Electrical resistivity of ferromagnetic nickel and iron based alloys. J. Phys. F: Met. Phys. 6: 849–871.

Fert, A., & Jaffrès, H. 2001. Conditions for efficient spin injection from a ferromagnetic metal into a semiconductor. Phys. Rev. B. 64: 184420.

Fert, A., & Jaffrès, H. 2002. Conditions for efficient spin injection from a ferromagnetic metal into a semiconductor. Phys. Rev. B. 64: 184420.

Fert, A., Cros, V., & Sampaio, J. 2013. Skyrmions on the track. Nat. Nanotech. 8: 152–156.

Fiederling, R., Keim, M., Reuscher, G., Ossau, W., Schmidt, G., Waag, A., & Molenkamp, L. W. 1999. Injection and detection of a spin-polarized current in a light-emitting diode. Nature. 402: 787–790.

Filip, A. T., Hoving, B. H., Jedema, F. J., van Wees, B. J., Dutta, B., & Borghs, S. 2000. Experimental search for the electrical spin injection in a semiconductor. Phys. Rev. B. 62: 9996–9999.

Finazzi, M., Savoini, M., Khorsand, A. R., Tsukamoto, A., Itoh, A., Duò, L., Kirilyuk, A., Rasing, Th., & Ezawa, M. 2013. Laser-induced magnetic nanostructures with tunable topological properties. Phys. Rev. Lett. 110: 177205.

Flatté, M. E., Yu, Z. G., Johnston-Halperin, E., & Awschalom, D. D. 2003. Theory of semiconductor magnetic bipolar transistors. Appl. Phys. Lett. 82: 4740–4742.

Flipse, J., Dejene, F. K., Wagenaar, D., Bauer, G. E. W., Ben Youssef, J., & van Wees, B. J. 2014. Observation of the spin peltier effect for magnetic insulators. Phys. Rev. Lett. 113: 027601.

Fluckey, S. P., Tiwari, S., Hinkle, C. L., Vandenberghe, W. G. 2022. Phys. Rev. Applied 18: 064037.

Fong, X. Y., Kim, Y. S., Yogendra, K., Fan, D. L., Sengupta, A., Raghunathan, A., & Roy, K. 2016. Spin-Transfer torque devices for logic and memory: prospects and perspectives. IEEE Transactions on Comput.-Aided Des. Integr. Circuits Syst. 35: 1–22.

Fouladi, A. A. 2016. Effect of uniaxial strain on the tunnel magnetoresistance of T-shaped graphene nanoribbon based spin-valve. Superlattices Microstruct. 95: 108–114.

Fouladi, A. A. 2017. The next nearest neighbor effect on the Spin-Dependent transport properties of DNA molecule based spin valve. J. Superconduct. Novel Magn. 30: 179–186.

Francis, T. L., Mermer, O., Veeraraghavan, G., & Wohlgenannt, M. 2004. Large magnetoresistance at room temperature in semiconducting polymer sandwich devices. New J. Phys. 6: 185–192.

Frégnac, Y. 2023. Flagship Afterthoughts: Could the Human Brain Project (HBP) Have Done Better?. eNeuro 10: ENEURO.0428-23.2023.

Freitas, P. P., & Berger, L. 1985. Observation of s-d exchange force between domain walls and electric current in very thin permalloy films. J. Appl. Phys. 57: 1266.

Freitas, P. P., Ferreira, R., & Cardosa, S. 2016. Spintronic sensors. Proc. IEEE. 104: 1894–1918.

Fried, J. P., Fangohr, H., Kostylev, M., & Metaxas, P. J. 2016. Exchange-mediated, nonlinear, out-of-plane magnetic field dependence of the ferromagnetic vortex gyrotropic mode frequency driven by core deformation. Phys. Rev. B. 94: 224407.

Friedman, J. S., Ismail, Y. I., Memik, G., Sahakian, A. V., & Wessels, B. W. 2012. Emitter-coupled spin-transistor logic. Proceedings of the 2012 IEEE/ACM International Symposium on Nanoscale Architectures, NANOARCH 2012, 139–145.

Friedman, A. L., van't Erve, O. M. J., Li, C. H., Robinson, J. T., & Jonker, B. T. 2014. Homoepitaxial tunnel barriers with functionalized graphene-on-graphene for charge and spin transport. Nat. Commun. 5: 3161.

Friedman, J. S., Wessels, B. W., Memik, G., & Sahakian, A. V. 2015. Emitter-coupled spin-transistor logic: cascaded spintronic computing beyond 10 GHz. IEEE J. Emerging Sel. Top. Circuits Syst. 5: 17–27.

Friedman, J. S., Memik, G., & Wessels, B. W. 2017. Emitter-coupled spin-transistor logic. US Patent US9780791B2.

Frisenda, R., Gaudenzi, R., Franco, C., Mas-Torrent, M., Rovira, C., Veciana, J., Alcon, I., Bromley, S. T., Burzuri, E., & van der Zant, H. S. J. 2015. Kondo effect in a neutral and stable all organic radical single molecule break junction. Nano Lett. 15: 3109–3114.

Frisenda, R., Harzmann, G. D., Gil, J. A. C., Thijssen, J. M., Mayor, M., & van der Zant, H. S. J. 2016. Stretching-induced Conductance Increase in a spin-crossover molecule. Nano Lett. 16: 4733–4737.

Frustaglia, D., Hentschel, M., & Richter, K. 2004. Aharonov-Bohm physics with spin. II. Spin-flip effects in two-dimensional ballistic systems. Phys. Rev. B. 69: 155327.

Fuchs, G. D., Krivorotov, I. N., Braganca, P. M., Emley, N. C., Garcia, A. G. F., Ralph, D. C., & Buhrmann, R. A. 2005. Adjustable spin torque in magnetic tunnel junctions with two fixed layers. Appl. Phys. Lett. 86: 152509.

Fujitsu Press Release. 2006. Fujitsu and Tokyo Institute of Technology Announce the Development of New Material for 256Mbit FeRAM Using 65-nanometer Technology – Low Power and High Speed FeRAMs for New Mobile Electronic Products, August 8, 2006 (http://www.fujitsu.com/sg/news/pr/fmal_20060808.html, accessed March 2018).

Fukushima, A., Yakushiji, K., Kubota, H., & Yuasa, S. 2015. Spin dice (physical random number generator using spin torque switching) and its thermal response. 2015 IEEE Magnetics Conf. (INTERMAG).

Fulton, T. A., & Dolan, G. J. 1987. Observation of single-electron charging effects in small tunnel junctions. Phys. Rev. Lett. 59: 109.

Gaididei, Y. B., Kravchuk, V. P., Sheka, D. D., & Mertens, F. G. 2008. Switching phenomena in magnetic vortex dynamics. Low Temp. Phys. 34: 528–534.

Gajek, M., Nowak, J. J., Sun, J. Z., Trouilloud, P. L., O'Sullivan, E. J., Abraham, D. W., Gaidis, M. C., Hu, G., Brown, S., Zhu, Y., Robertazzi, R. P., Gallagher, W. J. & Worledge, D. C. 2012. Spin torque switching of 20 nm magnetic tunnel junctions with perpendicular anisotropy. Appl. Phys. Lett. 100: 132408.

Gamba, I. M. 2014. Alternative computational methods for Boltzmann and Wigner models in charged transport systems. IEEE Xplore, International Workshop on Computational Electronics (IWCE).

Gan, L., Chung, S. H., Aschenbach, K. H., Dreyer, M., & Gomez, R. D. 2000. Pulsed-current-induced domain wall propagation in permalloy patterns observed using magnetic force microscope. IEEE Trans. Magn. 36: 3047–3049.

Gao, W. S., Hao, N. N., Zhen, F.-W., Ning, W., Wu, M., Zhu, X. D., Zheng, G., Zhang, J. L., Lu, J. W., Zhang, H. W., Xi, C. Y., Yang, J. Y., Du, H. F., Zhang, P., Zhang, Y. H., & Tian, M. L. 2017. Extremely large magnetoresistance in a topological semimetal candidate pyrite $PtBi_2$. Phys. Rev. Lett. 118: 256601.

Gao, Y., Li, S., Zhao, Y. L., Zhu, Z. Z., Cao, L. Y., Xu, J. W., Zhou, Y., & Wang, S. G. 2024. Topological transformation of magnetic hopfion in confined geometries. Microstructures 4: 2024001.

Garanin, D. A. 1997. Fokker-Planck and Landau-Lifshitz-Bloch equations for classical ferromagnets. Phys. Rev. B. 55: 3050–3057.

Garanin, D. A., & Chubykalo-Fesenko, O. 2004. Thermal fluctuations and longitudinal relaxation of single-domain magnetic particles at elevated temperatures. Phys. Rev. B. 70: 212409.

Gardelis, S., Smith, C. G., Barnes, C. H. W., Linfield, E. H., & Ritchie, D. A. 1999. Spin-valve effects in a semiconductor field-effect transistor: a spintronic device. Phys. Rev. B. 60: 7764.

Gärditz, C., & Mückl, A. G. 2005. Influence of an external magnetic field on the singlet and triplet emissions of tris-(8-hydroxyquinoline)aluminum(III) (Alq3). J. Appl. Phys. 98: 104507.

Garland, P., Birk, F. T., Jiang, W., & Davidović, D. 2013. Giant electron-spin g factors in a ferromagnetic nanoparticle. Phys. Rev. B. 88: 075303.

Garzón, E., de Rose, R., Crupi, F., Trojman, L., Teman, A., & Lanuzza, M. 2022. Adjusting thermal stability in double-barrier MTJ for energy improvement in cryogenic STT-MRAMs. Solid-State Electronics 194: 108315.

George, J. M., Pereira, L. G., Barthélémy, A., Petroff, F., Steren, L., Duvail, J. L., Fert, A., Loloee, R., Holody, P., & Schroeder, P. A. 1994. Inverse spin-valve-type magnetoresistance in spin engineered multilayered structures. Phys. Rev. Lett. 72: 408–411.

Germano, J., Martins, V. C., Cardoso, F. A., Almeida, T. M., Sousa, L., Freitas, P. P., & Piedade, M. S. 2009. A portable and autonomous magnetic detection platform for biosensing. Sensors. 9: 4119–4137.

Gerrits, Th., H., van den Berg, A. M., Hohlfeld, J., Bär, L., & Rasing, Th. 2002. Ultrafast precessional magnetization reversal by picosecond magnetic field pulse shaping. Nature. 418: 509–512.

Gholipour, B., Bastock, P., Craig, C., Khan, K., Hewak, D., & Soci, C. 2015. Amorphous metal-sulphide microfibers enable photonic synapses for brain-like computing. Adv. Opt. Mater. 3: 635–641.

Ghosh, S., Stern, N., Maertz, B., Awschalom, D. D., Xiang, G., Zhu, M., & Samarth, N. 2006. Internal magnetic field in thin ZnSe epilayers. Appl. Phys. Lett. 89: 242116.

Ghosh, S. & Grytsiuk, S. 2020. Orbitronics with uniform and nonuniform magnetic structures. Solid State Physics 71: 1–38.

Gilbert, T. L. 2004. A phenomenological theory of damping in ferromagnetic materials. IEEE Trans. Magn. 40: 3443–3449.

Giorgioni, A., Paleari, S., Cecchi, S., Vitiello, E., Grilli, E., Isella, G., Jantsch, W., Fanciulli, M., & Pezzoli, F. 2016. Strong confinement-induced engineering of the g factor and lifetime of conduction electron spins in Ge quantum wells. Nat. Commun. 7: 13886.

IBM 2018b. GMR – The Giant Magnetoresistive Head: A giant leap for IBM Research (http://www.research. ibm.com/research/gmr.html, accessed March 2018).

Gobbi, M., Golmar, F., Llopis, R., Casanova, F., & Hueso, L. E. 2011. Room-temperature spin transport in C60-based spin valves. Adv. Mater. 23: 1609–1613.

Göbel, B., Ashu Akosa, C., Tatara, G., & Mertig, I. 2020. Topological Hall signatures of magnetic hopfions. Phys. Rev. Research 2: 013315.

Godfrey, M. D., & Hendry, D. F. 1993. The computer as Von Neumann planned it. IEEE Ann. Hist. Comput. 15: 11–21.

Goldhaber-Gordon, D., Shtrikman, H., Mahalu, D., Abusch-Magder, D., Meirav, U., & Kastner, M. A. 1998. Kondo effect in a single-electron transistor. Nature. 391: 156–159.

Goldhaber-Gordon, D., Göres, J., Kastner, M. A., Shtrikman, H., Mahalu, D., & Meirav, U. 1998a. From the Kondo Regime to the mixed-valence regime in a single-electron transistor. Phys. Rev. Lett. 81: 5225.

Gomonay, E. V., & Loktev, V. M. 2014. Spintronics of antiferromagnetic systems. Low Temp. Phys. 40: 17–35.

Goodenough, J. B. 1955. Theory of the role of covalence in the Perovskite-type manganites [La, M(II)] MnO_3. Phys. Rev. 100: 564.

Gopman, D. B., Sampath, V., Ahmad, H., Bandyopadhyay, S., & Atulasimha, J. 2017. Static and dynamic magnetic properties of sputtered Fe-Ga thin films. IEEE Trans. Magn. 53: 6101304.

Gorelik, L. Y., Isacsson, A., Voinova, M. V., Kasemo, B., Shekhter, R. I., & Jonson, M. 1998. Shuttle mechanism for charge transfer in coulomb blockade nanostructures. Phys. Rev. Lett. 80: 4526.

Goumrhar, F., Bahmad, L., Mounkachi, O., & Benyoussef, A. 2018. Ab-initio calculations for the electronic and magnetic properties of Cr doped ZnTe. Comput. Condens. Matt. 15: 15–20.

Govinden, V., Prokhorenko, S., Zhang, Q., Rijal, S. Y., Nahas, Y., Bellaiche, L., & Valanoor, N. 2023. Spherical ferroelectric solitons. Nature Materials 22: 553–561.

Grabert, H., Ingold, G.-L., Devoret, M. H., Estève, D., Pothier, H., & Urbina, C. 1991. Single electron tunneling rates in multijunctions circuits. Z. Phys. B – Condens. Matt. 84: 143–155.

Grabert, H. and M. Devoret, (Ed.). 1992. Single Charge Tunneling. Plenum, New York.

Graham, D. L., Ferreira, H. A., & Freitas, P. P. 2004. Magnetoresistive-based biosensors and biochips. Trends Biotechnol. 22: 455.

Greaves, S., Kanai, Y., & Muraoka, H. 2009. Shingled recording for 2–3 Tbit/in^2. IEEE Trans. Magn. 45, 3823–2829.

Grechnev, G. E., Savchenko, N. V., Svechkarev, I. V., Lee, M. J. G., & Perz, J. M. 1989. Conduction-electron g-factors in the noble metals. Phys. Rev. B. 39: 9865–9873.

Grimsditch, M., Vavassori, P., Novosad, V., Metlushko, V., Shima, H., Otani, Y., & Fukamichi, K. 2002. Vortex chirality in an array of ferromagnetic dots. Phys. Rev. B. 65: 172419.

Grochot, K., Ogrodnik, P., Mojsiejuk, J., mazalski, P., Guzowska, U., Skowronski, W., & Stobiecki, T. 2024. Influence of ferromagnetic interlayer exchange coupling on current-induced magnetization switching and Dzyaloshinskii-Moriya interaction in Co/Pt/Co multilayer system. Sci. Rep. 14: 9938.

Grollier, J., Lacour, D., Cros, V., Hamzic, A., Vaurès, A., Fert, A., Adam, D., & Faini, G. 2002. Switching the magnetic configuration of a spin valve by current-induced domain wall motion. J. Appl. Phys. 92: 4825.

Grollier, J., Querlioz, D., & Stiles, M. D. 2016. Spintronic nanodevices for bioinspired computing. Proceedings of the IEEE 104: 2024–2039.

Grytsiuk, S., Hanke, J.-P., Hoffmann, M., Bouaziz, J., Gomonay, O., Bihlmayer, G., Lounis, S., Mokrousov, Y., Blügel, S. & 2020. Topo-logical–chiral magnetic interactions driven by emergent orbital magnetism. Nature Comm. 11: 511.

Gu, K., Guan, Y. C., Hazra, B. K., Deniz, H., Migliorini, A., Zhang, W. J., Parkin, S. S. P. 2022. Three-dimensional racetrack memory devices designed from freestanding magnetic heterostructures. Nature Nanotechnology 17: 1065–1071.

Guarnieri, M. 2011. Trailblazers in solid-state electronics. IEEE Ind. Electron. M. 5: 46–47.

Guo, E.-J., Cramer, J., Kehlberger, A., Ferguson, C. A., MacLaren, D. A., Jakob, G., & Kläui, M. 2016. Influence of thickness and interface on the low-temperature enhancement of the spin Seebeck effect in YIG films. Phys. Rev. X. 6: 031012.

Guo, L. J., Leobandun, E., Chou, S. Y. 1997. A Silicon Single-Eletrcon Transistor Memory Operating at Room Temperature. Science 275: 649–651.

Gupta, A., Zhang, R., Kumar, P., Kumar, V., & Kumar, A. 2020. Magnetochemistry 6: 15.

Gurram, M., Omar, S., & van Wees, B. J. 2018. Electrical spin injection, transport, and detection in graphene-hexagonal boron nitride van der Waals heterostructures: progress and perspectives. 2D Mater. 5: 032004.

Guslienko, K. Y. 2018. Neel skyrmion stability in ultrathin circular magnetic nanodots. Appl. Phys. Expr. 11: 063007.

Guslienko, K. Y. 2023. Emergent magnetic field and vector potential of the toroidal magnetic hopfions. Chaos, Solitons & Fractals 174: 113840.

Guslienko, K. 2024. 3D Magnetization Textures: Toroidal Magnetic Hopfion Stability in Cylindrical Samples. Nanomaterials 14: 125.

Guthrie, F. 1873. On a relation between heat and static electricity. The London, Edinburgh, and Dublin Philosophical Magazine and Journal of Science. 4th series, 46: 257–266.

Gutmann, E. 2008. Maintaining Moore's law with new memristor circuits. Ars Technica. https://arstechnica.com/uncategorized/2008/05/maintaining-moores-law-with-new-memristor-circuits, accessed April 2018.

Hamaya, K., Fujita, Y., Yamada, M., Kawano, M., Yamada, S., & Sawano, K. 2018. Spin transport and relaxation in germanium. J. Phys. D Appl. Phys. 51: 393001.

Han, W., Pi, K., Bao, W., McCreary, K. M., Li, Y., Wang, W. H., Lau, C. N., & Kawakami, R. K. 2009. Electrical detection of spin precession in single layer graphene spin valves with transparent contacts. Appl. Phys. Lett. 94: 222109.

Harfah, H., Wicaksono, Y., Sunnardianto, G. K., Majidi, M. A., Kusakabe, K. 2022. High magnetoresistance of a hexagonal boron nitride–graphene heterostructure-based MTJ through excited-electron transmission. Nanoscale Adv. 4: 117–124.

Hastings, A. 2005. The Art of Analog Layout (2nd ed.). Prentice Hall. ISBN 0-13-146410-8, Upper Saddle River, New Jersey.

Hatami, M., Bauer, G. E. W., Zhang, Q. F., & Kelly, P. J. 2007. Thermal spin-transfer torque in magnetoelectronic devices. Phys. Rev. Lett. 99: 066603.

Hayakawa, J., Ikeda, S., Miura, K., Yamanouchi, M., Lee, Y. M., Sasaki, R., Ichimura, M., Ito, K., Kawahara, T., Takemura, R., Meguro, T., Matsukura, F., Takahashi, H., Matsuoka, H., & Ohno, H. 2008. Current-induced magnetization switching in MgO barrier magnetic tunnel junctions with CoFeB-based synthetic ferrimagnetic free layers. IEEE Trans. Magn. 44: 1962–1967.

Hayashi, M., Thomas, L., Moriya, R., Rettner, C., & Parkin, S. S. P. 2008. Current-controlled magnetic domain-wall nanowire shift register. Science. 320: 209–211.

Hayashi, M., Takahashi, Y. K., & Mitani, S. 2012. Microwave assisted resonant domain wall nucleation in permalloy nanowires. Appl. Phys. Lett. 101: 172406.

Hayward, T. J., & Omari, K. A. 2017. Beyond the quasi-particle: stochastic domain wall dynamics in soft ferromagnetic nanowires. J. Phys. D. 50: 084006.

He, W., Huang, K., Ning, N., Ramanathan, K., Li, G., Jiang, Y., Sze, J., Shi, L., Zhao, R., & Pei, J. 2014. Enabling an integrated rate-temporal learning scheme on memristor. Sci. Rep. 4: 4755.

He, B., Tomasello, R., Luo, X. m.; Zhang, R., Nie, Z. Y., Carpentieri, M., Han, X. F., Finochio, G.; Yu, G. Q. All-Electrical 9-Bit Skyrmion-Based Racetrack Memory Designed with Laser Irradiation. Nano Lett. 23: 94782–9490.

Heberle, A. P., Rühle, W. W., & Ploog, K. 1994. Quantum beats of electron Larmor precession in GaAs wells. Phys. Rev. Lett. 72: 3887.

Heil, O. 1934. Improvements in or relating to electrical amplifiers and other control arrangements and devices. Patent No. GB439457 (filed in Great Britain 1934-03-02, published December 6 1935; originally filed in Germany March 2, 1934).

Heindl, R., Rippard, W. H., Russek, S. E., Pufall, M. R., & Kos, A. B. 2011. Validity of the thermal activation model for spin-transfer torque switching in magnetic tunnel junctions. J. Appl. Phys. 109: 073910.

Heintze, E., El Hallak, F., Clauß, C., Rettori, A., Pini, M. G., Totti, F., Dressel, M., & Bogani, L. 2012. Dynamic control of magnetic nanowires by light-induced domain-wall kickoffs. Nat. Mater. 12: 202–206.

Heinze, S., von Bergmann, K., Menzel, M., Brede, J., Kubetzka, A., Wiesendanger, R., Bihlmayer, G., & Blügel, S. 2011. Spontaneous atomic-scale magnetic skyrmion lattice in two dimensions. Nat. Phys. 7: 713–718.

Heller, P. 1967. Experimental investigations of critical phenomena. Rep. Progr. Phys. 30 pII: 731–826.

Heron, J. T., Trassin, M., Ashraf, K., Gajek, M., He, Q., Yang, S. Y., Nikonov, D. E., Chu, Y.-H., Salahuddin, S., & Ramesh, R. 2011. Electric-field-induced magnetization reversal in a ferromagnet-multiferroic heterostructure. Phys. Rev. Lett. 107: 217202.

Herranen, T., & Laurson, L. 2017. Bloch-line dynamics within moving domain walls in 3D ferromagnets. Phys. Rev. B. 96: 144422.

Hertel, R., Wulfhekel, W., & Kirschner, J. 2004. Domain-wall induced phase shifts in spin waves. Phys. Rev. Lett. 93: 257202.

Hirohata, A., Sagar, J., Fleet, L. R., & Parkin, S. S. P. 2016. Heusler Alloy Films for Spintronic Devices. In: (C. Felser and A. Hirohata, eds) Heusler Alloys: Properties, Growth, Applications. Springer Series in Materials Science, Basel, vol. 222, pp. 219–248.

Hitachi 2010. Hitachi Global Storage Technologies – Hitachi Research and Technology – Overview (http://www1.hgst.com/hdd/research/, accessed March 2018).

Hoffer, B. & Kvatinsky, S. 2022. Performing Stateful Logic Using Spin-Orbit Torque (SOT) MRAM. 2022 IEEE 22nd International Conference on Nanotechnology (NANO), Palma de Mallorca, Spain, 2022, pp. 571–574.

Hofherr, M., Maldonado, P., Schmitt, O., Berritta, M., Bierbrauer, U., Sadashivaiah, S., Schellekens, A. J., Koopmans, B., Steil, D., Cinchetti, M., Stadtmüller, B., Oppeneer, P. M., Mathias, S., & Aeschlimann, M. 2017. Speed and efficiency of femtosecond spin current injection into a nonmagnetic material. Phys. Rev. B. 96: 100403.

Högl, P., Matos-Abiague, A., Žutić, I., & Fabian, J. 2015. Magnetoanisotropic Andreev reflection in ferromagnet-superconductor junctions. Phys. Rev. Lett. 115: 116601.

Hoi, B. D., & Yarmohammadi, M. 2018. The role of electronic dopant on full band in-plane RKKY coupling in armchair graphene nanoribbons-magnetic impurity system. J. Magn. Magn. Mater. 454: 362–367.

Holmqvist, C., Belzig, W., & Fogelstrom, M. 2018. Non-equilibrium charge and spin transport insuperconducting-ferromagnetic- superconducting point contacts. Philos. Trans. R. Soc. A – Math. Phys. Eng. Sci. 376: 20150229.

Hook, J. R., & Hall, H. E. 2001. Solid State Physics. John Wiley & Sons. Hoboken, New Jersey ISBN 0-471-92805-4.

Hopf, H. 1931. Über die Abbildungen der dreidimensionalen Sphäre auf die Kugelfläche. Mathematische Annalen 104, 637–665. Available online: https://resolver.sub.uni-goettingen.de/purl? PPN235181684_0104.

Hortensius, J. R., Afanasiev, D., Vistoli, L., Matthiesen, M., Biebes, M, & Caviglia, A. D. 2023. Ultrafast activation of the double-exchange interaction in antiferromagnetic manganites. APL Mater. 11: 071107.

Hou, Q. Y., Li, W. L., Jia, X. F., & Xu, Z. C. 2018. Effects of stress and point defect on the physical properties of ZnO:Nd. J. Magn. Magn. Mater. 461: 82–90.

Houshangh, A., Khymyn, R., Fulara, H., Gangwar, A., Haidar, M., Etesami, S. R., Ferreira, R., Freitas, P. P., Dvornik, M., Dumas, R. K., & Akerman, J. 2018. Spin transfer torque driven higher-order propagating spin waves in nano-contact magnetic tunnel junctions. Nat. Commun. 9: 4374.

Hromadová, L., Martonák, R., & Tosatti, E. 2013. Structure change, layer sliding, and metallization in high-pressure MoS₂. Phys. Rev. B. 87: 144105–144105.

Hsu, P.-J., Kubetzka, A., Finco, A., Romming, N., von Bergmann, K., & Wiesendanger, R. 2017. Electric-field-driven switching of individual magnetic skyrmions. Nat. Nanotech. 12: 123–126.

Hsu, P.-J., Rozsa, L., Finco, A., Schmidt, L., Palotas, K., Vedmedenko, E., Udvardi, L., Szunyogh, L., Kubetzka, A., von Bergmann, K., & Wiesendanger, R. 2018. Inducing skyrmions in ultrathin Fe films by hydrogen exposure. Nat. Commun. 9: 1571.

Hsu, W.-H. & Victora, R. H. 2022. Heat-assisted magnetic recording – micromagnetic modeling of recording media and areal density: a review. J. Magn. Magn. Mater. 563: 169973.

Hu, B., & Wu, Y. 2007. Tuning magnetoresistance between positive and negative values in organic semiconductors. Nat. Mater. 6: 985–991.

Hu, S., Wu, H., Liu, Y., Chen, T., Liu, Z., Yu, Q., Yin, Y., & Hosaka, S. 2013. Design of an electronic synapse with spike time dependent plasticity based on resistive memory device. J. Appl. Phys. 113: 114502.

Hu, S., Itoh, H., & Kimura, T. 2014. Efficient thermal spin injection using CoFeAl nanowire. NPG Asia Mater. 6: e127.

Hu, S., & Kimura, T. 2014. Significant modulation of electrical spin accumulation by efficient thermal spin injection. Phys. Rev. B. 90: 134412.

Hu, M., Chen, Y., Yang, J. J., Wang, Y., & Li, H. 2016. A compact memristor-based dynamic synapse for spiking neural networks. IEEE Trans. Comput.-Aided Des. Integr. Circuits Syst. 36: 1353–1366.

Huai, Y., Albert, F., Nguyen, P., Pakala, M., & Valet, T. 2004. Observation of spin-transfer switching in deep submicron-sized and low-resistance magnetic tunnel junctions. Appl. Phys. Lett. 84: 3118.

Huang, Y., Duan, X., Wei, Q., & Lieber, C. M. 2001. Directed assembly of one-dimensional nanostructures into functional networks. Science. 291: 630.

Huang, Y., Duan, X., Cui, Y., Lauhon, L. J., Kim, K.-H., & Lieber, C. M. 2001a. Logic gates and computation from assembled nanowire building blocks. Science. 294: 1313.

Huang, B. Q., Monsma, D. J., & Appelbaum, I. 2007. Coherent spin transport through a 350 micron thick silicon wafer. Phys. Rev. Lett. 99: 177209.

Huang, L., Schofield, M. A., & Zhu, Y. 2010. Control of double-vortex domain configurations in a shape-engineered trilayer nanomagnet system. Adv. Mater. 22: 492.

Huang, Y. Q., Polojärvi, V., Hiura, S., Höjer, P., Aho, A.; Isoaho, R., Hakkarainen, T., Guina, M., Sato, S.; Takayama, J., Murayama, A.; Buyanova, I. A., & Chen, W. M. M. 2021. Room-temperature electron spin polarization exceeding 90% in an opto-spintronic semiconductor nanostructure via remote spin filtering. Nature Photonics 15: 475–482.

Project, Human Brain. 2017. https://www.humanbrainproject.eu/en/, accessed April 2018.

Huminiuc, T., Whear, O., Takahashi, T., Kim, J. Y., Vick, A., Vallejo-Fernandez, G., O'Grady, K., & Hirohata, A. 2018. Growth and characterisation of ferromagnetic and antiferromagnetic $Fe_{2+x}VyAl$ Heusler alloy films. J. Phys. D. 51: 325003.

Hung, C.-Y., & Berger, L. 1988. Exchange forces between domain wall and electric current in permalloy films of variable thickness. J. Appl. Phys. 63: 4276.

Hussein, M. S. D. A., Daghofer, M., Dagotto, E., & Moreo, A. 2018. Phenomenological three-orbital spin-fermion model for cuprates. Phys. Rev. B. 98: 035124.

Hwang, D. G., Lee, S. S., & Park, C. M. 1998. Effect of roughness slope on exchange biasing in NiO spinvalves. Appl. Phys. Lett. 72: 2162.

Hwang, E. H., Adam, S., & Das Sarma, S. 2007. Carrier transport in two-dimensional graphene layers. Phys. Rev. Lett. 98: 186806.

IBM. 1955. 650 magnetic drum data-processing machine – manual of operation (http://www.bitsavers.org/pdf/ibm/650/22-6060-2_650_OperMan.pdf, accessed March 2018).

IBM 350 disk storage unit. 1956. IBM Archives (http://www-03.ibm.com/ibm/history/exhibits/storage/storage_350.html, accessed March 2018).

IBM 1996. Did you ever wonder how your hard disk drive works? IBM Research (http://www.research.ibm.com/research/gmr/basics.html, accessed March 2018).

IBM. 2012. IBM research determines atomic limits of magnetic memory. IBM News Room (https://www-03.ibm.com/press/us/en/pressrelease/36473.wss, accessed March 2018).

IBM. 2014. IBM research: neurosynaptic chips, http://www.research.ibm.com/cognitive-computing/neurosynaptic-chips.shtml, accessed April 2018.

IBM100 – Icons of Progress. 2018. Magnetic stripe technology (http://www-03.ibm.com/ibm/history/ibm100/us/en/icons/magnetic/, accessed March 2018).

IBM 100 – Icons of Progress. 2018a. The floppy disk (http://www-03.ibm.com/ibm/history/ibm100/us/en/icons/floppy/, accessed March 2018).

IEEE. 2009. Milestones: invention of the First Transistor at Bell Telephone Laboratories, Inc., 1947. IEEE Global History Network.

Ielmini, D., Nardi, F., & Cagli, C. 2011. Physical models of size-dependent nanofilament formation and rupture in NiO resistive switching memories. Nanotechnology. 22: 254022.

Ielmini, D., Cagli, C., Nardi, F., & Zhang, Y. 2013. Nanowire-based resistive switching memories: devices, operation and scaling. J. Phys. D: Appl. Phys. 46: 074006.

Ikeda, S., Hayakawa, J., Ashizawa, Y., Lee, Y. M., Miura, K., Hasegawa, H., Tsunoda, M., Matsukura, F., & Ohno, H. 2008. Tunnel magnetoresistance of 604% at 300 K by suppression of Ta diffusion in CoFeB/MgO/CoFeB pseudo-spin-valves annealed at high temperature. Appl. Phys. Lett. 93: 39–42.

Ikeda, S., Miura, K., Yamamoto, H., Mizunuma, K., Gan, H. D., Endo, M., Kanai, S., Hayakawa, J., Matsukura, F., & Ohno, H. 2010. A perpendicular-anisotropy CoFeB-MgO magnetic tunnel junction. Nat. Mater. 9: 721–724.

Ikegami, T., Kawayama, I., Tonouchi, M., Nakao, S., Yamashita, Y., & Tada, H. 2008. Planar-type spin valves based on low-molecular-weight organic materials with La0.67Sr0.33MnO3 electrodes. Appl. Phys. Lett. 9: 153304.

Ilinskaya, O. A., Kulinich, S. I., Krive, I. V., Shekhter, R. I., Park, H. C., & Jonson, M. 2018. Mechanically induced thermal breakdown in magnetic shuttle structures. New J. Phys. 20: 063036.

Imai, Y., Amasaki, R., Yanagibashi, Y., Suzuki, S., Shikura, R., & Yagi, S. 2024. Magnetically Induced Near-Infrared Circularly Polarized Electroluminescence from an Achiral Perovskite Light-Emitting Diode. Magnetochemistry 10: 39.

Imre, A., Csaba, G., Ji, L., Orlov, A., Bernstein, G. H., & Porod, W. 2006. Majority logic gate for magnetic quantum-dot cellular automata. Science. 311: 205–208.

Indolese, D., Zihlmann, S., Makk, P., Junger, C., Thodkar, K., & Schonenberger, C. 2018. Wideband and on-chip excitation for dynamical spin injection into graphene. Phys. Rev. Appl. 10: 044053.

Ingold, G.-L., & Grabert, H. 1991. Finite-temperature current-voltage characteristics of ultrasmall tunnel junctions. Europhys. Lett. 14: 371–376.

Isaacson, R. A. 1968. Electron spin resonance in n-type InSb. Phys. Rev. 169: 312–314.

Ishibashi, S., Seki, T., Nozaki, T., Kubota, H., Yakata, S., Fukushima, A., Yuasa, S., Maehara, H., Tsunekawa, K., Djayaprawira, D. D., & Suzuki, Y. 2010. Large diode sensitivity of CoFeB/MgO/CoFeB magnetic tunnel junctions. Appl. Phys. Express. 3, 073001.

Ishii, T., Yamakawa, H., Kanaki, T., Miyamoto, T., Kida, N., Okamoto, H., Tanaka, M., & Ohya, S. 2018. Ultrafast magnetization modulation induced by the electric field component of a terahertz pulse in a ferromagnetic-semiconductor thin film. Sci. Rep. 8: 6901.

Ishikawa, M., Oka, T., Fujita, Y., Sugiyama, H., Saito, Y., & Hamaya, K. 2017. Spin relaxation through lateral spin transport in heavily doped n-type silicon. Phys. Rev. B. 95: 115302.

Islam, S. K. F., Dutta, P., Jayannavar, A. M., & Saha, A. 2018. Probing decoupled edge states in a zigzag phosphorene nanoribbon via RKKY exchange interaction. Phys. Rev. B. 97: 235424.

Iwasaki, J., Mochizuki, M & Nagaosa, N. 2013. Current-induced skyrmion dynamics in constricted geometries. Nat. Nano. 8: 742–747.

Iwata-Harms, J. M., Jan, G., Liu, H. L., Serrano-Guisan, S., Zhu, J., Thomas, L., Tong, R. Y., Sundar, V. & Wang, P. K. 2018. High-temperature thermal stability driven by magnetization dilution in CoFeB free layers for spin-transfer-torque magnetic random access memory. Sci. Rep. 8: 14409.

Jackson, B. L., Rajendran, B., Corrado, G. S., Breitwisch, M., Burr, G. W., Cheek, R., Gopalakrishnan, K., Raoux, S., Rettner, C. T., Padilla, A., Schrott, A. G., Shenoy, R. S., Kurdi, B. N., Lam, C. H., & Modha, D. S. 2013. Nanoscale electronic synapses using phase change devices. ACM J. Emerging Technol. Comput. Syst. 9: 12.

Jacome, M., He, C., de Veciana, G., & Bijansky, S. 2004. Defect tolerant probabilistic design paradigm for nanotechnologies. Proceedings of the 41st annual Design Automation Conference, 596–601.

Jansen, R. 2003. The spin-valve transistor: a review and outlook. J. Phys. D. 36: R289.

Jedema, F. J., Filip, A. T., & van Wees, B. J. 2001. Electrical spin injection and accumulation at room temperature in an all-metal mesoscopic spin valve. Nature. 410: 345–348.

Jedema, F. J., Heersche, H. B., Filip, A. T., Baselmans, J. J. A., & van Wees, B. J. 2002. Electrical detection of spin precession in a metallic mesoscopic spin valve. Nature. 416: 713–716.

Jeon, K.-R., Min, B.-C., Park, S.-Y., Lee, K.-D., Song, H.-S., Park, Y.-H., Jo, Y.-H., & Shin, S.-C. 2012. Thermal spin injection and accumulation in CoFe/MgO/n-type Ge contacts. Sci. Rep. 2: 962.

Jiang, Q., & Zhong, Z. 2017. Research and development of Ce-containing $Nd_2Fe_{14}B$-type alloys and permanent magnetic materials. J. Mat. Sci. Technol. 33: 1087–1096.

Jin, S., Tiefel, T. H., McCormack, M., Fastnacht, R. A., Ramesh, R., & Chen, L. H. 1994. Thousandfold change in resistivity in magnetoresistive La-Ca-Mn-O films. Science. 264: 413–415.

Jin, C. D., Song, C. K., Wang, J. B., & Liu, Q. F. 2016. Dynamics of antiferromagnetic skyrmion driven by the spin Hall effect. Appl. Phys. Lett. 109: 182404.

Johari, P., & Shenoy, V. B. 2012. Tuning the electronic properties of semiconducting transition metal dichalcogenides by applying mechanical strains. ACS Nano. 6: 5449–5456.

Johnson, M. 2001. Spin injection and detection in a ferromagnetic metal/2DEG structure. Physica E. 10: 472–477.

Johnson, M., & Silsbee, R. H. 1985. Interfacial charge-spin coupling: injection and detection of spin magnetization in metals. Phys. Rev. Lett. 55: 1790.

Johnson, M., & Silsbee, R. H. 1987. Thermodynamic analysis of interfacial transport and of the thermomagnetoelectric system. Phys. Rev. B. 35: 4959.

Johnson, M., & Silsbee, R. H. 1988. Ferromagnet-nonferromagnet interface resistance. Phys. Rev. Lett. 60: 377.

Jones, J. R. 1976. Coincident Current Ferrite Core Memories. Byte Magazine, July 1976.

Jonietz, F., Mühlbauer, S., Pfleiderer, C., Neubauer, A., Münzer, W., Bauer, A., Adams, T., Georgii, R., Böni, P., Duine, R. A., Everschor, K., Garst, M., & Rosch, A. 2010. Spin transfer torques in MnSi at ultra-low current densities. Science. 330: 1648–1651.

Jourdan, M. 2014. Revival of Heusler compounds for spintronics. Mater. Today. 17: 362–363.

Jovanovic, B., Brum, R. M., & Torres, L. 2014. A hybrid magnetic/complementary metal oxide semiconductor three-context memory bit cell for non-volatile circuit design. J. Appl. Phys. 115: 134316.

Juge, R., Je, S. G., Chaves, D. D., Pizzini, S., Buda-Prejbeanu, L. D., Aballe, L., Foerster, M., Locatelli, A., Mentes, T. O., Sala, A., Maccherozzi, F., Dhesi, S. S., Auffret, S., Gautier, E., Gaudin, G., Vogel, J., & Boulle, O. 2018. Magnetic skyrmions in confined geometries: effect of the magnetic field and the disorder. J. Magn. Magn. Mater. 455: 3–8.

Julliere, M. 1975. Tunneling between ferromagnetic films. Phys. Lett. A. 54: 225–226.

Juma, I. G.; Kim, G. W.; Jariwala, D.; Behura, S. K. Direct growth of hexagonal boron nitride on non-metallic substrates and its heterostructures with graphene. Perspective 2021, 24(11), 103374.

Jung, Y., Kim, J., Ryu, K., Kim, J. P., Kang, S. H., & Jung, S. O. 2014. An MTJ-based non-volatile flip-flop for high-performance SoC. Int. J. Circuit Theor. Appl. 42: 394–406.

Jungfleisch, M. B., Sklenar, J., Ding, J. J., Park, J., Pearson, J. E., Novosad, V., Schiffer, P., & Hoffmann, A. 2017. High-frequency dynamics modulated by collective magnetization reversal in artificial spin ice. Phys. Rev. Appl. 8: 064026.

Jungfleisch, M. B., Ding, J. J., Zhang, W., Jiang, W. J., Pearson, J. E., Novosad, V., & Hoffmann, A. 2017a. Insulating nanomagnets driven by spin torque. Nano Lett. 17: 8–14.

Kakazei, G. N., Guslienko, K. Y., Verba, R. V., Ding, J., Liu, X. M., & Adeyeye, A. O. 2020. Non-uniform along thickness spin excitations in magnetic vortex-state nanodots. Low. Temp. Phys. 46: 863–868.

Kalinowski, J., Cocchi, M., Virgili, D., Di Marco, P., & Fattori, V. 2003. Magnetic field effects on emission and current in Alq3-based electroluminescent diodes. Chem. Phys. Lett. 380: 710–715.

Kalinowski, J., Cocchi, M., Virgili, D., Fattori, V., & Di Marco, P. 2004. Magnetic field effects on organic electrophosphorescence. Phys. Rev. B. 70: 205303.

Kamalakar, M. V., Dankert, A., & Dash, S. P. 2017. Spintronics with graphene and von der Waals heterostructures. Contemporary topics in semiconductor spintronics, World Scientific. 241–258.

Kanda, N., Higuchi, T., Shimizu, H., Konishi, K., Yoshioka, K., & Kuwata-Gonokami, M. 2011. The vectorial control of magnetization by light. Nat. Commun. 2: 362.

Kang, D.-H., Jun, H.-G., Ryoo, K.-C., Jeong, H., & Sohn, H. 2015. Emulation of spike-timing dependent plasticity in nano-scale phase change memory. Neurocomputing. 155: 153–158.

Kang, W., Huang, Y. Q., Zheng, C. T., Lv, W. F., Lei, N., Zhang, Y. G., Zhang, X. C., Zhou, Y., & Zhao, W. S. 2016. Voltage Controlled Magnetic Skyrmion Motion for Racetrack Memory. Sci. Rep. 6: 23164.

Kanomata, T., Chieda, Y., Endo, K., Okada, H., Nagasako, M., Kobayashi, K., Kainuma, R., Umetsu, R. Y., Takahashi, H., Furutani, Y., Nishihara, H., Abe, K., Miura, Y., & Shirai, M. 2010. Magnetic properties of the half-metallic Heusler alloys Co_2VAl and Co_2VGa under pressure. Phys. Rev. B. 82: 144415.

Kaplan, D. 2003. Hands-On Electronics. New York: Cambridge University Press. pp. 47–61. ISBN 978-0-511-07668-8.

Kar, U., Panda, J., & Nath, T. K. 2018. Low temperature electrical spin injection from highly spin polarized Co2CrAl Heusler alloy into p-Si. J. Nanosci. Nanotech. 18: 4135–4141.

Karolak, M., & Jako, D. 2016. Effects of valence, geometry and electronic correlations on transport in transition metal benzene sandwich molecules. J. Phys. Cond. Mat. 28: 1–16.

Karwacki, Ł., Trocha, P., & Barnaś, J. 2015. Magnon transport through a quantum dot: Conversion to electron spin and charge currennts. Phys. Rev. B 92: 235449.

Kasavajhala, V. 2011. SSD vs HDD Price and Performance Study, a Dell technical white paper. Dell PowerVault Technical Marketing.

Kastner, M. 1993. Artificial atoms. Phys. Today. 46: 24.

Kasuya, T. 1956. A theory of metallic ferro- and antiferromagnetism on Zener's model. Prog. Theor. Phys. 16: 45–57.

Katayama, R. & Sugiura, S. 2020. Simulation on near-field light on recording medium generated by semiconductor ring resonator with metal nano-antenna for heat-assisted magnetic recording. Optical Review 27: 432–440.

Katine, J. A., Albert, F. J., Buhrman, R. A., Myers, E. B., & Ralph, D. C. 2000. Current-driven magnetization reversal and spin-wave excitations in Co / Cu / Co pillars. Phys. Rev. Lett. 84: 3149.

Kato, Y., Myers, R. C., Gossard, A. C., & Awschalom, D. D. 2004. Coherent spin manipulation without magnetic fields in strained semiconductors. Nature. 427: 50–53.

Kato, Y., Myers, R. C., Gossard, A. C., & Awschalom, D. D. 2004a. Current-induced spin polarization in strained semiconductors. Phys. Rev. Lett. 93: 176601.

Kato, Y. K., Myers, R. C., Gossard, A. C., & Awschalom, D. D. 2005. Electrical initialization and manipulation of electron spins in an L-shaped strained n-InGaAs channel. Appl. Phys. Lett. 87: 022503.

Kaufmann, U., Schneider, J., & Räuber, A. 1976. ESR detection of antisite lattice defects in GaP, $CdSiP_2$, and $ZnGeP_2$. Appl. Phys. Lett. 29: 312–313.

Kawahara, T., Ito, K., Takemura, R., & Ohno, H. 2012. Spin-transfer torque RAM technology: review and prospect. Microelectron. Reliab. 52: 613–627.

Kawamura, H., Shirotani, I., & Tachikawa, K. 1985. Anomalous superconductivity and pressure induced phase transitions in black phosphorus. Solid State Commun. 54: 775–778.

Keller, J., Miltényi, P., Beschoten, B., Güntherodt, G., Nowak, U., & Usadel, K. D. 2002. Domain state model for exchange bias. II. Experiments. Phys. Rev. B. 66: 014431.

Kent, N., Reynolds, N., Raftrey, D., Campbell, I. T. G., Virasawmy, S., Dhuey, S., Chopdekar, R. V., Hierro-Rodriguez, A., Sorrentino, A., Pereiro, E., Ferrer, S., Hellman, F., Sutcliffe, P., & Fischer, P. 2021. Creation and observation of Hopfions in magnetic multilayer systems. Nature Communications 12: 1562.

Kersten, S. P., Schellekens, A. J., Koopmans, B., & Bobbert, P. A. 2011. Magnetic-field dependence of the electroluminescence of organic light-emitting diodes: a competition between exciton formation and spin mixing. Phys. Rev. Lett. 106: 197402.

Khan, U., Adeela, N., Naz, S., Irfan, M., Khan, K., Sagar, R. U. R., Aslam, S., & Wu, D. 2018. Synthesis and low temperature magnetic measurements of polycrystalline Gadolinium nanowires. Mater. Lett. 228: 266–269.

Khang, N. H. D., Ueda, Y., & Hai, P. N. 2018. A conductive topological insulator with large spin hall effect for ultralow power spin-orbit torque switching. Nat. Mater. 17: 808–813.

Khitun, A., & Wang, K. L. 2006. Nano logic circuits with spin wave bus. Proceedings. Third International Conference on Information Technology: New Generation.

Khitun, A., Bao, M.-Q., & Wang, K. L. 2008. Spin wave magnetic nanofabric: a new approach to spin-based logic circuitry. IEEE Trans. Magn. 44: 2121–2152.

Khodzhaev, Z. & Turgut, E. 2022. Hopfion dynamics in chiral magnets. J. Phys. Condens. Matter 34: 225805.

Khomskii, D. I. 2006. Multiferroics: different ways to combine magnetism and ferroelectricity. J. Magn. Magn. Mater. 306: 1–8.

Khvalkovskiy, A. V., Apalkov, D., Watts, S., Chepulskii, R., Beach, R. S., Ong, A., Tang, X., Driskill-Smith, A., Butler, W. H., Visscher, P. B., Lottis, D., Chen, E., Nikitin, V., & Krounbi, M. 2013. Basic principles of STT-MRAM cell operation in memory arrays. J. Phys. D: Appl. Phys. 46: 074001.

Khvalkovskiy, A. V., Mikhailov, A. P., Leshchiner, D. R., & Apalkov, D. 2018. Magnetic memory with a switchable reference layer. J. Appl. Phys. 124: 133902.

Khymyn, R., Lisenkov, I., Voorheis, J., Sulymenko, O., Prokopenko, O., Tiberkevich, V., Akerman, J. & Slavin, A. 2018. Ultra-fast artificial neuron: generation of picosecond-duration spikes in a current-driven antiferromagnetic auto-oscillator. Sci. Rep. 8: 15727.

Kikkawa, J. M., & Awschalom, D. D. 1998. Resonant spin amplification in n-Type GaAs. Phys. Rev. Lett. 80: 4313.

Kikkawa, J. M., & Awschalom, D. D. 1999. Lateral drag of spin coherence in gallium arsenide. Nature. 397: 139–141.

Kief, M. T., & Victora, R. H. 2018. Materials for heat-assisted magnetic recording. MRS Bulletin. 43: 87–92.

Kim, R., Holmes, S., Halle, S., Dai, V., Meiring, J., Dave, A., Colburn, M. E., & Levinson, H. J. 2010. 22nm technology node active layer patterning for planar transistor devices. J. Micro/Nanolith. MEMS MOEMS. 9: 013001.

Kim, P. D., Orlov, V. A., Prokopenko, V. S., Zamai, S. S., Prints, V. Y., Rudenko, R. Y., & Rudenko, T. V. 2015. On the low-frequency resonance of magnetic vortices in micro- and nanodots. Phys. Solid State. 57: 30–37.

Kimishima, Y. 2009. Room temperature magneto-resistance effects of half-metallic oxides. spintronics: materials, applications, and devices 1–47.

Kimling, J., Choi, G.-M., Brangham, J. T., Matalla-Wagner, T., Huebner, T., Kuschel, T., Yang, F., & Cahill, D. G. 2017. Picosecond spin Seebeck effect. Phys. Rev. Lett. 118: 057201.

Kiselev, N. S., Bogdanov, A. N., Schäfer, R., & Rößler, U. K. 2011. Chiral skyrmions in thin magnetic films: new objects for magnetic storage technologies? J. Phys. D: Appl. Phys. 44: 392001.

Kitahara, M., Hara, K., Suzuki, S., Iwasaki, H., Yagi, S., & Imai, Y. 2023. Red–green–blue–yellow (RGBY) magnetic circularly polarized electroluminescence from iridium(III)-magnetic circularly polarized organic light-emitting diodes. Organic Electronics 119: 106814.

Kittel, C. 1976. Introduction to Solid State Physics, Wiley, Hoboken, New Jersey, 5th Edition.

Kleemann, W. 1993. Random-field induced antiferromagnetic, ferroelectric and structural domain states. Int. J. Mod. Phys. B. 7: 2469.

Knobel, R. G., & Cleland, A. N. 2003. Nanometre-scale displacement sensing using a single electron transistor. Nature. 424: 291–293.

Koch, R. H., Katine, J. A., & Sun, J. Z. 2004. Time-Resolved Reversal of Spin-Transfer Switching in a Nanomagnet. Phys. Rev. Lett. 92: 088302.

Komeda, T., Isshiki, H., Liu, J., Zhang, Y.-F., Lorente, N., Katoh, K., Breedlove, B. K., & Yamashita, M. 2011. Observation and electric current control of a local spin in a single-molecule magnet. Nat. Commun. 2: 217.

Kon, J.-I., Maruyama, T., Kojima, Y., Takahashi, Y., Sugatani, S., Ogino, K., Hoshino, H., Isobe, H., Kurokawa, M., & Yamada, A. 2012. Optimization of chemically amplified resist for high-volume manufacturing by electron-beam direct writing toward 14nm node and beyond. Proc. SPIE 8323, Alternative Lithographic Technologies IV: 832324.

Kondo, J. 1964. Resistance Minimum in Dilute Magnetic Alloys. Prog. Theor. Phys. 32: 37.

Kondo, K., Choi, J. Y., Baek, J. U., Jun, H. S., Jung, S., Shim, T. H., & Park, J. G. 2018. A two-terminal perpendicular spin-transfer torque based artificial neuron. J. Phys. D. Appl. Phys. 51: 504002.

Konopka, J. Conduction electron spin resonance in InAs. 1967. Phys. Lett. 26 A: 29–31.

Koopmans, B., Malinowski, G., Dalla Longa, F., Steiauf, D., Fähnle, M., Roth, T., Cinchetti, M., & Aeschlimann, M. 2010. Explaining the paradoxical diversity of ultrafast laser-induced demagnetization. Nat. Mater. 9: 259–265.

Korotkov, A. N., Chen, R. H., & Likharev, K. K. 1995. Possible performance of capacitively coupled single-electron transistors in digital circuits. J. Appl. Phys. 78: 2520.

Koshihara, S., Oiwa, A., Hirasawa, M., Katsumoto, S., Iye, Y., Urano, C., Takagi, H., & Munekata, H. 1997. Ferromagnetic Order Induced by Photogenerated Carriers in Magnetic III-V Semiconductor Heterostructures of (In,Mn)As/GaSb. Phys. Rev. Lett. 78: 4617.

Kostylev, M. P., Serga, A. A., Schneider, T., Leven, B., & Hillebrands, B. 2005. Spinwave logical gates. Appl. Phys. Lett. 87: 153501.

Kouwenhoven, L. P., & Marcus, C. M. 1998. Quantum dots. Phys. World June. 1998: 35–39.

Kouwenhoven, L., & Glazman, L. 2001. Revival of the Kondo effect. Phys. World January. 2001, 33–38.

Kozicki, M. N., Yun, M., Hilt, L., & Singh, A. 1999. Applications of Programmable Resistance Changes in Metal-Doped Chalcogenides. Electrochem. Soc. Proc. 99–13: 298–309.

Krainov, I. V., Klier, J., Dmietriev, A. P., Klyatskaya, S., Ruben, M., Wernsdorfer, W., & Gornyi, I. V. 2017. Giant Magnetoresistance in Carbon Nanotubes with Single-Molecule Magnets TbPc$_2$. ACS Nano. 11: 6868–6880.

Krasheninnikov, A. V., Lehtinen, P. O., Foster, A. S., Pyykkö, P., & Nieminen, R. M. 2009. Embedding transition-metal atoms in graphene: structure, bonding, and magnetism. Phys. Rev. Lett. 102: 126807.

Kriegner, D., Výborný, K., Olejník, K., Reichlová, H., Novák, V., Marti, X., Gazquez, J., Saidl, V., Nemec, P., Volobuev, V. V., Springholz, G., Holý, V., & Jungwirth, T. 2016. Multiple-stable anisotropic magnetoresistance memory in antiferromagnetic MnTe. Nat. Commun. 7: 11623.

Krivorotov, I. N., Emley, N. C., Sankey, J. C., Kiselev, S. I., Ralph, D. C., & Buhrman, R. A. 2005. Time-Domain Measurements of Nanomagnet Dynamics Driven by Spin-Transfer Torques. Science. 307: 228–231.

Kroemer, H. 1982. Heterostructure Bipolar Transistors and Integrated Circuits. Proceedings of the IEEE 70: 13–25.

Kubatkin, S., Danilov, A., Hjort, M., Cornil, J., Brédas, J.-L., Stuhr-Hansen, N., Hedegard, P., & Bjornholm, T. 2003. Single-electron transistor of a single organic molecule with access to several redox states. Nature. 425: 698–701.

Kuchibhotla, M., Talapatra, A.; Haldar, A., & Adeyeye, A. O. 2021. Magnetization dynamics of single and trilayer permalloy nanodots. J. Appl. Phys. 130: 083906.

Kuchkin, V. M., Kiselev, N. S., Rybakov, F. N., Lobanov, I. S., Blügel, S., & Uzdin, V. M. 2023. Heliknoton in a film of cubic chiral magnet. Front. Phys. 11: 1201018.

Kuerbanjiang, B., Fujita, Y., Yamada, M., Yamada, S., Sanchez, A. M., Hasnip, P. J., Ghasemi, A., Kepaptsoglou, D., Bell, G., Sawano, K., Hamaya, K., & Lazarov, V. K. 2018. Correlation between spin transport signal and Hensler/semiconductor interface quality in lateral spin-valve devices. Phys. Rev. B. 98: 115304.

Kulinich, S. I., Gorelik, L. Y., Kalinenko, A. N., Krive, I. V., Shekhter, R. I., Park, Y. W., & Jonson, M. 2014. Single-Electron Shuttle Based on Electron Spin. Phys. Rev. Lett. 112: 117206.

Kumar, A., Mondal, P. C., & Fontanesi, C. 2018. Chiral Magneto-Electrochemistry. Magnetochemistry 4: 36.

Kumar, A., & Choudhary, S. 2018. Enhanced Magnetoresistance in In-Plane Monolayer MoS$_2$ with CrO$_2$ Electrodes. J. Supercond. Novel Magn. 31: 3245–3250.

Kuncser, A., Antohe, S., & Kuncser, V. 2017. A general perspective on the magnetization reversal in cylindrical soft magnetic nanowires with dominant shape anisotropy. J. Magn. Magn. Mater. 423: 34–38.

Lack, W., Jenkins, S., Meo, A., Chantrell, R. W., McKenna, K. M., & Evans, R. F. L. 2024. Thermodynamic properties and switching dynamics of perpendicular shape anisotropy MRAM. J. Phys. Condens. Matter 36: 145801.

Lacour, D., Jaffrès, H., Nguyen Van Dau, F., Petroff, F., Vaurès, A., & Humbert, J. 2002. Field sensing using the magnetoresistance of IrMn exchange-biased tunnel junctions. J. Appl. Phys. 91: 4655–4658.

Ladak, S., Fernández-Pacheco, A., & Fischer, P. 2022. Science and technology of 3D magnetic nanostructures. APL. Mater. 10: 120401.

Lajavardi, M., Kenney, D. J., & Lin, S. H. 2000. Time-Resolved Phase Transitions of the Nanocrystalline Cubic to Submicron Monoclinic Phase in Mn_2O_3-ZrO_2. J. Chin. Chem. Soc. 47: 1055–1063.

Landau, L., & Lifshitz, E. 1935. On the theory of magnetic permeability in ferromagnetic bodies. Physik. Z. Sowjetunion. 8: 153–169.

Landau, L., & Lifshitz, E. 1980. Statistical Physics, vol. 5 of Course of Theoretical Physics. Butterworth-Heinenmann, Oxford, 3rd Ed.

Lau, K., & Lee, S. 1997. A programmable CMOS Gaussian synapse for analogue VLSI neural networks. Int. J. Electron. 83: 91–98.

Laukhin, V., Skumryev, V., Martí, X., Hrabovsky, D., Sánchez, F., García-Cuenca, M. V., Ferrater, C., Varela, M., Lüders, U., Bobo, J. F., & Fontcuberta, J. 2006. Electric-Field Control of Exchange Bias in Multiferroic Epitaxial Heterostructures. Phys. Rev. Lett. 97: 227201.

Lebedeva, N., Varpula, A., Novikov, S., & Kuivalainen, P. 2012. Stability of magnetic polarons in magnetic semiconductor single electron transistors: the effect of Coulomb interaction and external magnetic field. Phys. Stat. Sol. B – Basic Sol. State Phys. 249: 2244–2249.

Lebeugle, D., Colson, D., Forget, A., Viret, M., Bataille, A. M., & Gukasov, A. 2008. Electric-Field-Induced Spin Flop in $BiFeO_3$ Single Crystals at Room Temperature. Phys. Rev. Lett. 100: 227602.

Lebeugle, D., Mougin, A., Viret, M., Colson, D., & Ranno, L. 2009. Electric Field Switching of the Magnetic Anisotropy of a Ferromagnetic Layer Exchange Coupled to the Multiferroic Compound $BiFeO_3$. Phys. Rev. Lett. 103: 257601.

Lee, J.-C., & Sheu, B. J. 1990. Parallel digital image restoration using adaptive vlsi neural chips. Computer Design: VLSI in Computers and Processors. Proceedings of 1990 IEEE International Conference on ICCD'. 90: 126–129.

Lee, Y. M., Hayakawa, J., Ikeda, S., Matsukura, F., & Ohno, H. 2007. Effect of electrode composition on the tunnel magnetoresistance of pseudospin-valve magnetic tunnel junction with a MgO tunnel barrier. Appl. Phys. Lett. 90: 212507.

Lee, K.-S., & Kim, S.-K. 2008. Conceptual design of spin wave logic gates based on a Mach-Zehnder-type spin wave interferometer for universal logic functions. J. Appl. Phys. 104: 053909.

Lee, G.-H., Yu, Y.-J., Lee, C., Dean, C., Shepard, K. L., Kim, P., & Hone, J. 2011. Electron tunneling through atomically flat and ultrathin hexagonal boron nitride. Appl. Phys. Lett. 99: 243114.

Lee, T. H., Loke, D., Huang, K.-J., Wang, W.-J., & Elliott, S. R. 2014. Tailoring transient-amorphous states: towards fast and power-efficient phase-change memory and neuromorphic computing. Adv. Mater. 26: 7493–7498.

Lee, W. K., Whitener, K. E., Robinson, J. T., & Sheehan, P. E. 2015. Patterning magnetic regions in hydrogenated graphene via E-beam irradiation. Adv. Mater. 27: 1774–1778.

Lee, S.-E., Baek, J.-U., & Park, J.-G. 2017. Highly Enhanced TMR Ratio and Δ for Double MgO-based p-MTJ Spin-Valves with Top Co2Fe6B2 Free Layer by Nanoscale-thick Iron Diffusion-barrier. Sci. Rep. 7: 11907.

Lee, A., Lee, H., Ebrahimi, F., Lam, B., Chen, W. H., Chang, M. F., Amiri, P. K., & Wang, K. L. 2018. A Dual-Data Line Read Scheme for High-Speed Low-Energy Resistive Nonvolatile Memories. IEEE Trans. Very Large Scale Integr. (VLSI) Syst. 26: 272–279.

Lehndorff, T. L., Bürgler, D. E., Kakay, A., Hertel, R., & Schneider, C. M. 2008. Spin-Transfer Induced Dynamic Modes in Single-Crystalline Fe–Ag–Fe Nanopillars, Spin-Transfer Induced Dynamic Modes in Single-Crystalline Fe–Ag–Fe Nanopillars. IEEE Trans. Magn. 44: 1951–1956.

Lei, Y. L., Zhang, Y., Liu, R., Chen, P., Song, Q. L., & Xiong, Z. H. 2009. Driving current and temperature dependent magnetic-field modulated electroluminescence in Alq3-based organic light emitting diode. Org. Electron. 10: 889–894.

Leighton, C., Nogués, J., Suhl, H., & Schuller, Ivan K. 1999. Competing interfacial exchange and Zeeman energies in exchange biases bilayers. Phys. Rev. B. 60: 12837.

Leitão, U. A., Kleemann, W., & Ferreira, I. B. 1988. Metastability of the uniform magnetization in three-dimensional random-field Ising model systems. I. Fe0,7Mg0,3Cl2. Phys. Rev. B. 38: 4765.

Leitao, D. C., Silva, A. V., Paz, E., Ferreira, R., Cardoso, S., & Freitas, P. P. 2016. Magnetoresistive nanosensors: controlling magnetism at the nanoscale. Nanotechnology. 27: 045501.

Leong, P. H., & Jabri, M. A. 1992. A VLSI neural network for morphology classification. IJCNN – International Joint Conference on Neural Networks. 2: 678–683.

Leonov, A. O. 2023. Swirling of Horizontal Skyrmions into Hopfions in Bulk Cubic Helimagnets. Magnetism 3: 297–307.

Leutenantsmeyer, J. C., Liu, T., Gurram, M., Kaverzin, A. A., & van Wees, B. J. 2018. Bias-dependent spin injection into graphene on YIG through bilayer hBN tunnel barriers. Phys. Rev. B. 98: 125422.

Li, H., & Chen, Y. 2009. An overview of non-volatile memory technology and the implication for tools and architectures. Design, Automation & Test in Europe Conference & Exhibition '09.

Li, B., Kao, C. Y., Yoo, J. W., Prigodin, V. N., & Epstein, A. J. 2011. Magnetoresistance in an all-organic-based spin valve. Adv. Mater. 23: 3382–3386.

Li, B. X., Zheng, J., & Chi, F. 2012. Spin-Selective Transport of Electron in a Quantum Dot under Magnetic Field. Chin. Phys. Lett. 29: 107302.

Li, D., Vetter, J. S., Marin, G., McCurdy, C., Cira, C., Liu, Z., & Yu, W. 2012. Identifying Opportunities for Byte-Addressable Non-Volatile Memory in Extreme-Scale Scientific Applications. In: IPDPS, 945–956, IEEE Computer Society.

Li, Y., Ma, Z. Y., Song, X. J., Yang, Z., Xu, L. C., Liu, R. P., Li, X. Y., Liu, X. G., & Hu, D. Y. 2017. The magnetoresistance effect and spin-polarized photocurrent of zigzag graphene-graphyne nanoribbon heterojunctions. Comp. Mat. Sci. 136: 1–11.

Li, M., Wang, J. B., & Lu, J. 2017a. General planar transverse domain walls realized by optimized transverse magnetic field pulses in magnetic biaxial nanowires. Sci. Rep. 7: 43065.

Li, Y., Ngo, A. T., DiLullo, A., Latt, K. Z., Kersell, H., Fisher, B., Zapol, P., Ulloa, S. E., & Hla, S. W. 2017b. Anomalous Kondo resonance mediated by semiconducting graphene nanoribbons in a molecular heterostructure. Nat. Commun. 8: 946.

Li, X. Y., & Melcher, C. 2018. Stability of axisymmetric chiral skyrmions. J. Funct. Anal. 275: 2817–2844.

Li, Z., Zhang, Y. G., Huang, Y. Q., Wang, C. X., Zhang, X. C., Liu, Y., Zhou, Y., Kang, W., Koli, S. C., & Lei, N. 2018. Strain-controlled skyrmion creation and propagation in ferroelectric/ferromagnetic hybrid wires. J. Magn. Magn. Mater. 455: 19–24.

Li, J., Sansos, J. T., & Sillanpää, M. A. 2018a. High-Precision Displacement Sensing of Monolithic Piezoelectric Disk Resonators Using a Single-Electron Transistor. J. Low Temp. Phys. 191: 316–329.

Li, C. H., van't Erve, O.M. J., Yan, C., Li, L., & Jonker, B. T. 2018b. Electrical Detection of Charge-to-Spin and Spin-to-Charge Conversion in a Topological Insulator Bi_2Te_3 Using BN/Al_2O_3 Hybrid Tunnel Barrier. Sci. Rep. 8: 10265.

Li, X., Song, M., Xu, N., Luo, S. J., Zou, Q. M., Zhang, S., Hong, J. M., Yang, X. F., Min, T., Han, X. F., Zou, X. C., Zhu, J. G., Salahuddin, S., & You, L. 2018c. Novel cascadable magnetic majority gates for implementing comprehensive logic functions. IEEE Trans. Electron Devices. 65: 4687–4693.

Li, R., Jiang, J. W.; Shi, X. H., Mi, W. B.; & Bai, H. L. 2021. Two-dimensional Janus FeXY (X, Y = Cl, Br, and I, X ≠ Y) monolayers: half-metallic ferromagnets with tunable magnetic properties under strain. ACS Appl. Mater. Interfaces 13: 38897–38905.

Li, S., Xia, J., Shen, L. C., Zhang, X. C., Ezawa, M., & Zhou, Y. 2022. Mutual conversion between a magnetic Néel hopfion and a Néel toron. Phys. Rev. B 105: 174407.

Li, N., Liu, X., Chen, W.; Pan, W.; & Yu, Z. 2024. Dynamic Time-Domain Sensing Scheme for Spin-Orbit Torque MRAM, in IEEE Transactions on Electron Devices. Online first, DOI: 10.1109/TED.2024.3397239.

Li, D., & Yu, G. 2021. Innovation of Materials, Devices, and Functionalized Interfaces in Organic Spintronics. Adv. Funct. Mater. 31: 2100550.

Liang, S., Geng, R., Yang, B., Zhao, W., Subedi, R. Chandra, Li, X., Han, X., & Nguyen, T. D. 2016. Curvature-enhanced Spin-orbit Coupling and Spinterface Effect in Fullerene-based Spin Valves. Sci. Rep. 6: 19461.

Liang, J. J., Yu, J.H., Chen, J., Qin, M. H., Zeng, M., Lu, X. B., Gao, X. S., & Liu, J. M. 2018. Magnetic field gradient driven dynamics of isolated skyrmions and antiskyrmions in frustrated magnets. New J. Phys. 20: 053037.

Liang, S., Wu, K., & Wang, J.-P. 2024. Magnetoresistive (MR) biosensor. In Magnetic Nanoparticles in Nanomedicine, Woodhead Publishing, pp. 289–322.

Lilienfeld, J. E. 1926. Method and apparatus for controlling electric currents. U. S. Patent No. 1,745,175 (Filed October 8, 1926; issued January 18, 1930).

Likharev, K. K. 1987. Single-electron transistors: electrostatic analogs of the DC SQUIDS. IEEE Trans. Magn. 23: 1142.

Likharev, K. K. 1988. Correlated discrete transfer of single electrons in ultrasmall tunnel junctions. IBM J. Res. Dev. 32: 144.

Likharev, K. K. 1999. Single-electron devices and their applications. Proc. IEEE. 87: 606–632.

Likharev, K. K., & Strukov, D. B. 2005. CMOL: Devices, Circuits, and Architectures. Introducing Molecular Electronics. Springer, Basel, 447–477.

Lin, T. H., & Dunn, K. J. 1986. High-pressure and low-temperature study of electrical resistance of lithium. Phys. Rev. B. 33: 807–811.

Lin, I. C., Law, Y. K., & Xie, Y. 2018. Mitigating BTI-Induced Degradation in STT-MRAM Sensing Schemes. IEEE Trans. Very Large Scale Integr. (VLSI) Syst. 26: 50–62.

Lisenkov, I., Tyberkevych, V., Nikitov, S., & Slavin, A. 2016. Theoretical formalism for collective spin-wave edge excitations in arrays of dipolarly interacting magnetic nanodots. Phys. Rev. B. 93: 214441.

Liu, L., Moriyama, T., Ralph, D. C., & Buhrman, R. A. 2011. Spin-torque ferromagnetic resonance induced by the spin Hall effect. Phys. Rev. Lett. 106: 036601.

Liu, L., Pai, C.-F., Li, Y., Tseng, H. W., Ralph, D. C., & Buhrman, R. A. 2012. Spin-torque switching with the giant spin Hall effect of tantalum. Science. 336: 555–558.

Liu, J. Y., Guo, J., Yang, M. M., Zeng, X. Y., & Yan, M. 2017. Nonrecessive and Sustained Precessional Domain Wall Motion Driven by Perpendicular Field Pulses in Thin Magnetic Nanocylinders. Spin. 7: 1740008.

Liu, Y. H., Lake, R. K., & Zang, J. D. 2018. Shape dependent resonant modes of skyrmions in magnetic nanodisks. J. Magn. Magn. Mater. 455: 9–13.

Liu, Z. Y., Huang, P. W., Hernandez, S., Ju, G. P., & Rausch, T. 2018a. Systematic Evaluation of Microwave-Assisted Magnetic Recording. IEEE Trans. Magn. 54: 3200905.

Liu, Y. Z., Lake, R. K., & Zang, J. D. 2018. Binding a hopfion in a chiral magnet nanodisk. Phys. Rev. B 98: 174437.

Liu, Y. Z., Hou, W. T., Han, X. F., & Zang, J. D. 2020. Three-Dimensional Dynamics of a Magnetic Hopfion Driven by Spin Transfer Torque. Phys. Rev. Lett. 124: 127204.

Liu, Y. Z., Watanabe, H., & Nagaosa, N. 2022. Emergent Magnetomultipoles and Nonlinear Responses of a Magnetic Hopfion. Phys. Rev. Lett. 129: 267201.

Liu, Y. Z., Watanabe, H., & Nagaosa, N. 2023. Current-induced nonreciprocal dynamics and nonlinear Hall effect of a magnetic hopfion. 2023 IEEE International Magnetic Conference – Short Papers (INTERMAG Short Papers), Sendai, Japan, pp. 1–2.

Lobanov, I. S. & Uzdin, V. M. 2023. Lifetime, collapse, and escape paths for hopfions in bulk magnets with competing exchange interactions. Phys. Rev. B 107: 104405.

Lobo T., Neto, M. A., da Silva, M. G., & Salmon O. D. R. 2020. Carbon nanotube with pressure inducing pseudogaps: Kondo effect study. J. Appl. Phys. 127:155102.

Locatelli, N., Vincent, A. F., Mizrahi, A., Friedman, J. S., Vodenicarevic, D., Kim, J. V., Klein, J. O., Zhao, W. S., Grollier, J., & Querlioz, D. 2015. Spintronic Devices as Key Elements for Energy-Efficient Neuroinspired Architectures. 2015 Design, Automation & Test in Europe Conference & Exhibition, 994–999.

Lottermoser, T., Lonkai, T., Amann, U., Hohlwein, D., Ihringer, J., & Fiebig, M. 2004. Magnetic phase control by an electric field. Nature. 430: 541–544.

Lou, X., Adelmann, C., Furis, M., Crooker, S. A., Palmstrom, C. J., & Crowell, P. A. 2006. Electrical Detection of Spin Accumulation at a Ferromagnet-Semiconductor Interface. Phys. Rev. Lett. 96: 176603.

Lou, P. C., & Kumar, S. 2018. Spin-Hall effect and emergent antiferromagnetic phase transition in n-Si. J. Magn. Magn. Mater. 452: 129–133.

LTO Generation 8. 2018 (https://www.lto.org/technology/lto-generation-8/, accessed March 2018).

Lu, Y., Altman, R. A., Marley, A., Rishton, S. A., Trouilloud, P. L., Xiao, G., Gallagher, W. J., & Parkin, S. S. P. 1997. Shape-anisotropy-controlled magnetoresistive response in magnetic tunnel junctions. Appl. Phys. Lett. 70: 2610–2612.

Lu, Y., Betto, D., Fursich, K., Suzuki, H., Kim, H. H., Cristiani, G., Logvenov, G., Brookes, N. B., Benckiser, E., Haverkort, M. W., Khaliullin, G., le Tacon, M., Minola, M., & Keimer, B. 2018. Site-Selective Probe of Magnetic Excitations in Rare-Earth Nickelates Using Resonant Inelastic X-ray Scattering. Phys. Rev. X. 8: 031014.

Luong, V. S., Nguyen, A. T., & Nguyen, A. T. 2018. Exchange biased spin valve-based gating flux sensor. Measurement. 115: 173–177.

Luong, V. S., Su, Y. H., Lu, C. C., Jeng, J. T., Hsu, J. H., Liao, M. H., Wu, J. C., Lai, M. H., & Chang, C. R. 2018a. Planarization, Fabrication, and Characterization of Three-Dimensional Magnetic Field Sensors. IEEE Trans. Nanotechnol. 17: 11–25.

Lupo, P., Haghshenasfard, Z., Cottam, M. G., & Adeyeye, A. O. 2016. Ferromagnetic resonance study of interface coupling for spin waves in narrow NiFe/Ru/NiFe multilayer nanowires. Phys. Rev. B. 94: 214431.

Lutwyche, M. T., & Wada, Y., 1994. Estimate of the ultimate performance of the single-electron transistor. J. Appl. Phys. 75: 3654.

Lv, W. X., Zou, J. T., Li, X., Wang, J., & Zhao, Z. J. 2018. Magnetostatic interaction in multi-shell $Ni_{80}Fe_{20}$/Cu composite wires. J. Magn. Magn. Mater. 460: 1–5.

Lyberatos, A., Komineas, S., & Papanicolaou, N. 2011. Precessing vortices and antivortices in ferromagnetic elements. J. Appl. Phys. 109: 023911.

Ma, Q. Y., Wan, W. H., Li, Y. M., & Liu, Y. 2022. First principles study of 2D half-metallic ferromagnetism in Janus Mn_2XSb (X = As, P) monolayers. Appl. Phys. Lett. 120: 112402.

MacDonald, A. H., & Tsoi, M. 2011. Antiferromagnetic metal spintronics. Philos. Trans. A. Math. Phys. Eng. Sci. 369: 3098–3114.

Maciejko, J., Kim, E.-A., & Qi, X.-L. 2010. Spin Aharonov-Bohm effect and topological spin transistor. Phys. Rev. B. 82: 195409.

Madami, M., Gubbiotti, G., Tacchi, S., & Carlotti, G. 2018. Magnetization dynamics of single-domain nanodots and minimum energy dissipation during either irreversible or reversible switching. J. Phys. D. 50: 453002.

Mahato, B. K., Rana, B., Kumar, D., Barman, S., Sugimoto, S., Otani, Y., & Barman, A. 2014. Tunable spin wave dynamics in two-dimensional $Ni_{80}Fe_{20}$ nanodot lattices by varying dot shape. Appl. Phys. Lett. 105: 012406.

Mahato, B. K., Choudhury, S., Mandal, R., Barman, S., Otani, Y., & Barman, A. 2015. Tunable configurational anisotropy in collective magnetization dynamics of $Ni_{80}Fe_{20}$ nanodot arrays with varying dot shapes. J. Appl. Phys. 117: 213909.

Mahmoudi, H., Windbacher, T., Sverdlov, V., & Selberherr, S. 2013. Implication logic gates using spin-transfer-torque-operated magnetic tunnel junctions for intrinsic logic-in-memory. Solid-State Electron. 84: 191–197.

Maier, P., Hartmann, F., Emmerling, M., Schneider, C., Kamp, M., Höfling, S., & Worschech, L. 2016. Electro-photo-sensitive memristor for neuromorphic and arithmetic computing. Phys. Rev. Appl. 5: 054011.

Majhi, J. & Maiti, S. K. 2023. Generation and manipulation of pure spin current in a conducting loop coupled to an Aharonov–Bohm ring. J. Phys. Condens. Matter 35: 195301.

Maji, N., & Nath, T. K. 2018. Rectifying Magnetic Tunnel Diode Like Behavior in $Co_2MnSi/ZnO/p-Si$ Heterostructure. AIP Conf. Proc. 1942: 130021.

Majumdar, S., Majumdar, H. S., Laiho, R., & Österbacka, R. 2006. Comparing small molecules and polymer for future organic spin-valves. J. Alloys Compd. 423, 169–171.

Majumdar, S., Das, A. K., & Ray, S. K. 2009. Magnetic semiconducting diode of $p-Ge_{1-x}Mn_x/n-Ge$ layers on silicon substrate. Appl. Phys. Lett. 94: 122505.

Makarov, A., Sverdlov, V., Osintsev, D., & Selberherr, S. 2011. Reduction of switching time in pentalayer magnetic tunnel junctions with a composite-free layer. Phys. Status Solidi RRL. 5: 1–3.

Makarov, A., Sverdlov, V., Osintsev, D., & Selberherr, S. 2011a. Fast switching in magnetic tunnel junctions with double barrier layer. Proc. of 2011 SSDM: 456–457.

Makarov, A., Sverdlov, V., & Selberherr, S. 2012. Emerging memory technologies: trends, challenges, and modeling methods. Microelectron. Reliab. 52: 628–634.

Makarov, A., Windbacher, T., Sverdlov, V., & Selberherr, S. 2016. CMOS-compatible spintronics devices: a review. Semicond. Sci Technol. 31: 113006.

Malajovich, I., Berry, J. J., Samarth, N., & Awschalom, D. D. 2001. Persistent sourcing of coherent spins for multifunctional semiconductor spintronics. Nature. 411: 770–772.

Malozemoff, A. P. 1987. Random-field model of exchange bias anisotropy at rough ferromagnetic-antiferromagnetic interfaces. Phys. Rev. B. 35: 3679.

Malozemoff, A. P. 1988. Mechanisms of exchange anisotropy. J. Appl. Phys. 63: 3874.

Malozemoff, A. P. 1988a. Heisenberg-to-Ising crossover in a random-field model with uniaxial anisotropy. Phys. Rev. B. 37: 7673.

Mamouni, N., Vijaya, J. J., Benyoussef, A., El Kenz, A., & Bououdina, M. 2018. Electronic structure and magnetic studies of V-doped ZnO: ab initio and experimental investigations. Bull. Mater. Sci. 41: 87.

Mani, R. G., Smet, J. H., von Klitzing, K., Narayanamurti, V., Johnson, W. B., & Umansky, V. 2004. Radiation-induced oscillatory magnetoresistance as a sensitive probe of the zero-field spin-splitting in high-mobility $GaAs/Al_xGa_{1-x}As$ devices. Phys. Rev. B. 69: 193304.

Mann, J. R., & Gilbert, S. 1989. An analog self-organizing neural network chip. Adv. Neural Inf. Process. Syst. 1: 739–747.

Mantsevich, V. N. & Smirnov, D. S. 2023. Kondo enhancement of current-induced spin accumulation in a quantum dot. Phys. Rev. B 108: 035409.

Markram, H. 2006. The Blue Brain Project. Nat. Rev. Neurosci. 7: 153–160.

Marti, X., Fina, I., Frontera, C., Liu, J., Wadley, P., He, Q., Paull, R. J., Clarkson, J. D., Kudrnovský, J., Turek, I., Kunes, J., Yi, D., Chu, J.-H., Nelson, C. T., You, L., Arenholz, E., Salahuddin, S., Fontcuberta, J., Jungwirth, T., & Ramesh, R. 2014. Room-temperature antiferromagnetic memory resistor. Nat. Mater. 13: 367–374.

Martin, J., Akerman, N., Ulbricht, G., Lohmann, T., Smet, J. H., von Klitzing, K., & Yacoby, A. 2008. Observation of electro-hole puddles in graphene using a scanning single-electron transistor. Nat. Phys. 4: 144–148.

Martinek, J., Borda, L., Utsumi, Y., König, J., von Delft, J., Ralph, D. C., Schon, G., & Maekawa, S. 2007. Kondo effect in single-molecule spintronic devices. J. Magn. Magn. Mater. 310: E343–E345.

Maruyama, T., Shiota, Y., Nozaki, T., Ohta, K., Toda, N., Mizuguchi, M., Tulapurkar, A. A., Shinjo, T., Shiraishi, M., Mizukami, S., Ando, Y., & Suzuki, Y. 2009. Large voltage-induced magnetic anisotropy change in a few atomic layers of iron. Nat. Nanotechnol. 4: 158–161.

Maruyama, T., Machida, Y., Sugatani, S., Takita, H., Hoshino, H., Hino, T., Ito, M., Yamada, A., Iizuka, T., Komatsu, S., Iked, M., & Asada, K. 2012. CP element based design for 14 nm node EBDW high volume manufacturing. Proc. SPIE 8323, Alternative Lithographic Technologies IV: 832314.

Md, Masud, Md, A., Islam, S., & Khosru, Q. D. M. 2015. Modified Ebers-Moll model of magnetic bipolar transistor. 2015 IEEE International Conference on Electron Devices and Solid-State Circuits (EDSSC), 812–815.

Masell, J. & Everschor-Sitte, K. 2021. Current-Induced Dynamics of Chiral Magnetic Structures: Creation, Motion, and Applications. In Kamenetskii, E. (Ed.), Chirality, Magnetism and Magnetoelectricity, Topics in Applied Physics, vol. 138, Springer, Cham, pp. 147–181.

Mathon, J., & Umerski, A. 2001. Theory of tunneling magnetoresistance of an epitaxial Fe/MgO/Fe(001) junction. Phys. Rev. B. 63: 220403(R).

Matityahu, S., Aharony, A., Entin-Wohlman, O., & Tarucha, S. 2013. Spin filtering in a Rashba-Dresselhaus-Aharonov-Bohm double-dot interferometer. New J. Phys. 15: 125017.

Matlak, J., Rismaniyazdi, E., & Komvopoulos, K. 2018. Nanostructure, structural stability, and diffusion characteristics of layered coatings for heat-assisted magnetic recording head media. Sci. Rep. 8: 9807.

Matlak, J., & Komvopoulos, K. 2018. Ultrathin amorphous carbon films synthesized by filtered cathodic vacuum arc used as protective overcoats of heat-assisted magnetic recording heads. Sci. Rep. 8: 9647.

Matsumoto, K., Takahashi, S., Ishii, M., Hoshi, M., Kurokawa, A., Ichimura, S., & Ando, A. 1995. Application of STM nanometer-size oxidation process to planar-type MIM diode. Jpn. J. Appl. Phys. 34: 1387.

Matsumoto, K., Ishii, M., Segawa, K., Oka, Y., Vartanian, B. J., & Harris, J. S. 1996. Room temperature operation of a single electron transistor made by the scanning tunneling microscope nanooxidation process for the TiO_x/Ti system. Appl. Phys. Lett. 68: 34–36.

Matsumoto, R., Chanthbouala, A., Grollier, J., Cros, V., Fert, A., Nishimura, K., Nagamine, Y., Maehara, H., Tsunekawa, K., & Fukushima, A. 2011. Spin-torque diode measurements of MgO-based magnetic tunnel junctions with asymmetric electrodes. Appl. Phys. Expr. 4: 063001.

Matsuo, S., Ueda, K., Baba, S., Kamata, H., Tateno, M., Shabani, J., Palmstrom, C. J., & Tarucha, S. 2018. Equal-spin andreev reflection on junctions of spin-resolved quantum hall bulk state and spin-singlet superconductor. Sci. Rep. 8: 3454.

Mauri, D., Siegmann, H. C., Bagus, P. S., & Kay, E. 1987. Simple model for thin ferromagnetic films exchange coupled to an antiferromagnetic substrate. J. Appl. Phys. 62: 3047.

Maxwell, J. C. 1861. On physical lines of force. Philosophical Magazine 90: 11–23; reprinted from Philosophical Magazine Series 4: 12–24.

Mayergoyz, I., Bertotti, G., Serpico, C., Liu, Z., & Lee, A. 2012. Random magnetization dynamics at elevated temperatures. J. Appl. Phys. 111: 07D501.

Mazin, I. I., Papaconstantopoulos, D. A., & Mehl, M. J. 2002. Superconductivity in compressed iron: role of spin fluctuations. Phys. Rev. B. 65: 100511(R).

McDaniel, T. 2012. Application of Landafshitz-Bloch dynamics to grain switching in heat-asisted magnetic recording. J. Appl. Phys. 112: 013914.

Mead, C. 1990. Neuromorphic Electronic Systems, Proceedings of the IEEE 78: 1629–1636.

Medapalli, R., Afanasiev, D., Kim, D. K., Quessab, Y., Manna, S., Montoya, S. A., Kirilyuk, A., Rasing, T., Kimel, A. V., & Fullerton, E. E. 2017. Multiscale dynamics of helicity-dependent all-optical magnetization reversal in ferromagnetic Co/Pt multilayers. Phys. Rev. B. 96: 224421.

Medlej, I., Wang, J. L., Hu, C. Y., & Yu, K. L. 2024. Hopfion based magnonic crystal. J. Magn. Magn. Mater. 591: 171726.

Mehlin, A., Gross, B., Wyss, M., Schefer, T., Tutuncuoglu, G., Heimbach, F., Morral, A. F. I., Grundler, D., & Poggio, M. 2018. Observation of end-vortex nucleation in individual ferromagnetic nanotubes. Phys. Rev. B. 97: 134422.

Meier, G., Bolte, M., Eiselt, R., Krüger, B., Kim, D.-H., & Fischer, P. 2007. Direct imaging of stochastic domain-wall motion driven by nanosecond current pulses. Phys. Rev. Lett. 98: 187202.

Meiklejohn, W. H., & Bean, C. P. 1956. New magnetic anisotropy. Phys. Rev. 102: 1413.

Meiklejohn, W. H., & Bean, C. P. 1957. New magnetic anisotropy. Phys. Rev. 105: 904.

Meiklejohn, W. H. 1962. Exchange bias – a review. J. Appl. Phys. 33: 1328.

Meinken, R. H., & Stammerjohn, L. W. 1965. Memory devices. Bell Lab. Rec. 43: 229–235.

Meir, Y., Wingreen, N. S., & Lee, P. A. 1993. Low-temperature transport through a quantum dot: the Anderson model out of equilibrium. Phys. Rev. Lett. 70: 2601.

Meirav, U., & Foxman, E. B. 1995. Single-electron phenomena in semiconductors. Semiconduc. Sci. Technol. 10: 255.

Meisenheimer, P. B., Novakov, S., Vu, N. M., & Heron, J. T. 2018. Perspective: magnetoelectric switching in thin film multiferroic heterostructures. J. Appl. Phys. 123: 240901.

Mesoraca, S., Knudde, S., Leitao, D. C., Cardosa, S., & Blamire, M. G. 2018. All-spinel oxide Josephson junctions for high-efficiency spin filtering. J. Phys. Cond. Matter. 30: 015804.

Metlov, K. L. 2023. Two types of metastable hopfions in bulk magnets. Physica A: Nonlinear Phenomena 443: 133561.

Miller, J. B., Zumbühl, D. M., Marcus, C. M., Lyanda-Geller, Y. B., Goldhaber-Gordon, D., Campman, K., & Gossard, A. C. 2003. Gate-controlled spin-orbit quantum interference effects in lateral transport. Phys. Rev. Lett. 90: 076807.

Miltat, J., Rohart, S., & Thiaville, A. 2018. Brownian motion of magnetic domain walls and skyrmions, and their diffusion constants. Phys. Rev. B. 97: 214426.

Miltényi, P. 2000. Mikroskopischer Ursprung der Austauschkopplung an ferromagnetischen/ antiferromagnetischen Schichten, dissertation thesis, RWTH Aachen University, Aachen.

Miltényi, P., Gierlings, M., Keller, J., Beschoten, B., Güntherodt, G., Nowak, U., & Usadel, K. D. 2000. Diluted antiferromagnets in exchange bias: proof of the domain state model. Phys. Rev. Lett. 84: 4224.

Miron, I. M., Gaudin, G., Auffret, S., Rodmacq, B., Schuhl, A., Pizzini, S., Vogel, J., & Gambardella, P. 2010. Current-driven spin torque induced by the Rashba effect in a ferromagnetic metal layer. Nat. Mater. 9: 230–233.

Miron, I. M., Garello, K., Gaudin, G., Zermatten, P.-J., Costache, M. V., Auffret, S., Bandiera, S., Rodmacq, B., Schuhl, A., & Gambardella, P. 2011. Perpendicular switching of a single ferromagnetic layer induced by in-plane current injection. Nature. 476: 189–193.

Mishra, D., Markus, T. Z., Naaman, R., Kettner, M., Göhler, B., Zacharias, H., Friedman, N., Sheves, M., & Fontanesi, C. 2013. Spin-dependent electron transmission through bacteriorhodopsin embedded in purple membrane. Proc. Natl. Acad. Sci. U.S.A. 110: 14872–14876.

Misiorny, M., Weymann, I., & Barnaś, J. 2012. Temperature dependence of electroni transport through molecular magnets in the Kondo regime. Phys. Rev. B 86: 035417.

Miyazaki, T., Yaoi, T., & Ishio, S. 1991. Large magnetoresistance effect in 82Ni-Fe/Al-Al_2O_3/Co magnetic tunneling junction. J. Magn. Magn. Mater. 98: L7–L9.

Miyazaki, T., & Tezuka, N. 1995. Giant magnetic tunneling effect in Fe/Al_2O_3/Fe junction. J. Magn. Magn. Mater. 139: L231–L234.

Miyazaki, T., & Tezuka, N. 1995a. Spin polarized tunneling in ferromagnet/insulator/ferromagnet junctions. J. Magn. Magn. Mater. 151: 403–410.

Mizukami, S., Watanabe, D., Oogane, M., Ando, Y., Miura, Y., Shirai, M., & Miyazaki, T. 2009. Low damping constant for Co_2FeAl Heusler alloy films and its correlation with density of states. J. Appl. Phys. 105: 07D306.

Modak, A. R., Parkin, S. S. P., & Smith, D. J. 1994. Microstructural characterization of Co/Cu multilayers. J. Magn. Magn. Mat. 129: 415–422.

Modak, R., Srinivasu, V., & Srinivasu, A. 2018. Effect of Cu/Fe/Co substitution on static and dynamic magnetic properties of Ni-Mn-Sn alloy thin films. J. Magn. Magn. Mater. 464: 50–55.

Modha, D. S. 2018. Introducing a Brain-inspired Computer – TrueNorth's neurons to revolutionize system architecture. IBM Research: Brain-inspired Chip, http://www.research.ibm.com/articles/brain-chip.shtml, accessed April 2018.

Mohanan, V. P., & Kumar, P. S. A. 2017. Chirality dependent pinning and depinning of magnetic vortex domain walls at nano-constrictions. J. Magn. Magn. Mater. 422: 419–424.

Mondal, S., Choudhury, S., Barman, S., Otani, Y., & Barman, A. 2016. Transition from strongly collective to completely isolated ultrafast magnetization dynamics in two-dimensional hexagonal arrays of nanodots with varying inter-dot separation. RSC Adv. 6: 110393–110399.

Mondal, S., Barman, S., Choudhury, S., Otani, Y., & Barman, A. 2018. Influence of anisotropic dipolar interaction on the spin dynamics of Ni80Fe20 nanodot arrays arranged in honeycomb and octagonal lattices. J. Magn. Magn. Mater. 458: 95–104.

Monson, F. G., & Roukes, M. 1999. Spin injection and the local Hall effect in InAs quantum wells. J. Magn. Magn. Mater. 198–199: 632–335.

Moodera, J. S., Kinder, L. R., Wong, T. M., & Meservey, R. 1995. Large magnetoresistance at room temperature in ferromagnetic thin film tunnel junctions. Phys. Rev. Lett. 74: 3273–3276.

Moon, J.-H., Lee, T. Y., & You, C.-Y. 2018. Relation between switching time distribution and damping constant in magnetic nanostructure. Sci. Rep. 8: 13288.

Moore, G. E. 1965. Cramming more components onto integrated circuits. Electronics Magazine April 1965, p. 4.

Moran, T. J., Gallego, J. M., & Schuller, Ivan K. 1995. Increased exchange anisotropy due to disorder at permalloy/CoO interfaces. J. Appl. Phys. 78: 1887.

Moran, T. J., & Schuller, Ivan K. 1996. Effects of cooling field strength on exchange anisotropy at permalloy/CoO interfaces. J. Appl. Phys. 79: 5109.

Moreau-Luchaire, C., Moutafis, C., Reyren, N., Sampaio, J., Vaz, C. A. F., van Horne, N., Bouzehouane, K., Garcia, K., Deranlot, C., Warnicke, P., Wohlhüter, P., George, J.-M., Weigand, M., Raabe, J., Cross, V., & Fert, A. 2016. Additive interfacial chiral interaction in multilayers for stabilization of small individual skyrmions at room temperature. Nat. Nanotechnol. 11: 444–448.

Moreno, J. A., Bran, C., Vazquez, M., & Kosel, J. 2021. Cylindrical Magnetic Nanowires Applications. IEEE Transactions on Magnetics 57: 800317.

Moritz, J., Vinai, G., Auffret, S., & Dieny, B. 2011. Two-bit-per-dot patterned media combining in-plane and perpendicular-to-plane magnetized thin films. J. Appl. Phys. 109: 083902.

Moriya, T. 1960. Anisotropic superexchange interaction and weak ferromagnetism. Phys. Rev. 120: 91–98.

Moriyama, T., Cao, R., Xiao, J. Q., Lu, J., Wang, X. R., Wen, Q., & Zhang, H. W. 2007. Microwave-assisted magnetization switching of Ni80Fe20 in magnetic tunnel junctions. Appl. Phys. Lett. 90: 152503.

Moriyama, T., Matsuzaki, N., Kim, K.-J., Suzuki, I., Taniyama, T., & Ono, T. 2015. Sequential write-read operations in FeRh antiferromagnetic memory. Appl. Phys. Lett. 107: 122403.

Mosse, I. S., Sodisetti, V. R., Coleman, C., Ncube, S., de Sousa, A. S., Erasmus, R. M., Flahaut, E., Blon, T.; Lassagne, B., Samoril, T., & Bhattacharyya, S. Tuning Magnetic Properties of a Carbon Nanotube-Lanthanide Hybrid Molecular Complex through Controlled Functionalization. Molecules 26: 563.

Motsnyi, V. F., Van Dorpe, P., Van Roy, W., Goovaerts, E., Safarov, V. I., Borghs, G., & De Boeck, J. 2003. Optical investigation of electrical spin injection into semiconductors. Phys. Rev. B. 68: 245319.

Mühlbauer, S., Binz, B., Jonietz, F., Pfleiderer, C., Rosch, A., Neubauer, A., Georgii, R., & Böni, P. 2009. Skyrmion lattice in a chiral magnet. Science. 323: 915–919.

Mukhtar, A., Wu, K. M., Cao, X. M., & Gu, L. Y. 2020. Magnetic nanowires in biomedical applications. Nanotechnology 31: 433001.

Müller, G., Nagel, N., Pinnow, C.-U. U., & Röhr, T. 2003. Emerging non-volatile memory technologies. Proc. of the 29th European Solid-State Circuits Conference, ESSCIRC '03.

Muneeb, M., Akram, I., & Nazir, A. 2006. Smart Electronic Materials – Non-Volatile Random Access Memory Technologies, Lecture, KTH – School of information and communication technology, Sweden (http://citeseerx.ist.psu.edu/viewdoc/summary?doi=10.1.1.98.2239, accessed March 2018).

Muneeswaran, M., Gopiraman, M., Dhanabalan, S. S., Giridharan, N. V., & Akbari-Fakhrabadi, A. 2020. Multiferroic Properties of Rare Earth-Doped BiFeO$_3$ and Their Spintronic Applications. In Rajendran,

S., Qin, J., Gracia, F., Lichtfouse, E. (Eds.) Metal and Metal Oxides for Energy and Electronics, pp. 375–395, Springer, Cham.

Murray, A. F., & Smith, A. V. 1988. Asynchronous vlsi neural networks using pulse-stream arithmetic. IEEE J. Solid-State Circuits. 23: 688–697.

Murray, G., White, Ch. V., & Weise, W. 2007. Introduction to Engineering Materials. CRC Press, Tyler&Francis Group, Boca Roton, London, New York, p. 339.

Naaman, R., & Waldeck, D. H. J. 2012. Chiral-induced spin selectivity effect. Phys. Chem. Lett. 3: 2178–2187.

Nagaosa, N., & Tokura, Y. 2013. Topological properties and dynamics of magnetic skyrmions. Nat. Nanotechnol. 8: 899–911.

Nakagawa, K., Kimura, K., Hayashi, Y., Taruma, K., Ashizawa, Y., & Ohnuki, S. 2018. Design for cooling near-field transducer using surface plasmon polariton waveguide for heat-assisted magnetic recording. Jap. J. Appl. Phys. 57: 09TB01.

Nakamoto, K., Hoshiya, H., Katada, H., Hoshino, K., Yoshida, N., Shiimoto, M., Takei, H., Sato, Y., Hatatani, M., Watanabe, K., Carey, M., Maat, S., & Childress, J. 2007. CPP-GMR heads with a current screen layer for 300 Gb/in^2 Recording. IEEE Trans. Magn. 44: 95–99.

Nakayama, M., Kai, T., Shimomura, N., Amano, M., Kitagawa, E., Nagase, T., Yoshikawa, M., Kishi, T., Ikegawa, S., & Yoda, H. 2008. Spin transfer switching in TbCoFe/CoFeB/MgO/CoFeB/TbCoFe magnetic tunnel junctions with perpendicular magnetic anisotropy. J. Appl. Phys. 103: 07A710.

Nakagawa, Y., Takagishi, M., Narita, N., Nagasawa, T., Koizumi, G., Chen, W. Y., Kawasaki, S., Roppongi, T., Takeo, A., & Maeda, T. Spin-torque oscillator with coupled out-of-plane oscillation layers for microwave-assisted magnetic recording: experimental, analytical, and numerical studies. Appl. Phys. Lett. 122: 042403.

Narayanapillai, K., Qui, X. P., Rhensius, J., & Yang, H. 2014. Thermally assisted domain wall nucleation in perpendicular anisotropy trilayer nanowires. J. Phys. D. Appl. Phys. 47: 105005.

Navel Education and Training Command. 1978. Thin Film. In: Digital Computer Basics. Cited from: E. Thelen: Facts and stories about Antique (lonesome) Computers (http://ed-thelen.org/comp-hist/navy-thin-film-memory-desc.html, accessed March 2018) p. 106 ff., US Government Printing Office, Washington, D. C., USA.

Nawrocki, P., Kanak, J., Wojcik, M., & Sobiecki, T. 2018. ^{59}Co NMR analysis of CoFeB-MgO based magnetic tunnel junctions. J. Alloys Compd. 741: 775–780.

Nayak, A. P., Bhattacharyya, S., Zhu, J., Liu, J., Wu, X., Pandey, T., Jin, C., Singh, A. K., Akinwande, D., & Lin, J.-F. 2014. Pressure-induced semiconducting to metallic transition in multilayered molybdenum disulphide. Nature Comm. 5: 3731.

Nayak, A. K., Kumar, V., Ma, T. P., Werner, P., Pippel, E., Sahoo, R., Damay, F., Rößler, U. K., Felser, C., & Parkin, S. S. P. 2017. Magnetic antiskyrmions above room temperature in tetragonal Heusler materials. Nature. 548: 561–566.

NCR. 1962. NCR CRAM: card random access memory, product brochure. The National Cash Register Co. (http://s3data.computerhistory.org/brochures/ncr.cram.1960.102646240.pdf, accessed March 2018).

Ncube, S., Coleman, C., de Sousa, A. S., Nie, C., Lonchambon, P., Flahaut, E., Strydom, A., & Bhattacharyya, S. 2018. Observation of strong Kondo like features and co-tunnelling in superparamagnetic GdCl3 filled 1D nanomagnets. J. Appl. Phys. 123: 213901.

Ncube, S., Coleman, C., Strydom, A., Flahaut, E., de Sousa, A. S., & Bhattacharyya, S. 2018a. Kondo effect and enhanced magnetic properties in gadolinium functionalized carbon nanotube supramolecular complex. Sci. Rep. 8: 8057.

Nguyen, T. D., Hukic-Markosian, G., Wang, F., Wojcik, L., Li, X.-G., Ehrenfreund, E., & Vardeny, Z. V. 2010. Isotope effect in spin response of π-conjugated polymer films and devices. Nat. Mater. 9: 345.

Nishioka, K., Honjo, H., Ikeda, S., Watanabe, T., Miura, S., et al. 2020. Novel Quad-Interface MTJ Technology and its First Demonstration With High Thermal Stability Factor and Switching Efficiency for STT-MRAM Beyond 2X nm. IEEE Transactions on Electron Devices 67: 995–1000.

Nogués, J., Lederman, D., Moran, T. J., & Schuller, Ivan K. 1996. Positive exchange bias in FeFe$_2$-Fe bilayers. Phys. Rev. Lett. 76: 624.

Nogués, J., & Schuller, Ivan K. 1999. Exchange bias. J. Magn. Magn. Mat. 192: 203.

Nogués, J., Moran, T. J., Lederman, D., Schuller, Ivan K., & Rao, K. V. 1999. Role of interfacial structure on exchange-biased FeF$_2$-Fe. Phys. Rev. B. 59: 6984.

Nogués, J., Leighton, C., & Schuller, Ivan K. 2000. Correlation between antiferromagnetic interface coupling and positive exchange bias. Phys. Rev. B. 61: 1315.

Noori, F., Ramazani, A., & Kashi, M. A. 2018. Controlling structural and magnetic properties in CoNi and CoNiFe nanowire arrays by fine-tuning of Fe content. J. Alloys Compd. 756: 193–201.

Noske, M., Stoll, H., Fahnle, M., Gangwar, A., Woltersdorf, G., Slavin, A., Weigang, M., Dieterle, G., Forster, J., Back, C. H., & Schütz, G. 2016. Three-dimensional character of the magnetization dynamics in magnetic vortex structures: hybridization of flexure gyromodes with spin waves. Phys. Rev. Lett. 117: 037208.

Nossa, J. F., Islam, M. F., Canali, C. M., & Pederson, M. R. 2013. Electric control of a {Fe4} single-molecule magnet in a single-electron transistor. Phys. Rev. B. 88: 224423.

Nowak, U., Usadel, K. D., Keller, J., Miltényi, P., Beschoten, B., & Güntherodt, G. 2002. Domain state model for exchange bias. I. Theor. Phys. Rev. B. 66: 014430.

O'Brian, L., Read, D. E., Zeng, H. T., Lewis, E. R., Petit, D., & Cowburn, R. P. 2009. Bidirectional magnetic nanowire shift register. Appl. Phys. Lett. 95: 232502.

Octavio, M., Tinkham, M., Blonder, G. E., & Klapwijk, T. M. 1983. Subharmonic energy-gap structure in superconducting constrictions. Phys. Rev. B. 27: 6739.

Ogino, K., Hoshino, H., Maruyama, T., Machida, Y., & Sugatani, S. 2012. Proximity effect correction using multilevel area density maps for character projection based electron beam direct writing toward 14 nm node and beyond. Proc. SPIE 8323, Alternative Lithographic Technologies IV: 832328.

Ogut, E., Yanik, C., Kaya, I. I., Ow-Yang, C., & Sendur, K. 2018. Focusing short-wavelength surface plasmons by a plasmonic mirror. Opt. Lett. 43: 2208–2211.

O'Hara, D. J., Zhu, T. C., Trout, A. H., Ahmed, A. S., Luo, Y. K., Lee, C. H., Brenner, M. R., Rajan, S., Gupta, J. A., McComb, D. W., & Kawakami, R. K. 2018. Room temperature intrinsic ferromagnetism in epitaxial manganese selenide films in the monolayer limit. Nano Lett. 18: 3125–3131.

Ohldag, H., Tyliszczak, T., Höhne, R., Spemann, D., Esquinazi, P., Ungureanu, M., & Butz, T. 2007. π-electron ferromagnetism in metal-free carbon probed by soft X-ray dichroism. Phys. Rev. Lett. 98: 187204.

Ohmichi, E., Komatsu, E., & Osada, T. 2004. Application of a tunnel diode oscillator to noncontact resistivity measurement in pulsed magnetic fields. Rev. Sci. Instr. 75: 2094.

Ohno, H. 1998. Making nonmagnetic semiconductors ferromagnetic. Science. 281: 951–956.

Ohno, H., Chiba, D., Matsukura, F., Omiya, T., Abe, E., Dietl, T., Ohno, Y., & Ohtani, K. 2000. Electric-field control of ferromagnetism. Nature. 408: 944–946.

Ohno, H. 2011. Magnetoresistive random access memory with spin transfer torque write (Spin RAM): present and future. Proc. of 2011 SSDM: 957–958.

Oiwa, A., Mitsumori, Y., Moriya, R., Slupinski, T., & Munekata, H. 2002. Effect of optical spin injection on ferromagnetically coupled Mn spins in the III-V magnetic alloy semiconductor (Ga,Mn) As. Phys. Rev. Lett. 88: 137202.

Olsen, K. H., & Best, R. L. 1959. Magnetic core memory. US Patent 3161861.

Ono, T., Miyajima, H., Shigeto, K., Mibu, K., Hosoito, N., & Shinjo, T. 1999. Propagation of a magnetic domain wall in a submicrometer magnetic wire. Science. 284: 468–470.

Ono, A., & Ishihara, S. 2017. Double-exchange interaction in optically induced nonequilibrium state: a conversion from ferromagnetic to antiferromagnetic structure. Phys. Rev. Lett. 119: 207202.

Ormaza, M., Robles, R., Bachellier, N., Abufager, P., Lorente, N., & Limot, L. 2016. On-surface engineering of a magnetic organometallic nanowire. Nano Lett. 16: 588–593.

Ostanin, S., Brubitsin, V., Staunton, J. B., & Savrasov, S. Y. 2003. Density functional study of the phase diagram and pressure-induced superconductivity in P: implication for spintronics. Phys. Rev. Lett. 91: 087002.

Ostanin, S., Ernst, A., Sandratskii, L. M., Bruno, P., Däne, M., Hughes, I. D., Staunton, J. B., Hergert, W., Mertig, I., & Kudrnovský, J. 2007. Mn-stabilized zirconia: from imitation diamonds to a new potential high-TC ferromagnetic spintronics material. Phys. Rev. Lett. 98: 016101.

Ouardi, S., Fecher, G. H., Felser, C., & Keubler, J. 2013. Realization of spin gapless semiconductors: the Heusler compound Mn_2CoAl. Phys. Rev. Lett. 110: 100401.

Oujja, M., Martin-Garcia, L., Robollar, E., Quesada, A., Garcia, M. A., Fernandez, J. F., Marco, J. F., de la Figuera, J., & Castillejo, M. 2018. Effect of wavelength, deposition temperature and substrate type on cobalt ferrite thin films grown by pulsed laser deposition. Appl. Surf. Sci. 452: 19–31.

Padrón-Hernández, E., Azevedo, A., & Rezende, S. M. 2011. Amplification of spin waves by thermal spin-transfer torque. Phys. Rev. Lett. 107: 197203.

Palii, A., Clemente-Juan, J. M., Aldoshin, S., Korchagin, D., Golosov, E., Zilberg, S., & Tsukerblat, B. 2020. Can the Double Exchange Cause Antiferromagnetic Spin Alignment?, Magnetochemistry 6: 36.

Pan, H., & Lue, R. 2008. Effects of the spin-orbit interaction on the spin polarization of quantum dots in the presence of Andreev reflection. J. Phys. D. – Cond. Matt. 20: 195220.

Pan, H., Xu, H.-Z., & Lue, R. 2010. Charge and spin pumping effects in a single-dot Aharonov-Bohm ring with ferromagnetic leads. Physica E – Low-Dimension. Syst. Nanostruct. 43: 85–88.

Pan, H., & Zhang, Y.-W. 2012. Tuning the electronic and magnetic properties of MoS_2 nanoribbons by strain engineering. J. Phys. Chem. C. 116: 11752–11757.

Pan, L., Song, B., Sun, J., Zhang, L., Hofer, W., Du, S. X., & Gao, H.-J. 2013. The origin of half-metallicity in conjugated electron systems – a study on transition-metal-doped graphyne. J. Phys.: Condens. Matt. 25: 505502.

Pan, W., Lu, P., Ihlefeld, J. F., Lee, S. R., Choi, E. S., Jiang, Y., & Jia, Q. X. 2018. Electrical-current-induced magnetic hysteresis in self-assembled vertically aligned $La_{2/3}Sr_{1/3}MnO_3$: ZnOnanopillar composites. Phys. Rev. Mater. 2: 021401.

Panda, J., Benerjee, P., & Nath, T. K. 2014. Electrical spin extraction and giant positive junction magnetoresistance in a Fe_3O_4/MgO/n-Si magnetic diode like heterostructure. J. Phys. D: Appl. Phys. 47, 415103.

Pandey, N., Kumar, A., & Chakrabarti, S. 2020. Tunnelling magnetoresistance and spin filtration effect in functionalized graphene sheet with CrO_2 as electrode: an ab-initio study. J. Magn. Magn. Mater. 497: 166073.

Pandey, E., Sharangi, P., Sahoo, A., Mahanta, S. P, Mallik, S., & Bedanta, S. 2023. A Perspective on multifunctional ferromagnet/organic molecule spinterface. Appl. Phys. Lett. 123: 040501.

Paulo, V. I. M., Neves-Araujo, J., Revoredo, F. A., & Padrón-Hernández, E. 2018. Magnetization curves of electrodeposited Ni, Fe and Co nanotubes. Mater. Lett. 223: 78–81.

Park, B. G., Wunderlich, J., Martí, X., Holý, V., Kurosaki, Y., Yamada, M., Yamamoto, H., Nishide, A., Hayakawa, J., Takahashi, H., Shick, A. B., & Jungwirth, T. 2011. A spin-valve-like magnetoresistance of an antiferromagnet-based tunnel junction. Nat. Mater. 10: 347–351.

Parkin, S. S. P., & Mauri, D. 1991. Spin engineering: direct determination of the Ruderman-Kittel-Kasuya-Yosida far-field range function in ruthenium. Phys. Rev. B. 44: 7131(R).

Parkin, S. S. P., Moon, K. S., Pettit, K. E., Smith, D. J., Dunin-Borkowski, R. E., & McCartney, M. R. 1999. Magnetic tunnel junctions thermally stable to above 300 °C. Appl. Phys. Lett. 75: 543–545.

Parkin, S. S. P., Kaiser, C., Panchula, A., Rice, P. M., Hughes, B., Samant, M., & Yang, S.-H. 2004. Giant tunnelling magnetoresistance at room temperature with MgO (100) tunnel barriers. Nat. Mater. 3: 862–867.

Parkin, S. S. P., Hayashi, M., & Thomas, L. 2008. Magnetic domain-wall racetrack memory. Science. 320: 190–194.

Parvin, M. S., Oogane, M., Kubota, M., Tsunoda, M., & Ando, Y. 2018. Epitaxial L1(0)-MnAl thin films with high perpendicular magnetic anisotropy and small surface roughness. IEEE Trans. Magn. 54: 3401704.

Pati, S. P., Al-Mahdawi, M., Shiokawa, Y., Sahashi, M., & Endo, Y. 2017. Effect of a platinum buffer layer on the magnetization dynamics of sputter deposited YIG polycrystalline thin films. IEEE Trans. Magn. 53: 6101105.

Pauli, W. 1926. Zur Quantenmechanik des magnetischen Elektrons. Zeitschrift für Physik. 43: 601–623.

PC guide. 2001. Function of the Read/Write Heads (http://www.pcguide.com/ref/hdd/op/heads/opFunc tion-c.html, accessed March 2018).

Pearton, S. J., Abernathy, C. R., Overberg, M. E., Thaler, G. T., Norton, D. P., Theodoropoulou, N., Hebard, A. F., Park, Y. D., Ren, F., Kim, J., & Boatner, L. A. 2003. Wide band gap ferromagnetic semiconductors and oxides. J. Appl. Phys. 93: 1.

Peltier, P. M. 1834. Nouvelles expériences sur la caloricité des courants électrique. Annales de Chimie et de Physique. 56: 371–386.

Pershin, Y. V., & Privman, V. 2003. Spin relaxation of conduction electrons in semiconductors due to interaction with nuclear spins. Nano Lett. 3: 695.

Pershoguba, S. S., Andreoli, D., & Zang, J. D. 2021. Electronic scattering off a magnetic hopfion. Phys. Rev. B 104: 075102.

Peters, R., Pruschke, T., & Anders, F. B. 2006. Numerical renormalization group approach to Green's functions for quantum impurity models. Phys. Rev. B 74: 245114.

Pettersson, J., Wahlgren, P., Delsing, P., Haviland, D. B., Claeson, T., Rorsman, N., & Zirath, H. 1996. Extending the high-frequency limit of a single-electron transistor by on-chip impedance transformation. Phys. Rev. B. 53: R13272.

Piedade, M., Sousa, L. A., de Almeida, T. M., Germano, J., da Costa, B. D., Lemos, J. M., Freitas, P. P., Ferreira, H. A., & Cardoso, F. A. 2006. A new hand-held microsystem architecture for biological analysis. IEEE Trans. Circuits Syst. I: Regul. Pap. 53: 2384.

Pinna, D., Araujo, F. A., Kim, J. V., Cros, V., Querlioz, D., Bessiere, P., Droulez, J., & Grollier, J. 2018. Skyrmion gas manipulation for probabilistic computing. Phys. Rev. Appl. 9: 064018.

Piquemal-Banci, M., Galceran, R., Godel, F., Caneva, S., Martin, M.-B., Weatherup, R. S., Kidambi, P. R., Bouzehouane, K., Xavier, S., Anane, A., Petroff, F., Fert, A., Dubois, S. M. M., Charlier, J.-C., Robertson, J., Hofmann, S., Dlubak, B., & Seneor, P. 2018. Insulator-to-metallic spin-filtering in 2D-magnetic tunnel junctions based on hexagonal boron nitride. ACS Nano. 12: 4712–4718.

Piramanayagam, S. N. 2007. Perpendicular recording media for hard disk drives. J. Appl. Phys. 102: 011301.

Ploog, K. 1980. Molecular Beam Epitaxy of III-V Compounds. In: Chrystals: Growth, Properties, and Applications, H. C. Freyhardt (Ed.), Springer, New York, vol. 3, pp. 73–162.

Pohm, A. V., Huang, J. S. T., Daughton, J. M., Krahn, D. R., & Mehra, V. 1988. The design of a one megabit nonvolatile M-R memory chip using 1.5×5 µm cells. IEEE Trans. Mag. 24: 3117–3119.

Polley, D., Pancaldi, M., Hudl, M., Vavassori, P., Urazhdin, S., & Bonetti, S. 2018. THz-driven demagnetization with perpendicular magnetic anisotropy: towards ultrafast ballistic switching. J. Phys. D. 51: 084001.

Polyakov, O. P., Korobova, J. G., Stepanyuk, O.V., & Bazhanov, D. I. 2017. Impact of surface strain on the spin dynamics of deposited Co nanowires. J. Appl. Phys. 121: 014306.

Pontinen, R. E., & Sanders, T. M. 1960. New electron spin resonance spectrum in antimony-doped germanium. Phys. Rev. Lett. 5: 311–313.

Popadiuk, D., Tartakovskaya, E., Krawczyk, M., & Guslienko, K. 2023. Emergent Magnetic Field and Nonzero Gyrovector of the Toroidal Magnetic Hopfion. Phys. Stat. Sol. RRL 17: 2300131.

Poulsen, V., & Pedersen, P. O. 1907. Telegraphone. US Patent US873083A.

Prakash, A., Brangham, J., Yang, F., & Heremans, J. P. 2016. Spin Seebeck effect through antiferromagnetic NiO. Phys. Rev. B. 94: 014427.

Pramanik, S., Stefanita, C. G., Patibandla, S., Bandyopadhyay, S., Garre, K., Harth, N., & Cahay, M. 2007. Observation of extremely long spin relaxation times in an organic nanowire spin valve. Nat. Nanotech. 2: 216–219.

Pramanik, T., Roy, U., Jadaun, P., Register, L. F., & Banerjee, S. K. 2018. Write error rates of in-plane spin-transfer-torque random access memory calculated from rare-event enhanced micromagnetic simulations. J. Magn. Magn. Mater. 467: 96–107.

Prejbeanu, I. L., Kerekes, M., Sousa, R. C., Sibuet, H., Redon, O., Dieny, B., & Nozières, J. P. 2007. Thermally assisted MRAM. J. Phys.: Condens. Matt. 19: 165218.

Prigodin, V., Bergeson, J. D., Lincoln, D. M., & Epstein, A. J. 2006. Anomalous room temperature magnetoresistance in organic semiconductors. Synth. Met. 156: 757.

Privitera, A., Righetto, M., Cacialli, F., & Riede, M. K. 2021. Perspectives of Organic and Perovskite-Based Spintronics. Adv. Optical Mater. 9: 2100215.

Psaroudaki, C., & Loss, D. 2018. Skyrmions driven by intrinsic magnons. Phys. Rev. Lett. 120: 237203.

Quantum LTO Drives. 2018 (https://iq.quantum.com/exLink.asp?10444458OP44N16I37407297, accessed March 2018).

Quantum LTO-5 Tape Drive. 2009. (http://downloads.quantum.com/lto5/6-66786-01_RevA.pdf, accessed March 2018).

Qin, P.-X., Ya, H., Wang, X.-N., Feng, Z.-X., Guo, H.-X. Zho, X.-R., Wu, H.-J., Zhang, X., Leng, Z. G. G., Chen, H.-Y., & Liu, Z.-Q. 2020. Noncollinear spintronics and electric-field control: a review. Rare Metals 39: 95–112.

Raftrey, D. & Fischer, P. 2021. Field-Driven Dynamics of Magnetic Hopfions. Phys. Rev. Lett. 127: 257201.

Rahaman, S. Z., Wang, I. J., Chen, T. Y., Pai, C. F., Wang, D. Y., Wie, J.H., Lee, H. H., Hsin, Y. C., Chang, Y. J., Yang, S. Y., Kuo, Y. C., Su, Y. H., Chen, Y. S., Huang, K. C., Wu, C. I., & Deng, D. L. 2018. Pulse-width and temperature effect on the switching behavior of an etch-stop-on-MgO-barrier spin-orbit torque MRAM cell. IEEE Electron Device Lett. 39: 1306–1309.

Ralph, D. C., & Stiles, M. D. 2008. Spin transfer torques. J. Magn. Magn. Mater. 320: 1190–1216.

Ramakrishnan, S., Hasler, P. E., & Gordon, C. 2011. Floating gate synapses with spike-time-dependent plasticity. IEEE Trans. Biomed. Circuits Syst. 5: 244–252.

Ramirez, A. P. 1997. Colossal magnetoresistance. J. Phys.: Condens. Matt. 9: 8171.

Ramos, R., Kikkawa, T., Uchida, K., Adachi, H., Lucas, I., Aguirre, M. H., Algarabel, P., Morellón, L., Maekawa, S., Saitoh, E., & Ibarra, M. R. 2013. Observation of the spin Seebeck effect in epitaxial Fe_3O_4 thin films. Appl. Phys. Lett. 102: 072413.

Rana, B., Kumar, D., Barman, S., Pal, S., Fukuma, Y., Otani, Y., & Barman, A. 2011. Detection of Picosecond magnetization dynamics of 50 nm magnetic dots down to the single dot Regime. ACS Nano. 5: 9559–9565.

Rangaraju, N., Peters, J. A., & Wessels, B. W. 2010. Magnetoamplification in a bipolar magnetic junction transistor, Phys. Rev. Lett. 105: 117202.

Rao, S., Couet, S., van Beek, S., Kundu, S., Sharifi, S. H., Jossart, N., & Kar, G. S. 2021. A Systematic Assessment of W-Doped CoFeB Single Free Layers for Low Power STT-MRAM Applications. Electronics 10: 2384.

Rashba, E. I. 2000. Theory of electrical spin injection: tunnel contacts as a solution of the conductivity mismatch problem. Phys. Rev. B. 62: R16267–R16270.

Ravelosona, D., Lacour, D., Katine, J. A., Terris, B. D., & Chappert, C. 2005. Nanometer scale observation of high efficiency thermally assisted current-driven domain wall depinning. Phys. Rev. Lett. 95: 117203.

Reine, M., Aggarwal, R. L., & Lax, B. 1972. Stress-modulated magnetoreflectivity of gallium antimonide and gallium arsenide. Phys. Rev. B 5: 3033–3049.

Reiss, G., & Meyners, D. 2006. Reliability of field programmable magnetic logic gate arrays. Appl. Phys. Lett. 88: 043505.

Ren, S. X., Tang, L. Z., Sun, Q., Li, Z. H., Yang, H. F., & Zhao, J. J. 2018. The structure, oxygen vacancies and magnetic properties of TiO_x (0 < x < 2) synthesized by plasma assisted chemical vapor deposition and reduction. Mater. Lett. 228: 212–215.

Ren, H. W., Wu, S.-Y., Sun, J. Z., Fullerton, E. E. 2021. Ion beam etching dependence of spin–orbit torque memory devices with switching current densities reduced by Hf interlayers. APL Mater. 9: 091101.

Ren, H. T., Zhong, J., & Xiang, G. 2023. The Progress on Magnetic Material Thin Films Prepared Using Polymer-Assisted Deposition. Molecules 28: 5004.

Rezende, S. M., Rodriguez-Suarez, R. L., Cunha, R. O., Rodrigues, A. R., Machado, F. L. A., Guerra, G. A. F., Ortiz, J. C. L., & Azevedo, A. 2014. Magnon spin-current theory for the longitudinal spin-Seebeck effect. Phys. Rev. B. 89: 014416.

Rezende, S. M., Rodríguez-Suárez, R. L., Cunha, R. O., López Ortiz, J. C., & Azevedo, A. 2016. Bulk magnon spin current theory for the longitudinal spin Seebeck effect. J. Magn. Magn. Mater. 400: 171–177.

Rial, J. & Proenca, M. P. 2020. A Novel Design of a 3D Racetrack Memory Based on Functional Segments in Cylindrical Nanowire Arrays. Nanomaterials, 10: 2403.

Richardson, O. W. 1929. Thermionic phenomena and the laws which govern them. Nobel Lecture, Stockholm.

Richardson, C. F., & Ashcroft, N. W. 1997. Effective electron-electron interactions and the theory of superconductivity. Phys. Rev. B. 55: 15130–15145.

Richter, H., Dobin, A., Heinonen, O., Gao, K., Veerdonk, R., Lynch, R., Xue, J., Weller, D., Asselin, P., Erden, M., & Brockie, R. 2006. Recording on bit-patterned media at densities of 1 Tb/in^2 and beyond. IEEE Trans. Magn. 42: 2255–2260.

Rietman, E. A., Frye, R., Wong, C., & Kornfeld, C. 1989. Amorphous silicon photoconductive arrays for artificial neural networks. Appl. Opt. 28: 3474–3478.

Rippard, W. H., Pufall, M. R., Kaka, S., Russek, S. E., & Silva, T. J. 2004. Direct-current induced dynamics in $Co_{90}Fe_{10}/Ni_{80}Fe_{20}$ point contacts. Phys. Rev. Lett. 92: 027201.

Rodmacq, B., Manchon, A., Ducruet, C., Auffret, S., & Dieny, B. 2009. Influence of thermal annealing on the perpendicular magnetic anisotropy of Pt/Co/AlOx trilayers. Phys. Rev. B. 79: 024423.

Rodriguez, L., Miramond, B., & Granado, B. 2015. Toward a sparse self-organizing map for neuromorphic architectures. ACM J. Emerging Technol. Comput. Syst. 11: 33.

Romeira, B., Avó, R., Figueiredo, J. M. L., Barland, S., Javaloyes, J. 2016. Regenerative memory in time-delayed neuromorphic photonic resonators. Scientific Reports 6: 19510.

Romming, N., Hanneken, C., Menzel, M., Bickel, J. E., Wolter, B., von Bermann, K., Kubetzka, A., & Wiesendanger, R. 2013. Writing and deleting single magnetic skyrmions. Science. 341, 36–639.

Rossella, F., Bertoni, A., Ercolani, D., Rontani, M., Sorba, L., Beltram, F., & Roddaro, S. 2014. Nanoscale spin rectifiers controlled by the Stark effect. Nat. Nanotechnol. 9: 997–1001.

Rostky, G. 2007. Bubbles: the better memory. 2000 – The Century of the Engineer – Misunderstood Milestones, EETimes.com (http://web.archive.org/web/20070930031657/http://www.eetonline.com/special/special_issues/millennium/milestones/bobeck.html, accessed March 2018).

Rottmayer, R. E., Batra, S., Buechel, D., Challener, W. A., Hohlfeld, J., Kubota, Y., Li, L., Lu, B., Mihalcea, C., Mountfield, K., Pelhos, K., Peng, C., Rausch, T., Seigler, M. A., Weller, D., & Yang, X.-M. 2006. Heat-assisted magnetic recording. IEEE Trans. Magn. 42: 2417–2421.

Roy, K., Sharad, M., Fan, D. L., & Yogendra, K. 2013. Beyond charge-based computation: Boolean and non-Boolean computing with spin torque devices. 2013 ACM/IEEE International Symposium on Low Power Electronics and Design, 139–142.

Roy, K., Sharad, M., Fan, D. L., & Yogendra, K. 2013a. Exploring boolean and non-boolean computing with spin torque devices. 2013 IEEE/ACM International Conference on Computer-Aided Design, 576–580.

Roy, K., Sharad, M., Fan, D. L., & Yogendra, K. 2014. Brain-inspired computing with spin torque devices. 2014 Design, Automation and Text in Europe Conference and Exhibition, 14253493.

Roy, K., Fan, D. L., Fong, X. Y., Kim, Y. S., Sharad, M., Paul, S., Chatterjee, S., Bhunia, S., & Mukhopadhyay, S. 2015. Exploring spin transfer torque devices for unconventional computing. IEEE J. emerging Sel. Top. Circuits syst. 5: 5–16.

Ruderman, M. A., & Kittel, C. 1954. Indirect exchange coupling of nuclear magnetic moments by conduction electrons. Phys. Rev. 96: 99.

Rybakov, F. N., Kiselev, N. S., Borisov, A. B., Döring, L., Melcher, C., & Blügel, S. 2022. Magnetic hopfions in solids. APL Mater. 10: 111113.

Ryu, K., Kim, J., Jung, J., Kim, J. P., Kang, S. H., & Jung, S. O. 2012. A magnetic tunnel junction based zero standby leakage current retention flip-flop. IEEE Trans. Very Large Integr. (VLSI) Syst. 20: 2044–2053.

Rzeszut, O., Skowroński, W., Ziętek, S., Wrona, J., & Stobiecki, T. 2019. Multi-bit MRAM storage cells utilizing serially connected perpendicular magnetic tunnel junctions. J. Appl. Phys. 125: 223907.

Rzeszut, O., Checinski, J., Brzozowski, I., Zietek, S., Skowronski, W., & Stobiecki, T. 2022. Multi-state MRAM cells for hardware neuromorphic computing. Sci. Rep. 12: 7178.

Safranski, C., Barsukov, I., Lee, H. K., Schneider, T., Jara, A. A., Smith, A., Chang, H., Lenz, K., Lindner, J., Tserkovnyak, Y., Wu, M., & Krivorotov, I. N. 2017. Spin caloritronic nano-oscillator. Nat. Commun. 8: 117.

Saha, I., Kanazawa, K., Nitani, H., & Kuroda, S. 2022. Impact of growth conditions on the structural and magnetic properties of (Zn,Fe)Te thin films grown by molecular beam epitaxy (MBE). J. Crystal Growth 580: 126492.

Sahoo, S., Kontos, T., Furer, J., Hoffmann, C., Gräber, M., Cottet, A., & Schönenberger, C. 2005. Electric field control of spin transport. Nat. Phys. 1: 99–102.

Sahoo, S., Kontos, T., Schönenberger, C., & Sürgers, C. 2005a. Electrical spin injection in multiwall carbon nanotubes with transparent ferromagnetic contacts. Appl. Phys. Lett. 86: 112109.

Saikin, S. 2004. A drift-diffusion model for spin-polarized transport in a two-dimensional non-degenerate electron gas controlled by spin–orbit interaction. J. Phys.: Condens. Matt. 16: 5071.

Saikin, S., Shen, M., & Cheng, M.-C. 2004. Study of spin-polarized transport properties for spin-FET design optimization. IEEE Trans. Nanotechnol. 3: 173.

Saito, Y., Inokuchi, T., Ishikawa, M., Sugiyama, H., Marukame, T., & Tanamoto, T. 2011. Spin-based MOSFET and its applications. J. Electrochem. Soc. 158: H1068–H1076.

Saji, C., Troncoso, R. E., Carvalho-Santos, V. L., Altbir, D., & Nunez, A. S. 2023. Hopfion-Driven Magnonic Hall Effect and Magnonic Focusing. Phys. Rev. Lett. 131: 166702.

Sakr, G. B., & Yahia, I. S. 2010. Effect of illumination and frequency on the capacitance spectroscopy and the relaxation process of p-ZnTe/n-CdMnTe/GaAs magnetic diode for photocapacitance applications. J. Alloys Compd. 503: 213–219.

Sallermann, M.; Jónsson, H.; & Blügel, S. 2023. Stability of hopfions in bulk magnets with competing exchange interactions. Phys. Rev. B 107: 104404.

Samoilenka, A. & Shnir, Ya. 2018. Magnetic Hopfions in the Faddeev-Skyrme-Maxwell model. Phys. Rev. D, 97: 125014.

Sanchez, D., Lopez, R., & Choi, M. S. 2005. Spintronic transport and Kondo effect in quantum dots. J. Supercond. 18: 251–260.

Sando, D., Agbelele, A., Rahmedov, D., Liu, J., Rovillain, P., Toulouse, C., Infante, I. C., Pyatakov, A. P., Fusil, S., Jacquet, E., Carrétéro, C., Deranlot, C., Lisenkov, S., Wang, D., Le Breton, J.-M., Cazayous, M., Sacuto, A., Juraszek, J., Zvezdin, A. K., Bellaiche, L., Dkhil, B., Barthélémy, A., & Bibes, M. 2013. Crafting the magnonic and spintronic response of $BiFeO_3$ films by epitaxial strain. Nat. Mater. 12: 641–646.

Sankey, J. C., Braganca, P. M., Garcia, A. G. F., Krivorotov, I. N., Buhrman, R. A., & Ralph, D. C. 2006. Spin-transfer-driven ferromagnetic resonance of individual nanomagnets. Phys. Rev. Lett. 96: 227601.

Sanvito, S. 2011. Molecular spintronics. Chem. Soc. Rev. 40: 3336–3355.

Sapozhnik, A. A., Abrudan, R., Skourski, Y., Jourdan, M., Zabel, H., Kläui, M., & Elmers, H.-J. 2017. Manipulation of antiferromagnetic domain distribution in Mn2Au by ultrahigh magnetic fields and by strain. Phys. Stat. Sol. Rap. Res. Lett. 11: 1600438.

Satoh, T., Terui, Y., Moriya, R., Ivanov, B. A., Ando, K., Saitoh, E., Shimura, T., & Kuroda, K. 2012. Directional control of spin-wave emission by spatially shaped light. Nat. Photonics. 6: 662–666.

Sbiaa, R., Lua, S. Y. H., Law, R., Meng, H., Lye, R., & Tan, H. K. 2011. Reduction of switching current by spin transfer torque effect in perpendicular anisotropy magnetoresistive devices (invited). J. Appl. Phys. 109: 07C707.

Sbiaa, R., Meng, H., & Piramanayagam, S. N. 2011a. Materials with perpendicular magnetic anisotropy for magnetic random access memory. Phys. Status Solidi RRL. 5: 413–419.

Sbiaa, R. 2013. Magnetization switching by spin-torque effect in off-aligned structure with perpendicular anisotropy. J. Phys. D. Appl. Phys. 46: 395001.

Sbiaa, R., Piramanayagam, S. N., & Liew, T. 2013. High speed in spin-torque-based magnetic memory using magnetic nanocontacts. Phys. Status Solidi. 7: 332–335.

Sbiaa, R., Al Bahri, M., & Piramanayagam, S. N. 2018. Domain wall oscillation in magnetic nanowire with a geometrically confined region. J. Magn. Magn. Mater. 456: 324–328.

Schäpers, Th., Nitta, J., Heersche, H. B., & Takayanagi, H. 2000. Interference ferromagnet/semiconductor/ferromagnet spin field-effect transistor. Phys. Rev. B. 64: 125314.

Schemmel, J., Bruderle, D., Grubl, A., Hock, M., Meier, K., & Millner, S. 2010. A wafer-scale neuromorphic hardware system for large-scale neural modeling. Proceedings of 2010 IEEE International Symposium on Circuits and Systems (ISCAS), 1947–1950.

Schleicher, F., Halisdemir, U., Lacour, D., Gallart, M., Boukari, S., Schmerber, G., Davesne, V., Panissod, P., Halley, D., Majjad, H., Henry, Y., Leconte, B., Boulard, A., Spor, D., Beyer, N., Kieber, C., Sternitzky, E., Cregut, O., Ziegler, M., Montaigne, F., Beaurepaire, E., Gilliot, P., Hehn, M., & Bowen, M. 2014. Localized states in advanced dielectrics from the vantage of spin- and symmetry-polarized tunnelling across MgO. Nat. Commun. 5: 4547.

Schliemann, J., Egues, J. C., & Loss, D. 2003. Nonballistic spin-field-effect transistor. Phys. Rev. Lett. 90: 146801.

Schmidt, G., Ferrand, D., Molenkamp, L. W., Filip, A. T., & van Wees, B. J. 2000. Fundamental obstacle for electrical spin injection from a ferromagnetic metal into a diffusive semiconductor. Phys. Rev. B. 62: R4790(R).

Schmidt, R., Wilken, F., Nunnber, T. S., & Brouwer, P. W. 2018. Boltzmann approach to the longitudinal spin Seebeck effect. Phys. Rev. B. 98: 134421.

Schneider, V., Reinholdt, A., Kreibig, U., Weirich, T., Güntherodt, G., Beschoten, B., Tillmanns, A., Krenn, H., Rumpf, K., & Granitzer, P. 2006. Structural and magnetic properties of Ni/NiOxide and Co/CoOxide Core/Shell nanoparticles and their possible use for ferrofluids. Z. Phys. Chem. 220, 173–187.

Schneider, T., Serga, A. A., Leven, B., Hillebrands, B., Stamps, R. L., & Kostylev, M. P. 2008. Realization of spin-wave logic gates. Appl. Phys. Lett. 92: 022505.

Schöbitz, M., De Riz, A., Martin, S., Bochmann, S., Thirion, C., Vogel, J., Foerster, M., Aballe, L., Mentes, T. O., Locatelli, A., Genuzio, F., Le-Denmat, S., Cagnon, L., Toussaint, J. C., Gusakova, D., Bachmann, J., & Fruchart, O. 2019. Fast Domain Wall Motion Governed by Topology and Œrsted Fields in Cylindrical Magnetic Nanowires. Phys. Rev. Lett. 123: 217201.

Schoelkopf, R. J., Wahlgren, P., Kozhevnikov, A. A., Delsing, P., & Prober, D. E. 1998. The radio-frequency single-electron transistor (RF-SET): a fast and ultrasensitive electrometer. Science. 280: 1238–1242.

Scholz, W., Fidler, J., Schrefl, T., Suess, D., Dittrich, R., Forster, H., & Tsiantos, V. 2003. Scalable parallel micromagnetic solvers for magnetic nanostructures. Comp. Mat. Sci. 28: 366–383.

Schroeder, D. 2013. Modelling of Interface Carrier Transport for Device Simulation. Springer-Verlag Wien GmbH. ISBN 978-3-7091-6644-4 (e-book).

Schulthess, T. C., & Butler, W. H. 1998. Consequences of spin-flop coupling in exchange biased films. Phys. Rev. Lett. 81: 4516.

Schulthess, T. C., & Butler, W. H. 1999. Coupling mechanisms in exchange biased films. J. Appl. Phys. 85: 5510.

Schuman, C. D., Potok, T. E., Patton, R. M., Birdwell, J. D., Dean, M. E., Rose, G. S., & Plank, J. S. 2017. A survey of neuromorphic computing and neural networks in hardware. arXiv:1705.06963.

Schwee, L. J., Hunter, P. E., Restorff, K. A., & Shephard, M. T. 1982. The concept and initial studies of a crosstie random access memory (CRAM). J. Appl. Phys. 53: 2762–24764.

Seagate. 2017. Multi-Tier Caching Technology™ (https://www.seagate.com/files/www-content/product-content/barracuda-fam/barracuda-new/files/multi-tier-caching-technology-white-paper-2017.pdf, accessed March 2018).

Seebeck, T. J. 1822. Magnetische Polarisation der Metalle und Erze durch Temperatur-Differenz. Abhandlungen der Königlichen Akademie der Wissenschaften zu Berlin: Königlichen Akademie der Wissenschaften, Berlin. 265–373.

Seiler, D. G., Bajaj, B. D., & Stephens, A. E. 1977. Inversion-asymmetry splitting of the conduction band in InSb. Phys. Rev. B. 16: 2822.

Seki, T., Mitani, S., Yakushiji, K., & Takanashi, K. 2006. Magnetization reversal by spin-transfer torque in 90° configuration with a perpendicular spin polarizer. Appl. Phys. Lett. 89: 172504.

Seleznev, J. E., Burkin, J. A., & Kuzmin, S. V. 1977. Ferrite core memory. US Patent 4161037.

Sengupta, A., Panda, P., Raghunathan, A., & Roy, K. 2016. Neuromorphic computing enabled by spin-transfer torque devices. 2016 29th International Conference on VLSI Design and 2016 15th International Conference on Embedded Systems (VLSID), International Conference on VLSI Design, 32–37.

Sengupta, A., Shim, Y., & Roy, K. 2016a. Proposal for an all-spin artificial neural network: emulating neural and synaptic functionalities through domain wall motion in ferromagnets. IEEE Trans. Biomed. Circuits Syst. 10: 1152–1160.

Sengupta, A., & Roy, K. 2018. Neuromorphic computing enabled by physics of electron spins: prospects and perspectives. Appl. Phys. Express. 11: 030101.

Seo, Y. K.; Kwon, K.-W. 2021. Area Optimization Techniques for High-Density Spin-Orbit Torque MRAMs. Electronics 10: 792.

Seong, S., Lee, E., Kim, H. W., Min, B. I., Lee, S., Dho, J., Kim, Y., Kim, J.-Y., & Kang, J.-S. 2018. Experimental evidence for mixed-valent Cr ions in half-metallic CrO_2: temperature-dependent XMCD study. J. Magn. Magn. Mater. 452: 447–450.

Sharma, E., Rathi, R., Misharwa, J., Sinhmar, B., Kumari, S., Dalal, J., & Kumar, A. 2022. Evolution in lithography techniques: microlithography to nanolithography. Nanomaterials 12: 2754.

Shekhter, R. I., Gorelik, L. Y., Krive, I. V., Kiselev, M. N., Parafilo, A. V., & Jonson, M. 2013. Nanoelectromechanics of shuttle devices. Nanoelectromech. Syst. 1: 1–25.

Shen, W. F., Liu, X. Y., Mazumdar, D., & Xiao, G. 2005. In situ detection of single micron-sized magnetic beads using magnetic tunnel junction sensors. Appl. Phys. Lett. 86: 253901.

Shen, W. F., Schrag, B. D., Carter, M. J., Xie, J., Xu, C. J., Sun, S. H., & Xiao, G. 2008. Detection of DNA labeled with magnetic nanoparticles using MgO-based magnetic tunnel junction sensors. J. Appl. Phys. 103: 07A306.

Sheng, Y., Nguyen, T. D., Veeraraghavan, G., Mermer, Ö., Wohlgenannt, M., Qui, S., & Scherf, U. 2006. Hyperfine interaction and magnetoresistance in organic semiconductors. Phys. Rev. B. 74: 045213.

Shi, Y., & Wang, J. 2018. Stabilizing skyrmions by nonuniform strain in ferromagnetic thin films without a magnetic field. Phys. Rev. B. 97: 224428.

Shibata, H., Okano, M., & Watanabe, S. 2018. Ultrafast control of coherent spin precession in ferromagnetic thin films via thermal spin excitation processes induced by two-pulse laser excitation. Phys. Rev. B. 97: 014438.

Shick, A. B., Khmelevskyi, S., Mryasov, O. N., Wunderlich, J., & Jungwirth, T. 2010. Spin-orbit coupling induced anisotropy effects in bimetallic antiferromagnets: a route towards antiferromagnetic spintronics. Phys. Rev. B. 81: 212409.

Shigeta, I., Kubota, T., Sakuraba, Y., Molenaar, C. G., Beukers, J. N., Kimura, S., Golubov, A., Brinkman, A., Awaji, S., Takanashi, K., & Hiroi, M. 2018. Epitaxial contact Andreev reflection spectroscopy of NbN/Co2FeSi layered devices. Appl. Phys. Lett. 112: 072402.

Shim, J.H., Raman, K. V., Park, Y. J., Santos, T. S., Miao, G. X., Satpati, B., & Moodera, J. S. 2008. Large spin diffusion length in an amorphous organic semiconductor. Phys. Rev. Lett. 100: 226603.

Shim, Y., Chen, S. H., Sengupta, A., & Roy, K. 2017. Stochastic spin-orbit torque devices as elements for Bayesian inference. Sci. Rep. 7: 14101.

Shimizu, K., Kimura, T., Furomoto, S., Takeda, K., Kontani, K., Onuki, Y., & Amaya, K. 2001. Superconductivity in the non-magnetic state of iron under pressure. Nature. 412: 316–318.

Shimizu, K., Ishikawa, H., Takao, D., Yagi, T., & Amaya, K. 2002. Superconductivity in compressed lithium at 20 K. Nature. 419: 597–599.

Shockley, W., Bardeen, J., & Brattain, W. H. 1948. Electronic theory of the transistor. Science. 108: 678–679.

Shockley, W. 1976 The path to the conception of the junction transistor. IEEE Trans. Electron Dev. 23: 597–605.

Shur, M. 1987. GaAs Devices and Circuits. Plenum Press, New York, London.

Simmons, J. G. 1963. Generalized formula for the electric tunnel effect between similar electrodes separated by a thin insulating film. J. Appl. Phys. 34: 1793.

Sivasubramani, S., Debroy, S., Acharyya, S. G., & Acharyya, A. 2018. Tunable intrinsic magnetic phase transition in pristine single-layer graphene nanoribbons. Nanotechnology. 29: 455701.

Skumryev, V., Laukhin, V., Fina, I., Martí, X., Sánchez, F., Gospodinov, M., & Fontcuberta, J. 2011. Magnetization reversal by electric-field decoupling of magnetic and ferroelectric domain walls in multiferroic-based heterostructures. Phys. Rev. Lett. 106: 057206.

Skyrme, T. H. R. 1962. A unified field theory of mesons and baryons. Nucl. Phys. 31: 556–569.

Skyrme, T. H. R., & Brown, G. E. 1994. Selected Papers, with Commentary, of Tony Hilton Royle Skyrme. World Scientific Serie in 20th Century Physics – Vol. 3, World Scientific Publishing Co. Pte. Ltd., Singapore, New Jersey, London, Hong Kong.

Slachter, A., Bakker, F. L., Adam, J.-P., & van Wees, B. J. 2010. Thermally driven spin injection from a ferromagnet into a non-magnetic metal. Nat. Phys. 6: 879–882.

Slaughter, J. M. 2009. Materials for magnetoresistive random access memory. Annu. Rev. Mater. Res. 39: 277–296.

Slobodskyy, A., Gould, C., Slobodskyy, T., Becker, C. R., Schmit, G., & Molenkamp, L. W. 2003. Voltage-controlled spin selection in a magnetic resonant tunneling diode. Phys. Rev. Lett. 90: 246601.

Slonczewski, J. C. 1989. Conductance and exchange coupling of two ferromagnets separated by a tunneling barrier. Phys. Rev. B. 39, 6995.

Slonczewski, J. C. 1996. Current-driven excitation of magnetic multilayers. J. Magn. Magn. Mater. 159: L1–7.

Slonczewski, J. C. 2002. Currents and torques in metallic magnetic multilayers. J. Mag. Magn. Mat. 247: 324–338.

Slonczewski, J. C. 2005. Currents, torques, and polarization factors in magnetic tunnel junctions. Phys. Rev. B. 71, 024411.

Smith, T. 2005. Seagate pledges first 2.5in perpendicular HDD. The Register 8th June (http://www.theregister.co.uk/2005/06/08/seagate_hdd_roadmap/, accessed March 2018).

Sobucki, K., Krawczyk, M., Tartakivska, O., & Graczyk, P. 2022. Magnon spectrum of Bloch hopfion beyond ferromagnetic resonance. APL Mater. 10: 091103.

Sobucki, K., Krawczyk, M., Tartakivska, O., & Graczyk, P. 2023. Dynamic properties of magnetic hopfions. 2023 IEEE International Magnetic Conference – Short Papers (INTERMAG Short Papers), Sendai, Japan, pp. 1–2.

Solyom, A., Flansberry, Z., Tschudin, M. A., Leitao, N., Pioro-Ladriere, M., Sankey, J. C., & Childress, L. I. 2018. Probing a spin transfer controlled magnetic nanowire with a single nitrogen-vacancy spin in bulk diamond. Nano Lett. 18: 6494–6499.

Son, Y.-W., Cohen, M. L., & Louie, S. G. 2006. Half-metallic graphene nanoribbons. Nature. 444: 347.

Song, P. H., & Kim, K. M. 2002. Spin relaxation of conduction electrons in bulk III-V semiconductors. Phys. Rev. B. 66: 035207.

Song, Y. 2018. Electric-field-induced extremely large change in resistance in graphene ferromagnets. J. Phys. D. – Appl. Phys. 51: 025002.

Soulen Jr., R. J., Byers, J. M., Osofsky, M. S., Nadgorny, B., Ambrose, T., Cheng, S. F., Broussard, P. R., Tanaka, C. T., Nowak, J., Moodera, J. S., Barry, A., & Coey, J. M. D. 1998. Measuring the spin polarization of a metal with a superconducting point contact. Science. 282: 85–88.

Spiesser, A., Saito, H., Fujita, Y., Yamada, S., Hamaya, K., Yuasa, S., & Jansen, R. 2017. Giant spin accumulation in silicon nonlocal spin-transport devices. Phys. Rev. Appl. 8: 064023.

Srinivasan, G., Sengupta, A., & Roy, K. 2016. Magnetic tunnel junction based long-term short-term stochastic synapse for a spiking neural network with on-chip STDP learning. Sci. Rep. 6: 29545.

Stiles, M. D., & McMichael, R. D. 1999. Model for exchange bias in polycrystalline ferromagnet-antiferromagnet bilayers. Phys. Rev. B. 59: 3722.

Stoffel, A., Schneider, J., & Hartmann, F. 1970. Magnetic first-order phase transition in MnAs films. IEEE Trans. Magn. 6: 545–548.

Strukov, D. B., Snider, G. S., Stewart, D. R., & Williams, S. R. 2008. The missing memristor found. Nature. 453: 80–83.

Su, B., & Li, N. 2019. Electronic and magnetic properties of 5d transition metal atoms doped blue phosphorene: first-principles study. J. Magn. Magn. Mater. 469: 236–244.

Sukhostavets, O. V., Aranda, G. R., & Guslienko, K. Y. 2012. Magnetization configurations of a tri-layer nanopillar ferromagnet/nonmagnetic spacer/ferromagnet. J. Appl. Phys. 111: 093901.

Sun, Z. G., Mizuguchi, M., Manago, T., & Akinaga, H. 2004. Magnetic-field-controllable avalanche breakdown and giant magnetoresistive effects in Gold/semi-insulating-GaAs Schottky diode. Appl. Phys. Lett. 85: 5643–5645.

Sun, D., Ehrenfreund, E., & Vardeny, Z. V. 2014. The first decade of organic spintronics research. Chem. Commun. 50: 1781–1793.

Sun, M. L., Hao, Y. T., Ren, Q. Q., Zhao, Y. M., Du, Y. H., & Tang, W. C. 2016. Tuning electronic and magnetic properties of blue phosphorene by doping Al, Si, As and Sb atom: a DFT calculation. Solid State Commun. 242: 36–40.

Suri, M., Bichler, O., Querlioz, D., Traoré, B., Cueto, O., Perniola, L., Sousa, V., Vuillaume, D., Gamrat, C., & DeSalvo, B. 2012. Physical aspects of low power synapses based on phase change memory devices. J. Appl. Phys. 112: 054904.

Sutcliffe, P. 2018. Hopfions in chiral magnets. J. Phys. A. Math. Theor. 51, 375401.

Sverdlov, V., & Selberherr, S. 2015. Spin-based devices for future microelectronics. International Symposium on Next-Generation Electronics 2015.

Sverdlov, V., Makarov, A., Windbacher, T., & Selberherr, S. 2016. Magnetic field dependent tunneling magnetoresistance through a quantum well between ferromagnetic contacts. Int. Conf. Sim. Semicond. Proc. Dev. 2016: 315–318.

Sverdlov, V., Weinbub, J., & Selberherr, S. 2017. Spintronics as a non-volatile complement to modern microelectronics. Informacije Midem – J. Microelectron. Electron. Compon. Mater. 47: 195–210.

Sverdlov, V., & Selberherr, S. 2018. Demands for spin-based nonvolatility in emerging digital logic and memory devices for low power computing. Facta Universitatis – Ser. Electron. Energ. 31: 529–545.

Swartz, A. G., Inoue, H., & Hwang, H. Y. 2016. Electric polarization control of magnetoresistance in complex oxide hetero junctions. Proc. SPIE. 9931: 99310G.

Szczepański, T., Dugaev, V. K., Barnaś, J., Martinez, I., Cascales, J. P., Hong, J.-Y, Lin, M.-T., & Aliev, F. G. 2016. Shot noise in magnetic tunneling structures with two-level quantum dots. Phys. Rev. B 94: 235429.

Szulczewski, G., Tokuc, H., Oguz, K., & Coey, J. M. D. 2009. Magnetoresistance in magnetic tunnel junctions with an organic barrier and an MgO spin filter. Appl. Phys. Lett. 95: 202506.

Szulczewski, G., Sanvito, S., & Coey, M. A. 2009a. Spin of their own. Nat. Mater. 8: 693–695.

Tai, J.-S. B. & Smalyukh, I. I. 2018. Static Hopf Solitons and Knotted Emergent Fields in Solid-State Noncentrosymmetric Magnetic Nanostructures. Phys. Rev. Lett. 121: 187201.

Tai, J.-S. B., Wu, J.-S., & Smalyukh, I. I. 2022. Geometric transformation and three-dimensional hopping of Hopf solitons. Nature Comm. 13: 2986.

Takagishi, M., Yamada, K., Iwasaki, H., Fuke, H. N., & Hashimoto, S. 2010. Magnetoresistance ratio and resistance area design of CPP-MR film for 2–5 Tb/in^2 read sensors. IEEE Trans. Magn. 46: 2086–2089.

Takahashi, Y., Nagase, M., Namatsu, H., Kurihara, K., Iwdate, K., Nakajima, Y., Horiguchi, S., Murase, K., & Tabe, M. 1995. Fabrication technique for Si single-electron transistor operating at room temperature. IEEE Electron. Lett. 31: 136–137.

Takahashi, S., & Maekawa, S. 2008. Spin current, spin accumulation and spin Hall effect. Sci. Technol. Adv. Mater. 9: 014105.

Takano, K., Kodama, R. H., Berkowitz, A. E., Cao, W. & Thomas, G. 1997. Interfacial uncompensated antiferromagnetic spins: role in unidirectional anisotropy in polycrystalline $Ni_{81}Fe_{19}/CoO$ bilayers. Phys. Rev. Lett. 79: 1130.

Takano, K., Kodama, R. H., Berkowitz, A. E., Cao, W., & Thomas, G. 1998. Role of interfacial uncompensated antiferromagnetic spins in unidirectional anisotropy in $Ni_{81}Fe_{19}/CoO$ bilayers. J. Appl. Phys. 83: 6888.

Taniguchi, T., & Imamura, H. 2012. Theoretical study on dependence of thermal switching time of synthetic free layer on coupling field. J. Appl. Phys. 111: 07C901.

Taniguchi, T., & Imamura, H. 2012a. Spin torque assisted magnetization switching in thermally activated region. J. Korean Phys. Soc. 62: 1773–1777.

Taniguchi, T. 2018. Spin-current driven spontaneous coupling of ferromagnets. Phys. Rev. B. 98: 104417.

Tang, H. X., Monzon, F. G., Lifshitz, R., Cross, M. C., & Roukes, M. L. 2000. Ballistic spin transport in a two-dimensional electron gas. Phys. Rev. B. 61: 4437.

Tarassenko, L., Brownlow, M., Marshall, G., Tombs, J., & Murray, A. 1991. Real-time autonomous robot navigation using vlsi neural networks. Adv. neural inf. process. syst. 422–428.

Taylor, R., Pasternak, M., & Jeanloz, R. 1991. Hysteresis in the high pressure transformation of bcc- to hcp-iron. J. Appl. Phys. 69: 6126–6128.

Tchoe, Y., & Han, J. H. 2012. Skyrmion generation by current. Phys. Rev. B. 85: 174416.

TDK. 2013. Magnetic head technology paved the way for smaller hard disk drives with higher storage capacities. The Wonders of Electromagnetism Vol. 2, TDK Tech-Mag 2013.01.31 (http://www.global.tdk.com/techmag/inductive/vol2/index3.htm, accessed March 2018).

Tehrani, S. 2006. Status and outlook of MRAM memory technology. Electron Devices Meet. 1–4: 11–13.

Thangaraj, N., Echer, C., Krishnan, K. M., Farrow, R. F. C., Marks, R. F., & Parkin, S. S. P. 1994. Giant magnetoresistance and microstructural characteristics of epitaxial Fe-Ag and Co-Ag granular thin films. J. Appl. Phys. 75: 6900–6902.

Theil, S. 2015. Why the human brain project went wrong – and how to fix it. Scientific American, https://www.scientificamerican.com/article/why-the-human-brain-project-went-wrong-and-how-to-fix -it/, accessed April 2018.

Thevenard, L., Zeng, H. T., Petit, D., & Cowburn, R. P. 2010. Macrospin limit and configurational anisotropy in nanoscale permalloy triangles. J. Magn. Magn. Mater. 322: 2152.

Thorpe, B., Kalna, K., Langbein, F. C., & Schirmer, S. 2017. Monte Carlo simulations of spin transport in a strained nanoscale InGaAs field effect transistor. J. Appl. Phys. 122: 223903.

Thiyagarajah, N., Lee, K.-I., & Bae, S. 2012. Spin transfer switching characteristics in a $[Pd/Co]_m/Cu/[Co/Pd]_n$ pseudo spin-valve nanopillar with perpendicular anisotropy. J. Appl. Phys. 111: 07C910.

Tillmanns, A., Blachowicz, T., Fraune, M., Güntherodt, G., & Schuller, Ivan K. 2009. Anomalous magnetization reversal mechanism in unbiased Fe/FeF_2 investigated by means of the magneto-optic Kerr effect. J. Magn. Magn. Mater. 321: 2932–2935.

Tiwari, S., Rana, F., Hanafi, H., Hartstein, A., Crabbé, E. F., & Chan, K. 1996. A silicon nanocrystal based memory. Appl. Phys. Lett. 68: 1377.

Tomasello, R., Kimineas, S., Siracusano, G., Carpentieri, M., & Finocchio, G. 2018. Chiral skyrmions in an anisotropy gradient. Phys. Rev. B. 98: 024421.

Tombros, N., Józsa, C., Popinciuc, M., Jonkman, H. T., & van Wees, B. J. 2007. Electronic spin transport and spin precession in single graphene layers at room temperature. Nature. 448: 571–574.

Torrejon, J., Riou, M., Araujo, F. A., Tsunegi, S., Khalsa, G., Querlioz, D., Bortolotti, P., Cros, V., Yakushiji, K., Fukushima, A., Kubota, H., Uasa, S. Y., Stiles, M. D., & Grollier, J. 2017. Neuromorphic computing with nanoscale spintronic oscillators. Nature. 547: 428–431.

Toshiba. 2004. Toshiba Leads Industry in Bringing Perpendicular Data Recording to HDD – Sets New Record for Storage Capacity With Two New HDDs. 14 Dec., 2004 (http://www.toshiba.co.jp/about/press/2004_12/pr1401.htm, accessed March 2018).

Toshiba. 2010. Bit-Patterned Media for High-Density HDDs. TOSHIBA Technologies, LSI & Storage (http://www.toshiba.co.jp/rdc/rd/fields/11_e09_e.htm, accessed March 2018).

Tosi, L., & Aligia, A. A. 2011. Spin selective transport through Aharonov-Bohm and Aharonov-Casher triple quantum dot systems. Phys. Stat. Sol. B. – Basic Solid State Phys. 248: 732–740.

Tóth, A. I., Moca, C. P., Legeza, Ö., & Zaránd, G. 2008. Density matrix numerical renormalization group for non-Abelian symmetries. Phys. Rev. B 78: 245109.

Tralle, I. 1999. Quantum interference due to Larmor precession in mesoscopic loop structures. J. Phys.: Condens. Matt. 11: 8239.

Tralle, I., & Pasko, W. 2003. Coherent Spin Transport and Quantum Interference in Mesoscopic Loop Structures. In: T. Lulek, B. Lulek, and A. Wal (Ed.), Symmetry and Structural Properties of Condensed Matter 125–140.

Tralle, I., & Pasko, W. 2003a. On the quantum beat in mesoscopic loop structures. Phys. Lett. A. World Scientific, Singapore. 318: 463–472.

TROY. 2007. MICR – Magnetic Ink Character Recognition (http://www.whatismicr.com/index.html, accessed March 2018).

Tsang, C., Fontana, R. E., Lin, T., Heim, D. E., Speriosu, V. S., Gurney, B. A., & Williams, M. L. 1994. Design, fabrication and testing of spin-valve read heads for high density recording. IEEE Trans. Magn. 30: 3801–3806.

Tsoi, M., Jansen, A. G. M., Bass, J., Chiang, W.-C., Seck, M., Tsoi, V., & Wyder, P. 1998. Excitation of a magnetic multilayer by an electric current. Phys. Rev. Lett. 80: 4281–4284.

Tsoi, M., Fontana, R. E., & Parkin, S. S. P. 2003. Magnetic domain wall motion triggered by an electric current. Appl. Phys. Lett. 83: 2617.

Tu, B. D., Cuong, L. V., Hung, T. Q., Giang, D. T. H., Danh, T. M., Duc, N. H., & Kim, C. 2009. Optimization of spin-valve structure NiFe/Cu/NiFe/IrMn for Planar Hall effect based biochips. IEEE Trans. Magn. 45: 2378–2382.

Tucker, J. R. 1992. Complementary digital logic based on the "Coulomb blockade". J. Appl. Phys. 72: 4339.

Tulapurkar, A. A., Suzuku, Y., Fukushima, A., Kubota, H., Maehara, H., Tsunekawa, K., Djayaprawira, D. D., Watanabe, N., & Yuasa, S. 2005. Spin-torque diode effect in magnetic tunnel junctions. Nature. 438: 339–342.

Tyagi, P., Baker, C., & D'Angelo, C. 2014. Tunnel junction testbed based molecular devices. 2014 IEEE 14th International Conference on Nanotechnology 801–804.

Tyagi, P., Friebe, E., & Baker, C. 2015. Advantages of prefabricated tunnel junction-based molecular spintronics devices. Nano. 10: 1530002.

Tyryshkin, A. M., Lyon, S. A., Astashkin, A. V., & Raitsimring, A. M. 2003. Electron spin relaxation times of phosphorus donors in silicon. Phys. Rev. B. 68: 193207.

Uchida, K., Adachi, H., Ota, T., Nakayama, H., Maekawa, S., & Saitoh, E. 2010. Observation of longitudinal spin-Seebeck effect in magnetic insulators. Appl. Phys. Lett. 97: 172505–1172503.

Uchitomi, N., Hidaka, S., Saito, S., Asubar, J. T., & Toyota, H. 2018. Magnetic phase change in Mn-doped ZnSnAs2 thin films depending on Mn concentration. J. Appl. Phys. 123: 161566.

Urdampilleta, M., Klyatskaya, S., Cleuziou, J.-P., Ruben, M., & Wernsdorfer, W. 2011. Supramolecular spin valves. Nat. Mater. 10: 502–506.

Urdampilleta, M., Klyatskaya, S., Ruben, M., & Wernsdorfer, W. 2013. Landau-Zener tunneling of a single Tb3+ magnetic moment allowing the electronic read-out of a nuclear spin. Phys. Rev. B: Condens. Matter Mater. Phys. 87: 195412.

Upadhyaya, Y. K., Gupta, M. K., Hasan, M., & Maheshwari, S. 2016. High-density magnetic flash ADC using domain-wall motion and pre-charge sense amplifiers. IEEE Trans. Magn. 52: 4100110.

Uzdin, V. M., Potkina, M. N., Lobanov, I. S., Bessarab, P. F., & Jonsson, H. 2018. Energy surface and lifetime of magnetic skyrmions. J. Magn. Magn. Mater. 459: 236–240.

Vadimov, V. L., Sapozhnikov, M. V., & Mel'nikov, A. S. 2018. Magnetic skyrmions in ferromagnet-superconductor (F/S) heterostructures. Appl. Phys. Lett. 113: 032402.

Vaklinova, K., Polyudov, K., Burghard, M., & Kern, K. 2018. Spin filter effect of hBN/Co detector electrodes in a 3D topological insulator spin valve. J. Phys. Cond. Matt. 30: 105302.

Vandermeulen, J., Nasseri, S. A, van de Wiele, B., Durin, G., van Waeyenberge, B., & Dupre, L. 2016. The effect of Dzyaloshinskii-Moriya interaction on field-driven domain wall dynamics analysed by a semi-analytical approach. J. Phys. D. 49: 465003.

van Son, P. C., van Kempen, H., & Wyder, P. 1987. Boundary resistance of the ferromagnetic-nonferromagnetic metal interface. Phys. Rev. Lett. 58: 2271.

van't Erve, O. M. J., Friedman, A. L., Cobas, E., Li, C. H., Robinson, J. T., & Jonker, B. T. 2012. Low-resistance spin injection into silicon using graphene tunnel barriers. Nat. Nanotechnol. 7: 737–742.

Varade, V., Markus, T., Vankayala, K., Friedman, N., Sheves, M., Waldeck, D. H., & Naaman, R. 2018. Bacteriorhodopsin based non-magnetic spin filters for biomolecular spintronics. Phys. Chem. Chem. Phys. 20: 1091–1097.

Varentsova, A. S., Potkina, M. N., von Malottki, S., Heinze, S., & Bessarab, P. F. 2018. Interplay between size and stability of magnetic skyrmions. Nanosystems – Phys. Chem. Math. 9: 356–363.

Varghani, A., Peiravi, A., & Moradi, F. 2018. Perpendicular STT_RAM cell in 8 nm technology node using $Co_1/Ni_3(111)||Gr_2||Co_1/Ni_3(111)$ structure as magnetic tunnel junction. J. Magn. Magn. Mater. 452: 10–16.

Vasiliev, S. V., Kruglyak, V. V., Sokolovskii, M. L., & Kuchko, A.N. 2007. Spin wave interferometer employing a local nonuniformity of the effective magnetic field. J. Appl. Phys. 101: 113919.

Vatajelu, E. I., & Anghel, L. 2017. Fully-connected single-layer STT-MTJ-based spiking neural network under process variability. Proc. of the IEEE/ACM international symposium on nanoscale architectures (Nanoarch 2017), IEEE International Symposium on Nanoscale Architectures, 21–26.

Vavassori, P., Grimsditch, M., Novosad, V., Metlushko, V., & Ilic, B. 2003. Metastable states during magnetization reversal in square permalloy ring. Phys. Rev. B. 67: 134429.

Vavassori, P., Donzelli, O., Grimsditch, M., Metlushko, V., & Ilic, B. 2007. Chirality and stability of vortex state in permalloy triangular ring micromagnets. J. Appl. Phys. 101: 023902.

Veldhorst, M., Yang, C. H., Hwang, J. C. C., Huang, W., Dehollain, J. P., Muhonen, J. T., Simmons, S., Laucht, A., Hudson, F. E., Itoh, K. M., Morello, A., & Dzurak, A. S. 2015. A two-qubit logic gate in silicon. Nature 526: 410–414.

Venuthurumilli, P. K., Zeng Z., & Xu, X. 2021. Inverse Design of Near-Field Transducer for Heat-Assisted Magnetic Recording Using Topology Optimization, IEEE Transactions on Magnetics 57: 3101406.

Victora, R. H., & Chen, X. 2012. Exchange-assisted spin transfer torque switching. U. S. Patent 8, 134,864 (March 13, 2012).

Vila-Fungueirino, J. M., Rivas-Murias, B., Rubio-Zuazo, J., Carretero-Genevrier, A., Lazzari, M., & Rivadulla, F. 2018. Polymer assisted deposition of epitaxial oxide thin films. J. Mater. Chem. C. 6: 3834–3844.

Vinal, A. W., & Masnari, N. A. 1984. Operating principles of bipolar transistor magnetic sensors. IEEE Trans. Electron Devices. 31: 1486–1496.

Visscher, E. H., Lindeman, J., Verbrugh, S. M., Hadley, P., & Mooij, J. E. 1996. Broadband single-electron tunneling transistor. Appl. Phys. Lett. 68: 2014.

Voinescu, R., Tai, J.-S. B., & Smalyukh, I. I. 2020. Hopf Solitons in Helical and Conical Backgrounds of Chiral Magnetic Solids. Phys. Rev. Lett. 125: 057201.

Volmer, M., & Marioara, A. 2013. Signal dependence on magnetic nanoparticles position over a planar Hall effect biosensor. Microelectron. Eng. 108: 116–120.

von Korff Schmising, C., Giovannella, M., Weder, D., Schaffert, S., Webb, J., & Eisebitt, S. 2015. Nonlocal ultrafast demagnetization dynamics of Co/Pt multilayers by optical field enhancement. New J. Phys. 17: 033047.

von Neumann, J. 1982. First Draft of a Report on the EDVAC. In: Randell B. (eds) The Origins of Digital Computers. Texts and Monographs in Computer Science. Springer, Berlin, Heidelberg. 349–401.

Wagemans, W., & Koopmans, B. 2011. Spin transport and magnetoresistance in organic semiconductors. Phys. Status Solidi B. 248: 1029–1041.

Walewyns, T., Scheen, G., Tooten, E., & Francis, L. A. 2011. Synthesis of patterned freestanding Nickel nanowires by using ion track-etched polyimide. Proceedings of SPIE 8068: 806815.

Wang, X. F., Vasilopoulos, P., & Peeters, F. M. 2002. Ballistic spin transport through electronic stub tuners: spin precession, selection, and square-wave transmission. Appl. Phys. Lett. 80: 1400.

Wang, D., Nordman, C., Daughton, J. M., Qian, Z., & Fink, J. 2004. 70 % TMR at room temperature for SDT sandwich junctions with CoFeB as free and reference layers. IEEE Trans. Magn. 40: 2269–2271.

Wang, S. X., Bae, S. Y., Li, G. X., Sun, S. H., White, R. L., Kemp, J. T., & Webb, C. D. 2005. Towards a magnetic microarray for sensitive diagnostics. J. Magn. Magn. Mater. 293:731.

Wang, K. L., Khitun, A., & Flood, A. H. 2005a. Interconnects for nanoelectronics. Proceedings of the IEEE 2005 International Interconnect Technology Conference.

Wang, K. Y., Campion, R. P., Edmonds, K. W., Sawicki, M., Dietl, T., Foxon, C. T. & Gallagher, B. L. 2005b. Magnetism in (Ga,Mn)As Thin Films With TC Up To 173 K. Proc. 27th International Conference on Physics of Semiconductors, ed. J. Mendez and C. G. Van de Walle, Springer, New York, p. 333.

Wang, Q., Shang, D. S., Wu, Z. H., Chen, L. D., & Li, X. M. 2007. "Positive" and "negative" electric-pulse-induced reversible resistance switching effect in Pr0.7Ca0.3MnO3 films. Appl. Phys. A. 86: 357–360.

Wang, W. L., Meng, S., & Kaxiras, E. 2008. Graphene nanoflakes with large spin. Nano Lett. 8: 241–245.

Wang, W. H., Han, W., Pi, K., McCreary, K. M., Miao, F., Bao, W., Lau, C. N., & Kawakami, R. K. 2008a. Growth of atomically smooth MgO films on graphene by molecular beam epitaxy. Appl. Phys. Lett. 93: 183107.

Wang, W. L., Yazyev, O. V., Meng, S., & Kaxiras, E. 2009. Topological frustration in graphene nanoflakes: magnetic order and spin logic devices. Phys. Rev. Lett. 102: 157201.

Wang, R.-H., Jiang, J.-S., & Hu, M. 2009. Metallic cobalt microcrystals with flowerlike architectures: synthesis, growth mechanism and magnetic properties. Mater. Res. Bull. 44: 1468.

Wang, W.-G., Li, M., Hagemann, S., & Chien, C. L. 2012. Electric-field-assisted switching in magnetic tunnel junctions. Nat. Mater. 11: 64.

Wang, W.-G., & Chien, C. L. 2013. Voltage-induced switching in magnetic tunnel junctions with perpendicular magnetic anisotropy. J. Phys. D: Appl. Phys. 46: 074004.

Wang, X., Sun, G., Routh, P., Kim, D.-H., Huang, W., & Chen, P. 2014. Heteroatom-doped graphene materials: syntheses, properties and applications. Chem. Soc. Rev. 43: 7067–7098.

Wang, X.-G., Chotorlishvili, L., & Berakdar, J. 2017. Strain and thermally induced magnetic dynamics and spin current in magnetic insulators subject to transient optical grating. Front. Mater. 4: 19.

Wang, S., & Komvopoulos, K. 2018. Electromagnetic and thermomechanical analysis of near-field heat transfer in heat-assisted magnetic recording heads. IEEE Trans. Magn. 54: 3000906.

Wang, X. S., Yuan, H. Y., & Wang, X. R. 2018. A theory on skyrmion size. Commun. Phys. 1: 31.

Wang, Y., Ramaswamy, R., & Yant, H. 2018a. FMR-related phenomena in spintronic devices. J. Phys. D. Appl. Phys. 51: 273002.

Wang, S. H., Li, G., Wang, J. Y., & Jin, H. Y. K. X. 2018c. Laser-heating spin Seebeck effect of yttrium iron garnet/platinum heterostructure below room temperature. J. Magn. Magn. Mater. 468: 50–55.

Wang, Z. H., Wu, B., Li, Z. W., Lin, X. Y., Yang, J. L., Zhang, Y. G., & Zhao, W. S. 2018d. Evaluation of ultrahigh-speed magnetic memories using field-free spin-orbit torque. IEEE Trans. Magn. 54: 3401505.

Wang, X. S., Qaiumzadeh, A., & Brataas, A. 2019. Current-Driven Dynamics of Magnetic Hopfions. Phys. Rev. Lett. 123: 147203.

Wang, K., Bheemarasetty, V., & Xiao, G. Spin textures in synthetic antiferromagnets: challenges, opportunities, and future directions. APL Mater. 11: 070902.

Wang, J., Xu, Y. K.; Wang, S. H.; Dai, X. Z.; Yan, P. F.; Zhou, J., Wang, R. F.; Xu, Y. B., & He, L. 2024. Hole-Mediated RKKY Interaction in 2D Ferromagnetic $CrTe_2$ Ultra-Thin Films. Adv. Electron. Mater. 10: 2300646.

Waser, R., & Aono, M. 2007. Nanoionics-based resistive switching memories. Nat. Mater. 6: 833–840.

WD. 2017. WD Gold – Enterprise-class Hard Drives. Specification Sheet (https://www.wdc.com/content/dam/wdc/website/downloadable_assets/eng/spec_data_sheet/2879-800074.pdf, accessed March 2018).

Wei, J. S., Wang, R. Z., Yuan, R., You, J. Q., & Yan, H. 2005. Spin transport in quantum dot embedded in Aharonov-Bohm ring. Phys. Lett. A. 345: 211–217.

Weihe, H., & Güdel, H. U. 1997. Quantitative interpretation of the goodenough–Kanamori rules: a critical analysis. Inorg. Chem. 36: 3632–3639.

Wen, S., Zeng, Z., & Huang, T. 2013. Associative learning of integrate-and-fire neurons with memristor-based synapses. Neural process. lett. 38: 69–80.

Weymann, I., & Barnaś, J. 2013. Spin thermoelectric effects in Kondo quantum dots coupled to ferromagnetic leads. Phys. Rev. B 88: 085313.

White, A. M., Hinchliffe, I., & Dean, P. J. 1972. Zeeman spectra of the principal bound exciton in Sn-doped gallium arsenide. Solid State Commun. 10: 497–500.

Wiesendanger, R. 2016. Nanoscale magnetic skyrmions in metallic films and multilayers: a new twist for spintronics. Nat. Rev. Mater. 1: 16044.

Wigner, E.P. 1932. On the quantum correction for thermodynamic equilibrium. Phys. Rev. 40: 749–759.

Wilamowski, Z., Malissa, H., Schaeffler, F., & Jantsch, W. 2007. g-factor tuning and manipulation of spins by an electric current. Phys. Rev. Lett. 98: 187203.

Will, I., Ding, A., & Xu, Y. B. 2014. Shape and proximity effect of magnetic nanoelements used as biomolecular labels for magnetic notched nanowire biosensors. IEEE Trans. Magn. 50: 5200205.

Will, I., Ding, A., & Xu, Y. B. 2015. Proximity effect of magnetic permalloy nanoelements used to induce AMR changes in magnetic biosensor nanowires at specific receptor sites. J. Magn. Magn. Mater. 388: 5–9.

Wilson, K. G. 1975. The renormalization group: critical phenomena and the Kondo problem. Rev. Mod. Phys. 47: 773–840.

Windbacher, T., Mahmoudi, H., Sverdlov, V., & Selberherr, S. 2013. Spin torque magnetic integrated circuit. EU Patent EP 2784020.

Windbacher, T., Mahmoudi, H., Sverdlov, V., & Selberherr, S. 2013a. Rigorous simulation study of a novel non-volatile magnetic flip-flop. Proc. Int. Conf. Simul. Semiconduc. Proces. Devices (SISPAD) 2013, 368–371.

Windbacher, T., Mahmoudi, H., Sverdlov, V. & Selberherr, S. 2013b. Novel MTJ-based shift register for non-volatile logic applications. Proc. IEEE/ACM Int. Symp. Nanoscale Archit. (NANOARCH) 11: 36–37.

Windbacher, T., Makarov, A., Sverdlov, V., & Selberherr, S. 2015. Influence of magnetization variations in the free layer on a non-volatile magnetic flip flop. Solid-State Electron. 108: 2–7.

Windbacher, T., Ghosh, J., Makarov, A., Sverdlov, V., & Selberherr, S. 2015a. Modelling of multipurpose spintronic devices. Int. J. Nanotechnol. 12: 313–331.

Windbacher, T., Makarov, A., Sverdlov, V., & Selberherr, S. 2015b. Novel buffered magnetic logic gate grid. ECS Trans. 66: 295–303.

Wingreen, N. S., & Meir, Y. 1994. Anderson model out of equilibrium: noncrossing approximation approach to transport through a quantum dot. Phys. Rev. B. 49: 11040.

Woo, S., Litzius, K., Krüger, B., Im, M.-Y., Caretta, L., Richter, K., Mann, M., Krone, A., Reeve, R. M., Weigand, M., Agrawal, P., Lemesh, I., Mawass, M.-A., Fischer, P., Kläui, M., & Beach, G. S. D. 2016. Observation of room temperature magnetic skyrmions and their current-driven dynamics in ultrathin films. Nat. Mater. 15: 501–506.

Wright, E. P. G. 1951. Electric connecting device. US Patent 2667542.

Wu, Y., & Hu, B. 2006. Metal electrode effects on spin-orbital coupling and magnetoresistance in organic semiconductor devices. Appl. Phys. Lett. 89: 203510.

Wu, S.-Q., & Sun, W.-L. 2007. Fano versus Kondo resonances in a closed Aharonov-Bohm interferometer coupled to ferromagnetic electrodes. Chin. Phys. Lett. 24: 1054–1057.

Wu, Y., Xu, Z., Hu, B., & Howe, J. 2007. Tuning magnetoresistance and magnetic field-dependent electroluminescence through mixing strong-spin-orbital-coupling molecule and weak-spin-orbital-coupling polymer. Phys. Rev. B. 75: 035214.

Wu, S.-Q. 2009. Spin-dependent transport through a closed Aharonov-Bohm interferometer coupled to ferromagnetic electrodes. Act. Phys. Sin. 58: 4175–4182.

Wu, S.-Q., Hou, T., Zhao, G.-P., & Yu, W.-L. 2010. Magnetotransport through an Aharonov-Bohm ring with parallel double quantum dots coupled to ferromagnetic leads. Chinese Phys. B. 19: 047202.

Wu, T., Bur, A., Wong, K., Zhao, P., Lynch, Ch. S., Amiri, P. K., Wang, K. L., & Carman, G. P. 2011. Electrical control of reversible and permanent magnetization reorientation for magnetoelectric memory devices. Appl. Phys. Lett. 98: 262504.

Wu, S. M., Pearson, J. E., & Bhattacharya, A. 2015. Paramagnetic spin Seebeck effect. Phys. Rev. Lett. 114: 186602.

Wu, H., Wan, C. H., Yuan, Z. H., Zhang, X., Jiang, J., Zhang, Q. T., Wen, Z. C., & Han, X. F. 2015a. Observation of pure inverse spin Hall effect in ferromagnetic metals via ferromagnetic/antiferromagnetic exchange-bias structures. Phys. Rev. B. 92: 054404.

Wu, H., Wan, C. H., Zhang, X., Yuan, Z. H., Zhang, Q. T., Qin, J. Y., Wei, H. X., Han, X. F., & Zhang, S. 2016. Observation of magnon-mediated electric current drag at room temperature. Phys. Rev. B. 93: 060403.

Wu, X. Q., & Meng, H. 2016. Gate-voltage control of equal-spin Andreev reflection inhalf-metal/semiconductor/superconductor junctions. Phys. Lett. A. 380: 1672–1676.

Xiao, W. D., Liu, L. W., Yang, K., Zhang, L. Z., Song, B. Q., Du, S. X., & Gao, H. J. 2015. Tuning the spin, chirality, and adsorption site of metal-phthalocyanine on Au(111) surface with hydrogen atoms. Act. Phys. Sin. 64: 076802.

Xiong, Z.H., Di, W., Vardeny, Z. V., & Jing, S. 2004. Giant magnetoresistance in organic spin-valves. Nature. 427: 821.

Xiong, Y. C., Huang, H. M., Zhao, W. L., & Laref, A. 2017. Suppressed Kondo effect and Kosterlitz-Thouless-type phase transition induced by level difference in a triple dot device. J. Phy. Cond. Matt. 29: 405601.

Xiong, Y. C., Wang, W. Z., Luo, S. J., Yang, J. T., & Huang, H. M. 2017a. Screened spin-1 and -1/2 Kondo effect in a triangular quantum dot system with interdot Coulomb repulsion. Superlattices Microstruct. 103: 180–189.

Xiong, L. L., Kostylev, M., & Adeyeye, A. O. 2017b. Magnetization dynamics of $Ni_{80}Fe_{20}$ nanowires with continuous width modulation. Phys. Rev. B. 95: 224426.

Xu, Y., Liu, X. F., & Guo, W. L. 2014. Tensile strain induced switching of magnetic states in $NbSe_2$ and NbS_2 single layers. Nanoscale. 6: 12929–12933.

Xu, J., Li, Q., Zong, W. H., Zhang, Y. D., & Li, S. D. 2016. Ultra-wide detectable concentration range of GMR biosensors using Fe_3O_4 microspheres. J. Magn. Magn. Matter. 417: 25–29.

Xu, Q. Y., Dai, C. J., Han, Z. D., & Li, Q. 2018. The magnetic properties of $BaCo_{0.5}Ni_{0.5}F_4$. Journal of Magnetism and Magnetic Materials 453, 177–181.

Yacoby, A., Hess, H. F., Fulton, T. A., Pfeiffer, L. N., & West, K. W. 1999. Electrical imaging of the quantum hall state. Solid State Commun. 111: 1–13.

Yafet, Y. 1963. g factors and spin-lattice relaxation of conduction electrons. Solid State Phys. 14: 1–98.

Yagmur, A., Iguchi, R., Geprägs, S., Erb, A., Daimon, S., Saitoh, E., Gross, R., & Uchida, K. 2018. Lock-in thermography measurements of the spin Peltier effect in a compensated ferrimagnet and its comparison to the spin Seebeck effect. J. Phys. D. Appl. Phys. 51: 194002.

Yahia, I. S., Sakr, G. B., Wojtowicz, T., & Karczewski, G. 2010. p-ZnTe/n-CdMnTe/n-GaAs diluted magnetic diode for photovoltaic applications. Semiconduc. Sci. Technol. 25: 095001.

Yahia, I. S., Yakuphanoglu, F., Chusnutdinow, S., Wojtowicz, T., & Karczewski, G. 2013. Photovoltaic characterization of n-CdTe/p-CdMnTe/GaAs diluted magnetic diode. Curr. Appl. Phys. 13: 537–543.

Yahia, I. S., AlFaify, S., Yakuphanoglu, F., Chusnutdinow, S., Wojtowicz, T., & Karczewski, G. 2015. Impedeance Spectroscopy of n-CdTe/p-CdMnTe/p-GaAs Diluted Magnetic Diode. J. Electronic Mat. 44: 2768–2772.

Yakata, S., Kubota, H., Sugano, T., Seki, T., Yakushiji, K., Fukushima, A., Yuasa, S., & Ando, K. 2009. Thermal stability and spin-transfer switchings in MgO-based magnetic tunnel junctions with ferromagnetically and antiferromagnetically coupled synthetic free layers. Appl. Phys. Lett. 95: 242504.

Yakout, S. M. 2021. Spintronics and innovative memory devices: a review on advances in magnetoelectric $BiFeO_3$. J. Supercond. Novel Magn. 34: 317–338.

Yamada, H., Terao, K., Kondo, K., & Goto, T. 2002. Strong pressure dependences of the magnetization and Curie temperature for CrTe and MnAs with NiAs-type structure. J. Phys.: Condens. Matter. 14: 11785.

Yamada, K., Kasai, S., Nakatani, Y., Kobayashi, K., Kohno, H., Thiaville, A., & Ono, T. 2007. Electrical switching of the vortex core in a magnetic disk. Nat. Mater. 6: 270–273.

Yamamoto, K., Gomonay, O., Sinova, J., & Schwiete, G. 2018. Spin transfer torques and spin-dependent transport in a metallic F/AF/N tunneling junction. Phys. Rev. B. 98: 014406.

Yamanouchi, M., Chiba, D., Matsukura, F., & Ohno, H. 2004. Current-induced domain-wall switching in a ferromagnetic semiconductor structure. Nature. 428: 539–542.

Yang, Z., Ouyang, B., Lan, G. Q., Xu, L. C., Liu, R. P., & Liu, X. G. 2017. The tunneling magnetoresistance and spin-polarized optoelectronic properties of graphyne-based molecular magnetic tunnel junctions. J. Phys. D – Appl Phys. 50: 075103.

Yan, B. M., Li, X. Q., Zhao, J. J., Jia, Z. Z., Tang, F. D., Zhang, Z. H., Yu, D. P., Liu, K. H., Zhang, L. Y., & Wu, X. S. 2018. Gate tunable Kondo effect in magnetic molecule decorated graphene. Solid State Commun. 278: 24–30.

Yang, H. H., Wang, C., Wang, X. F., Wang, X. S., Cao, Y. S., & Yan, P. 2018. Twisted skyrmions at domain boundaries and the method of image skyrmions. Phys. Rev. B. 98: 014433.

Yano, K., Ishii, T., Hashimoto, T., Kobayashi, T., Murai, F., & Seki, K. 1993. A room-temperature single-electron memory device using fine-grain polycrystalline silicon. Technical Digest of 1993 International Electron Device Meeting, 541.

Yano, K., Ishii, T., Hashimoto, T., Kobayashi, T., Murai, F., & Seki, K. 1994. Room-temperature single-electron memory. IEEE Trans. Electron. Devices. 41: 1628–1638.

Yao, X. N., Duan, Q. Q., Tong, J. W., Chang, Y. F., Zhou, L. Q., Qin, G. W., & Zhang, X. M. 2018. Magnetoresistance effect and the applications for organic spin valves using molecular spacers. Materials. 11: 721.

Yi, G.-Y., Jiang, C.; Zhang, L.-L., Zhong, S.-R., Chu, H., & Gong, W.-J. 2020. Manipulability of the Kondo effect in a T-shaped triple-quantum-dot structure. Phys. Rev. B 102: 085418.

Yoo, M. J., Fulton, T. A., Hess, H. F., Willett, R. L., Dunkleberger, L. N., Chichester, R. J., Pfeiffer, L. N., & West, K. W. 1997. Scanning single-electron transistor microscopy: imaging individual charges. Science. 276: 579–582.

Yoon, I., & Raychowdhury, A. 2018. Modeling and analysis of magnetic field induced coupling on embedded STT-MRAM arrays. IEEE Trans. Comput.-Aided Des. Integr. Circuits Syst. 37: 337–349.

Yoshikawa, M., Kitagawa, E., Nagase, T., Daibou, T., Nagamine, K., Nishiyama, K., Kishi, T., & Yoda, H. 2008. Tunnel magnetoresistance over 100 % in MgO-based magnetic tunnel junction films with perpendicular magnetic L10-FePt electrodes. Proc. IEEE. 44:2573–76.

Yosida, K. 1957. Magnetic properties of Cu-Mn alloys. Phys. Rev. 106: 893.

You, C.-Y., & Bader, S. D. 2000. Voltage controlled spintronic devices for logic applications. J. Appl. Phys. 87: 5215–5217.

Yu, Z. G., & Flatté, M. E. 2002. Spin diffusion and injection in semiconductor structures: electric field effects. Phys. Rev. B. 66: 235302.

Yu, Z. G., & Flatté, M. E. 2002a. Electric-field dependent spin diffusion and spin injection into semiconductors. Phys. Rev. B. 66: 201202(R).

Yu, H. M., Granville, S., Yu, D. P., & Ansermet, J.-Ph. 2010. Evidence for thermal spin-transfer torque. Phys. Rev. Lett. 104: 146601.

Yu, W. Y., Zhu, Z., Niu, C. Y., Li, C., Cho, J.-H., & Jia, Y. 2016. Dilute magnetic semiconductor and half-metal behaviors in 3 d transition-metal doped black and blue phosphorenes: a first-principles study. Nanoscale Res. Lett. 11: 77–79.

Yu, X. Z., Morikawa, D., Yokouchi, T., Shibata, K., Kanazawa, N., Kagawa, F., Arima, T. H., & Tokura, Y. 2018. Aggregation and collapse dynamics of skyrmions in a non-equilibrium state. Nat. Phys. 14: 832–836.

Yu, X. Z., Liu, Y. Z., Iakoubovskii, K. V., Nakajima, K., Kanazawa, N., Nagaosa, N., & Tokura, T. 2023. Realization and Current-Driven Dynamics of Fractional Hopfions and Their Ensembles in a Helimagnet FeGe. Adv. Mater. 35: 2210646.

Yuan, Z., Wang, S., & Xia, K. 2010. Thermal spin-transfer torques on magnetic domain walls. Solid State Commun. 150: 548–551.

Yuan, R. Y. 2012. Resonant spin transport through a double quantum-dot system with ferromagnetic electrodes and tunable magnetic field. Przeglad Elektrotechniczny 88: 88–92.

Yüksel, Y. 2018. Exchange bias mechanism in FM/FM/AF spin valve systems in the presence of random unidirectional anisotropy field at the AF interface: the role played by the interface roughness due to randomness. Phys. Lett. A. 19: 1298–1304.

Yuasa, S., Nagahama, T., Fukushima, A., Suzuki, Y., & Ando, K. 2004. Giant room-temperature magnetoresistance in single-crystal Fe/MgO/Fe magnetic tunnel junctions. Nat. Mater. 3: 868–871.

Yuasa, S., Fukushima, A., Kubota, H., Suzuki, Y., & Ando, K. 2006. Giant tunneling magnetoresistance up to 410 % at room temperature in fully epitaxial Co/MgO/Co magnetic tunnel junctions with bcc Co(001) electrodes. Appl. Phys. Lett. 89: 042505.

Yuasa, S., Fukushima, A., Yakushiji, K., Nozaki, T., Konoto, M., Maehara, H., Kubota, H., Taiguchi, T., Arai, H., Imamura, H., Ando, K., Shiota, Y., Bonell, F., Suzuki, Y., Shimomura, N., Kitagawa, E., Ito, J., Fujita, S., Abe, K., Nomura, K., Noguchi, H., & Yoda, H. 2013. Future prospects of MRAM technologies. Proc. IEEE Int. Electron devices meeting (IEDM) 3.1.1–3.1.4.

Yung, K. W., Landecker, P. B., & Villani, D. D. 1998. An analytical solution for the force between two magnetic dipoles. Magn. Electr. Sep. 9: 39–52.

Zare, M., Parhizgar, F., & Asgari, R. 2018. Strongly anisotropic RKKY interaction in monolayer black phosphorus. J. Magn. Magn. Mater. 456: 307–315.

Zelezny, J., Wadley, P., Olejnik, K., Hoffmann, A., & Ohno, H. 2018. Spin transport and spin torque in antiferromagnetic devices. Nat. Phys. 14: 220–228.

Zener, C. 1951. Interaction between the d-shells in the transition metals. II. Ferromagnetic compounds of manganese with Perovskite structure. Phys. Rev. 82: 403.

Zeng, H. T., Read, D., O'Brian, L., Sampaio, J., Lewis, E. R., Petit, D., & Cowburn, R. P. 2010. Asymmetric magnetic NOT gate and shift registers for high density data storage. Appl. Phys. Lett. 96: 262510.

Zerrouati, K., Fabre, F., Bacquet, G., Bandet, J., Frandon, J., Lampel, G., & Paget, D. 1987. Spin-lattice relaxation in p-type gallium arsenide single crystals. Phys. Rev. B. 37: 1334.

Zhang, Y., & Blencowe, M. P. 2002. Intrinsic noise of a micro-mechanical displacement detector based on the radio-frequency single-electron transistor. J. Appl. Phys. 91: 4249–4255.

Zhang, Y. T., Guo, Y., & Li, Y. C. 2005. Persistent spin currents in a quantum ring with multiple arms in the presence of spin-orbit interaction. Phys. Rev. B. 72: 125334.

Zhang, G. L., Lu, M. W., Tang, Y., & Chen, S. Y. 2008. Giant magnetoresistance effect realized by depositing nanosized ferromagnetic and Schottky stripes on a semiconductor heterostructure. J. Phys. Cond. Matt. 20: 335221.

Zhang, W., & Haas, S. 2010. Phase diagram of magnetization reversal processes in nanorings. Phys. Rev. B. 81: 064433.

Zhang, X., Ezawa, M., & Zhou, Y. 2015. Magnetic skyrmion logic gates: conversion, dublication and merging of skyrmions. Sci. Rep. 5: 9400.

Zhang, X., Zhao, G. P., Fangor, H., Liu, J. P., Xia, W. X., Xia, J., & Morvan, F. J. 2015a. Skyrmion-skyrmion and skyrmion-edge repulsions in skyrmion-based racetrack memory. Sci. Rep. 5: 7643.

Zhang, X. C., Ezawa, M., Xiao, D., Zhao, G. P., Liu, Y. W., & Zhou, Y. 2015b. All-magnetic control of skyrmions in nanowires by a spin wave. Nanotechnology. 26: 225701.

Zhang, X. C., Zhou, Y., Ezawa, M., Zhao, G. P., & Zhao, W. S. 2015c. Magnetic skyrmion transistor: skyrmion motion in a voltage-gated nanotrack. Sci. Rep. 5: 11369.

Zhang, D. M., Zeng, L., Cao, K. H., Wang, M. X., Peng, S. Z., Zhang, Y., Zhang, Y. G., Klein, J.-O., Wang, Y., & Zhao, W. S. 2016. All spin artificial neural networks based on compound spintronic synapse and neuron. IEEE Trans. Biomed. Circuits yst 10: 828–836.

Zhang, D. M., Zeng, L., Zhang, Y. G., Zhao, W. S., & Klein, J. O. 2016a. Stochastic spintronic device based synapses and spiking neurons for neuromorphic computation. Proceedings of the 2016 IEEE/ACM International Symposium on Nanoscale Architectures (NANOARCH), IEEE International Symposium on Nanoscale Architectures, 173–178.

Zhang, G. P., Bai, Y. H., & George, T. F. 2017. Is perpendicular magnetic anisotropy essential to all-optical ultrafast spin reversal in ferromagnets? J. Phys. Cond. Mater. 29: 425801.

Zhang, X., Tong, J., Zhu, H., Wang, Z., Zhou, L., Wang, S., Miyashita, T., Mitsuishi, M., & Qin, G. 2017a. Room temperature magnetoresistance effects in ferroelectric poly(vinylidene fluoride) spin valves. J. Mater. Chem. C. 5: 5055–5062.

Zhang, L., Deng, Y., Yu, J. Y., Zhang, M., Huang, K., Li, Y. B., Li, P., & Zhu, X. B. 2018. Synthesis and properties of $CO_{5.47}N$ thin films by chemical solution deposition. Mater. Lett. 225: 145–148.

Zhang, S. L., Wang, W. W., Burn, D. M., Peng, H., Berger, H., Bauer, A., Pfleiderer, C., van der Laan, G., & Hesjedal, T. 2018a. Manipulation of skyrmion motion by magnetic field gradients. Nat. Commun. 9: 2115.

Zhang, S. L., van der Laan, G., Müller, J., Heinen, L., Garst, M., Bauer, A., Berger, H., Pfleiderer, C., & Hesjedal, T. 2018b. Reciprocal space tomography of 3D skyrmion lattice order in a chiral magnet. Proc. Natl. Acad. Sci. USA. 115: 6386–6391.

Zhang, L. S.; Zhou, J., Li, H., Shen, L., & Feng, Y. P. 2021. Recent progress and challenges in magnetic tunnel junctions with 2D materials for spintronic applications. Appl. Phys. Rev. 8: 021308.

Zhang, Z. Z.; Lin, K. L.; Zhang, Y.; Bournel, A.; Xia, K.; Kläui, M.; Zhao, W. S. Magnon scattering modulated by omnidirectional hopfion motion in antiferromagnets for meta-learning. *Sci. Adv.* 2023, *9*, eade7439.

Zhao, W. S., Belhaire, E., Chappert, C., Dieny, B., & Prenat, G. 2009. TAS-MRAM-based low-power high-speed runtime reconfiguration (RTR) FPGA. ACM T. Reconfig. Technol. 2: 8.

Zhao, W. S., Duval, J., Ravelosona, D., Klein, J. O., & Kim, J. V. 2011. A compact model of domain wall propagation for logic and memory design. J. Appl. Phys. 109: 07D501.

Zhao, X. F., Ren, R. Z., Xie, G., & Liu, Y. 2018. Single antiferromagnetic skyrmion transistor based on strain manipulation. Appl. Phys. Lett. 112: 252402.

Zheng, Y., & Wudl, F. 2014. Organic spin transporting materials: present and future. J. Mater. Chem. A 2, 48–57.

Zheng, F. S., Kiselev, N. S., Rybakov, F. N., Yang, L. Y., Shi, W., Blügel, S., & Dunin-Borkowski, R. E. 2023. Hopfion rings in a cubic chiral magnet. Nature 623: 718–723.

Zhong, Z., Wang, D., Cui, Y., Bockrath, M. W., & Lieber, C. M. 2003. Nanowire crossbar arrays as address decoders for integrated nanosystems. Science. 302: 1377.

Zhong, Y., Li, Y., Xu, L., & Miao, X. 2015. Simple square pulses for implementing spike-timing-dependent plasticity in phase-change memory. physica status solidi (RRL)-Rapid Res. Lett. 9: 414–419.

Zhou, J. & Chen, J. S. 2021. Prospect of Spintronics in Neuromorphic Computing. Adv. Electron. Mater. 7: 2100465.

Zhou, J., & Wu, M. W. 2008. Spin relaxation due to the Bir-Aronov-Pikus mechanism in intrinsic and p-type GaAs quantum wells from a fully microscopic approach. Phys. Rev. B. 77: 075318.

Zhou, Y., Han, W., Chan, L.-T., Xiu, F., Wang, M., Oehme, M., Fischer, I. A., Schulze, J., Kawakami, R. K., & Wang, K. L. 2011. Electrical spin injection and transport in germanium. Phys. Rev. B. 84: 125323.

Zhou, Y., & Ezawa, M. 2014. A reversible conversion between a skyrmion and a domain-wall pair in a junction geometry. Nat. Commun. 5: 4652.

Zhou, X., Ma, L., Shi, Z., Fan, W. J., Evans, R. F. L., Zheng, J.-G., Chantrell, R. W., Mangin, S., Zhang, H. W., & Zhou, S. M. 2015. Mapping motion of antiferromagnetic interfacial uncompensated magnetic moment in exchange-biased bilayers. Sci. Rep. 5: 9183.

Zhou, L., Wang, X., Zhang, H. M., Shen, X. D., Dong, S., & Long, Y. W. 2018. High pressure synthesis and physical properties of multiferroic materials with multiply-ordered perovskite structure. Act. Phys. Sin. 67: 157505.

Zhu, J.-G., Zhu, X., & Tang, Y. 2008. Microwave assisted magnetic recording. IEEE Trans. Magn. 44: 125–131.

Zhu, J., Zhang, J. L., Kong, P. P., Zhang, S. J., Yu, X. H., Zhu, J. L., Liu, Q. Q., Li, X., Yu, R. C., Ahuja, R., Yang, W. G., Shen, G. Y., Mao, H. K., Weng, H. M., Dai, X., Fang, Z., Zhao, Y. S., & Jin, C. Q. 2013. Superconductivity in topological insulator Sb_2Te_3 induced by pressure. Sci. Rep. 3: 2016.

Zhu, H., Zhao, E. H., Richter, C. A., & Li, Q. L. 2014. Topological insulator Bi_2Se_3 nanowire field effect transistors. ECS Trans. 64: 51–59.

Zhu, W.-H., Ding, G.-H., & Dong, B. 2014a. Negative differential conductance and hysteretic current switching of benzene molecular junction in a transverse electric field. Nanotechnology. 25: 465202.

Zhu, S., & Li, T. 2016. Strain-induced programmable half-metal and spin-gapless semiconductor in an edge-doped boron nitride nanoribbon. Phys. Rev. B. 93: 115401.

Zhu, T. C., Singh, S., Katoch, J., Wen, H., Belashchenko, K., Zutic, I., & Kawakami, R. K. 2018. Probing tunneling spin injection into graphene via bias dependence. Phys. Rev. B. 98: 054412.

Zhukova, V., Mino, J., del Val, J. J., Varga, R., Martinez, G., Baibich, M., Ipatov, M., & Zhukov, A. 2017. Kondo-like behavior and GMR effect in granular $Cu_{90}Co_{10}$ microwires. AIP Adv. 7: 055906.

Ziegler, M., & Kohlstedt, H. 2013. Mimic synaptic behavior with a single floating gate transistor: a memflash synapse. Journal of Applied Physics. 114: 194506.

Žitko, R., Pruschke, T. 2009. Energy resolution and discretization artifacts in the numerical renormalization group. Phys. Rev. B 79: 085106.

Žutić, I., Fabian, J. & Das Sarma, S., 2004. Spintronics: fundamentals and applications, Rev. Mod. Phys. 76: 323.

Žutić, I., Fabian, J., & Erwin, S. C. 2006. Spin injection and detection in silicon. Phys. Rev. Lett. 97: 026602.

Žutić, I., & Fabian, J. 2007. Spintronics: silicon twists. Nature. 447: 268–269.

Index

Also of interest

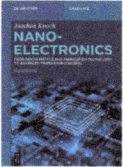

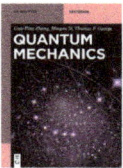

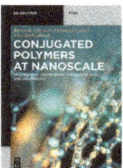

www.ingramcontent.com/pod-product-compliance
Lightning Source LLC
Chambersburg PA
CBHW080937220326
41598CB00034B/5809

* 9 7 8 3 1 1 1 3 8 2 4 9 4 *